I0478603

THE ANGELS

BOOK ONE, THE CIA AREA 51 CHRONICLES

The Complete Illustrated History of the CIA at Area 51

By

TD Barnes

TD Barnes Copyright 2024

Contents
Table of Contents

Dedication

The author dedicates this book to all who served the Central Intelligence Agency at its Groom Lake flight-test facility in Area 51, Nevada. I include the unknown participants who worked surreptitiously at Area 51's affiliated power projection locations throughout the world.

The author also dedicates this account of Cold War history to the many unsung heroes conducting post-WWII reconnaissance operations. It was their untold sacrifices and those of their families, which led to the CIA to become officially responsible for American overhead reconnaissance.

Acknowledgments

I first acknowledge my family, who for years dropped me off in a secured area at the McCarran Airport or Nellis AFB to catch a plane on Monday morning and picked me up when I returned Friday evening, not knowing where I went or what I did. I especially acknowledge my Area 51 CIA contemporaries referred to as the Roadrunners for whom I have served as the alumni president for many years. These were CIA, Air Force, and contractor personnel working at or directly affiliated with the CIA projects at Area 51, many of them named in this book.

Retired BGen Gerald E. McIlmoyle, while not serving with the CIA, did fly the U-2 for the Air Force. His book of collected memoirs quoted many times herein, provides for readers the best source of experiences and personal stories of U-2 pilots and crews.

Herbert I. Shingler, III, son of Col Herbert Shingler, provided considerably "never before told" information about his father, who served with the 4070th Support Wing as Director of Materiel and later as the Wing Commander, stationed at March Air Force Base, California. James L. Gibbs likewise provided a treasure trove of declassified information about his father, BGen Jack Gibbs, who served as deputy to the CIA U-2 AQUATONE boss, Richard "Dick" Bissell. Jim Rose, son of CIA U-2 pilot Wilburn S. Rose, likewise contributed to this book.

Lastly, I acknowledge all contemporaries who served at Area 51. They include unnamed agency, military, and civilian workers whose participation is unknown or lost beneath the shroud of compartmentalized secrecy. Together, we undertook an endeavor more highly classified than the Manhattan Project that developed the atomic bomb. Many have remained unknown and destined never to receive the acknowledgment deserved.

Declassification

The **CIA Area 51 Chronicles** are based on the now-declassified information. Sources include *CIA, Directorate of Science and Technology (DST), **History of the Office of Special Activities (OSA) From Inception to 1969*** by Kleyla and O'Hern. Also, see **The CIA and Overhead Reconnaissance: The U-2 and OXCART Programs, 1954-1974**, by Pedlow and Welzenbach. The CIA declassified these sources from Top-secret/Sensitive Compartmented status in 2016 and 2013, respectively.

During 1997, the National Air Intelligence Center declassified the formerly top-secret *HAVE DOUGHNUT*, *HAVE DRILL*, and *HAVE FERRY* reports that the author references.

In 2013, the CIA acknowledged its role in the MiG exploitation projects with the release of 'the Area 51 file' titled by the National Security Archive as, "The CIA Declassifies Area 51." The National Security Archive credited the author in this release:

https://nsarchive.wordpress.com/2013/08/16/the-cia-declassifies-area-51/

1950	★		1966	★ Louis A. O'Jibway
1951	★ Jerome P. Ginley		1967	★ Walter L. Ray
1952	★ Norman A. Schwartz		1968	★ Billy Jack Johnson
	★ Robert C. Snoddy			★ Jack W. Weeks
1956	★ Wilburn S. Rose			★ Wayne J. McNulty
	★ William P. Boteler			★ Richard M. Sisk
	★ Frank G. Grace, Jr.		1970	★
	★ Howard Carey		1971	★ Paul C. Davis
1960	★ Chiyoki Ikeda			★ David L. Konzelman
1961	★ Leo F. Baker		1972	★ Wilbur Murray Greene
	★ Wade C. Gray			★ Raymond L. Seaborg
	★ Thomas W. Ray			★ John Peterson
	★ Riley W. Shamburger, Jr.			★ John W. Kearns
	★ Nels L. Benson		1974	★
1964	★ John G. Merriman		1975	★ William E. Bennett
1965	★			★ Richard S. Welch
	★		1976	★ James A. Rawlings
	★ Buster E. Edens			★ Tucker Gougelmann
	★ John W. Waltz		1978	★
	★ Edward Johnson			★
	★ Michael M. Deuel			
	★ Michael A. Maloney			

CIA Book of Honor page for 1950-1978
"They labored in the shadows to shine the light on our adversaries."
– CIA Deputy Director Joan Dempsey

Foreword

Pilots Wilburn S. Rose, Frank G. Grace, and Howard Carey made the ultimate sacrifice for which the CIA carved stars into the marble of the CIA Memorial Wall. This wall stands as a silent, simple memorial to certain of those employees "who gave their lives in the service of their country." Their names became what are now the 6th, 7th, and 8th inscribed in the CIA Book of Honor.

Rose, Grace, and Carey each died in the crash of his U-2 aircraft during 1956. The CIA would later carve memorial stars and inscribe the names of Buster Eugene Edens, Walter L. Ray, and Jack W. Weeks, lost in 1965, 1967, and 1968 respectively, as their U-2 and A-12 aircraft suffered crashes. This book will work to bring you the story of these brave pilots and many other persons who heralded the Central Intelligence Agency into the world of overhead reconnaissance.

In the years to follow, the CIA added more stars to the wall, so that as of this writing in 2018, there are 129 stars. As of 2018, the Book of Honor has inscribed the names of 91 of these men and women who gave their lives. The names of the 38 other covert officers remain secret, even in death—represented only by a gold star followed by a blank space.

One can never overstate the risks assumed by these dedicated individuals or the sacrifices of their families. Of the brave men who piloted the U-2 and the A-12 at Groom Lake, each time one climbed into the cockpit, he headed to the very edge of the technical horizon—a place of known and unknown yet palpable danger.

All served in the CIA's Directorate of Plans, now known as the Directorate of Operations, the clandestine arm of the CIA and the national authority for the coordination, deconfliction, and evaluation of clandestine operations across the intelligence community of the United States.

These losses of America's elite occurred as the CIA protected the United States from an attack by communist Russia who had already killed more than 100 American military airmen flying ferret flights to see what the Soviet Union was up to behind its closed borders.

Like many of the military pilots lost earlier in the post-WWII reconnaissance missions flown in modified fighters and bombers, Wilburn Rose left behind a young wife, a son, and two daughters, the youngest 18 months of age. Frank Grace similarly left four children when he lost his life in 1956. All the families and loved ones felt the losses deeply.

These pilots were not the only casualties at Groom Lake facility called Watertown during the U-2 AQUATONE and A-12 OXCART projects at Groom Lake, now known as Area 51. On November 17, 1955, Skymaster, 44-9068 crashed while en route to Groom Lake, killing five military personnel and nine civilians—five CIA, two Hycon, and two Lockheed. Two F-101 pilots would die, one flying chase for the A-12 and one conducting a weather flight.

This book, *The Angels*, is the first book in *The CIA Area 51 Chronicles* series. It is not about the mythical Area 51 publicized by conspiracy theorists around the world. Instead, it is an insider's account of the recently declassified legacy of how the CIA became the world's leader in secret military aviation technology and aeronautical engineering. Because it is about military aircraft, logically, it should recognize the efforts of the United States Air Force, Navy, Marine Corps, Coast Guard, or the National Guard. However, while this book references these military services, it is primarily about the CIA's creation of a flight-test facility that in 1959 became known as Area 51, a venue for testing secret high-flying spy planes, development of stealth technology, and the exploitation of enemy assets. The recent declassification of CIA documents now allows the author, an Area 51 veteran, to answer the questions Who, What, Where, When, and Why about Agency activities there.

This first volume tells about the genesis of the CIA, the politics involved, and why it became engaged in aerial reconnaissance. It answers many of the above-stated questions as revealed in the March 2016 declassification by the CIA of the report titled: *CIA, Directorate of Science and Technology (DST), "History of the Office of Special Activities (OSA) From Inception to 1969,"* by Helen Kleyla and Robert O'Hern.

Introduction

The Need for Aerial Reconnaissance

Aerial reconnaissance originated with manned tethered balloons as platforms for gathering information during the French conflict with Austria in 1794. During the American Civil War of the 1860s, balloon-borne observers employed aerial photography for the first time. Twenty years later, aerial reconnaissance techniques expanded to include tethered kites and pigeons carrying small cameras with timers. A century later, armies lofted cameras into the sky using artillery projectiles and rockets.

The invention of the airplane, however, truly revolutionized both warfare and intelligence collection. During the war between the Kingdom of Italy and the Ottoman Empire in 1911, the Italian Air Force employed airplanes for scouting over enemy territory. The first such mission involved an Italian overflight of the Ottoman army in Thessaly, Greece. That same day, German mercenaries on the Thracian front flew a similar mission against the Bulgarians. A few days later, a Bulgarian aircraft performed one of Europe's first combat reconnaissance missions over the Turkish lines on the Balkan Peninsula. It didn't take long for both sides to recognize the airplane as a weapon. An Italian scout pilot became the first to use a plane as a platform for aerial bombardment when he tossed four grenades from the open cockpit of his monoplane.

The British made the significant discovery of stereoscopic photography in 1912. The Royal Air Force (RAF) then flew the dirigible Beta on a reconnaissance mission, taking vertical photos with a 60% overlap. Analysts found that this created a three-dimensional effect when viewed through a stereoscope. This advancement eventually inspired the RAF to form the first aerial reconnaissance unit of fixed-wing aircraft. Germany already used airborne cameras for scouting missions, and France soon followed with several squadrons of observation planes.

These developments proved significant during World War I, which devastated Europe from July 1914 to November 1918. German and British combat strategists alike used aerial photography for strategic targeting, tactical planning, and battle damage assessment. General Werner von Fritsch, commander-in-chief of the German forces, felt an army's knowledge of the enemy's forces as its best weapon. He said that the nation with the best aerial reconnaissance capabilities would win any future wars. The use of aerial photography matured rapidly during World War I with aircraft carrying cameras to record enemy movements and defenses. By 1918, both sides were photographing the entire front twice a day. Overall, the combatants took more than half a million photos during the conflict.

The United States formalized aerial reconnaissance during the Punitive Expedition into Mexico in 1916 when the US Army formed the 1st Aero Squadron. The unit's primary mission involved communications and observation that enabled Gen John J. Pershing to keep in touch with his thinly spread fast-moving troops during the first phase of the Punitive Expedition. The squadron also performed several reconnaissance missions but failed to locate the enemy forces. As an interesting side note, half a century later, the 1st Reconnaissance Squadron of the United States Air Force—a direct descendant of the 1st Aero Squadron—flew the Lockheed SR-71, the most advanced reconnaissance aircraft of the twentieth century.

Developments during World War II, which raged across the globe from 1939 to 1945, included aircraft that could fly higher and faster than their predecessors. But the most significant advances came during the post-war era that became known as the Cold War.

The surprise attacks of Nazi Germany against its West European and its Russian neighbors had shocked the invaded nations into reality. Japan's attack on Pearl Harbor served a similar lesson. The need for accurate information about one's opponents became even more acute during the Cold War. Nuclear brinksmanship and other factors raised higher the stakes and tension, as we will see.

Nuclear Brinksmanship

Consider the tension in Germany starting from late 1958, when the Soviets gave an ultimatum that American, British, and French forces must withdraw from Berlin within six months. The Western powers resisted this ultimatum, and both sides continued to test nuclear weapons and add them to their large arsenals. In his speeches of August 7 and 11, Khrushchev conjured up the specter of nuclear war.

On 12 August 1961, East Berlin ordered the border closed, and a Wall erected to stop the flood of East Germans escaping to the West. An estimated 3 to 4 million escaped, with the daily number rising after Khrushchev's threatening speeches.

On 13-14 August 1961, East German police and military sealed off all arteries leading to West Berlin. They pulled up train tracks and roads and erected barriers topped with barbed wire, completely isolating the Western sector. Also, during August, the Soviets moved three army divisions closer to Berlin. President Kennedy called up 148,000 US servicemen to duty. By the week of 25-31 October 1961, both sides brought up tanks to the Berlin Wall.

During 27-28 October, Soviet T-55 and American M-48 tanks faced each other at 100 yards (meters) across border checkpoints. Loaded with live ammunition, they had orders to fire if fired upon. As if to add an exclamation point, on 30 October 1961, the Soviets exploded the Tsar Bomb nuclear bomb, the largest ever built.

Dropped over uninhabited regions in the Soviet Arctic, the Tsar Bomb broke windows in Norway and Finland. Although detonated over 2½ miles above ground, the shock wave peaked around magnitude five and measured circling the crust of the earth three times by seismic sensors. The mushroom cloud went seven times higher than Mt. Everest. Analysis indicated that the heat released from this explosion would have been able to cause third-degree burns 100 km (60 miles) away.

However, these tensions of the times escalated higher—so that in less than a year, again Kennedy and Khrushchev faced off in Cuba, using offensive missiles and naval forces. In the cartoon, notice the fingers of governmental authority poised over the nuclear buttons. During the acute Cuban missile crisis of October 1962, people felt particularly at the brink of nuclear war and 'mutually assured destruction.'

By the time of these peaks of Cold War tension, the norm had come to involve propaganda battles, political struggles, KGB and CIA intrigues and armed contests in parts of Africa, Asia, and Latin America, and full-blown proxy wars. Berlin and Cuba were only the latest crisis areas 'blowing up' as the superpowers pushed and shoved with each other. How did such a Cold War develop?

Why the CIA came to be the US agency responsible for overhead reconnaissance goes back at least to WWII and its aftermath: "The shock of the 1941 Japanese attack on Pearl Harbor and the horrors of the war that followed were still fresh in the minds of American leaders. When the Soviet Union exploded its atomic bomb during September 1949, the sense of vulnerability increased, with the realization that the next surprise attack could destroy American cities and kill millions." - Suhler, pp. 1-2

Chapter 1 - Why the Need for Intelligence

The year 1947 found the battlefields silent at last and the world recovering from World War II, with over 400,000 US servicemen and women killed. Some estimate as many as 600,000 US WWII service deaths.

The devastating conflict left the world in a state of political and military tension between the dominant powers in the Western and the Eastern Blocs. Thus, the United States remained in a new kind of conflict, a war of ideals, culture, power plays, and espionage called the Cold War, called so because of no military forces actively engaging in battle. The aftermath of World War II found the United States facing the most powerful adversary ever:

"On VE Day in 1945, Soviet armies occupied Poland, Czechoslovakia, Hungary, Bulgaria, Romania, and a third of Germany, including the eastern half of Hitler's capital, Berlin. By the time of Japan's surrender on August 14, 1945, Moscow had added a military presence in Manchuria, the northern islands of Japan, and the northern part of Korea. But it all had come at great cost. No military victor had ever suffered as much as the Soviet Union did in World War II. The Nazis destroyed an estimated seventeen hundred Soviet towns and seventy thousand villages, in all about a third of the wealth of the Soviet Union. The human toll remains beyond reckoning. Battlefield deaths, twice those suffered by Nazi Germany, reached about seven million. The civilian loss was even higher, estimated somewhere between seventeen and twenty million people."

"Yet what might have turned other civilizations inward seemed to propel the Soviet Union onto the world stage. Within five years of the end of World War II, Russia would detonate its atomic device, threaten its neighbors, assist a Communist regime's rise to power in China, and participate in North Korea's invasion of South Korea. This spectacular case of imperial stretch inspired fear and concern that not only was the Soviet Union's political influence worldwide, but its military ambitions were boundless." - Fursenko & Naftali, pp. 5-8

Three years later, from 1947, the United States again found itself in a shooting war, a domestic conflict between North and South Korea that became a proxy war between superpowers in the East and West. The United Nations, with the US providing the principal force, came to the aid of South Korea. The Communist powers, which included the People's Republic of China and the Soviet Union, provided military assistance to the North.

President Harry S. Truman attempted to assuage the concerns of the war-weary American public by describing this undeclared war as an international "police action." Nonetheless, those who fought considered it a war, a bloody war that ended in a stalemate (without a formalized peace treaty) during July 1953, with some 36,574 Americans killed, 7,984 missing in action, 4,714 of them listed as prisoners of war, many believed taken to the Russian-dominated Soviet Union.

To a much greater extent than Communist China, the Soviet Union became the most significant threat to Western democracy and an existential threat to the United States. Time after time, the Soviets opposed the US, either as a proxy enemy combatant or as an adversary at the United Nations. Moreover, the Soviet Union, an ideological adversary, became a powerful military opponent with a nuclear arsenal.

US officials feared and prepared for a preemptive nuclear attack preceding a possible invasion. The Truman administration, therefore, established a nationwide emergency radio broadcast alert system to warn US citizens of a Soviet invasion, should it occur. In the event of an attack, after transmitting an alert message, all radio communication transmissions stopped except for the two designated low power AM frequencies (640 and 1240 kHz). This radio silence prevented enemy planes from using radio transmitters as navigation aids for direction finding.

From a 1952 news article: *"President Truman made a bid Saturday for more volunteers to man 6000 lookout stations in 27 states which will go on a round-the-clock watch Monday against air invaders, saying the project is 'a commonsense precaution in which Americans can serve proudly.'"* Observing posts later totaled 16,000, involving 600,000 civilian volunteers from ages 7 to 86 years old, working in shifts.

Air raid sirens sprouted up in cities and towns nationwide, and the warbling sounds of Civil Defense klaxons heralded monthly drills. The US established the National Emergency Alarm Repeater (NEAR) program to supplement the existing siren warning systems and radio broadcast in the event of a nuclear attack. American schools conducted "duck and cover" drills where students took cover beneath their school desks with instructions not to look at the blinding fireball.

High-ranking government officials had nuclear-proof bunkers in which to retreat. The military built its crucial military facilities deep inside mountains and sealed behind thick steel blast-proof doors. Everywhere citizens knew to look for the ominous placarded yellow and black trefoil warning signs that identified fallout shelters stocked with food and water.

A 1955 US newsreel about the Operation Alert air raid drill noted that President Eisenhower headed for "a secret retreat, the first time the government has abandoned the Capitol since it burned in the War of 1812." Thus, Americans prepared for nuclear attack or invasion by the Soviet Union.

In 1955, the United States also became embroiled in a military conflict in Southeast Asia. As with Korea, it started as a civil war in 1950. This time the area was Vietnam, then known as French Indochina. American involvement began with providing military advisors and support to the French against communist forces in North Vietnam. A decisive defeat in 1954 effectively ended a century of French colonial rule and split Vietnam in two.

In many ways, the civil war resembled the Korean conflict, wherein the North Vietnamese government and the Viet Cong (also known as the National Liberation Front) fought to reunify their country. Also, like what happened in Korea, it immediately became another Cold War proxy conflict between the superpowers. The US government viewed its involvement in the war—to prevent a communist takeover of South Vietnam—as part of a more comprehensive containment policy aimed at stopping the spread of communism by the Soviet Union, China, and other communist allies who did so by providing advisors and material support to the North. The war continued to escalate throughout the 1960s and ended with a North Vietnamese victory during April 1975. By that time, casualties included 58,315 Americans dead, 303,644 wounded, and more than 1,600 listed as missing in action. Estimates of Vietnamese military and civilian casualties vary from 966,000 to 3.1 million, plus thousands more in neighboring Cambodia and Laos.

As the Cold War heated up, it became increasingly vital for US military leaders to understand their adversaries' capabilities. The closed societies of the Communist nations made it nearly impossible to collect intelligence at ground level. Therefore, it became necessary to observe them from above, which the US Air Force and Navy attempted by flying reconnaissance missions over the Soviet Union only to lose many pilots and crews to Russian antiaircraft guns. President Dwight Eisenhower wanted to develop a spy plane that the Soviet Union could not shoot down. However, his Air Force chief of staff, Gen Curtis LeMay,

refused to support the procurement of an aircraft that did not shoot guns or drop bombs. LeMay also jealously protected his turf by opposing any agency that wished to conduct such operations other than the Air Force. Consequently, President Eisenhower saw no choice but to turn to the recently formed CIA to build such a plane secretly. The mere thought of the CIA's replacing the Air Force and Navy for such missions sparked a political battle in Washington that continued through 12 directors of Central Intelligence and in the Oval Office of six US presidents.

Why the OSS, Office of Strategic Services

Truman wasn't the first American president who sought to establish an intelligence agency. During July 1941, Franklin D. Roosevelt appointed William J. "Wild Bill" Donovan, a

World War I veteran, lawyer, intelligence officer, and diplomat as his Coordinator of Information (COI). As such an officer, Donovan headed the nation's first peacetime, non-departmental intelligence organization. However, this appointment occurred too late for Donovan to develop the necessary knowledge to prevent the Japanese sneak attack on Pearl Harbor that drew the United States into World War II.

The surprise attack by Japan prompted Roosevelt to reevaluate the role of the Coordinator of Information. In 1942, he placed the COI under the aegis of the Joint Chiefs of Staff and renamed the Office of Strategic Services (OSS, jokingly called the "Oh So Social."). Military services, Department of State, Federal Bureau of Investigation (FBI), and other agencies with separate intelligence units jealously guarded their turf. The president, therefore, tasked OSS specifically with collecting and analyzing strategic information required by the joint chiefs and as restricted to conduct special operations not assigned to other agencies.

One such operation involved the collection of radar signal data. In 1943, the crew of a specially equipped B-24 overflying Japanese-held Kiska Island in the Aleutians intercepted emanations from radar sites on the island. Throughout the war, such missions, known as "ferret" flights, enabled the OSS to supply policymakers with the intelligence needed to aid military campaigns.

Although the president dissolved the OSS immediately following the war, the fledgling agency helped shape the US intelligence hierarchy and created a blueprint for sharing authority over foreign intelligence activities with the FBI. Meanwhile, the president left it to the military branches to conduct intelligence operations within their areas of responsibility.

Why the Cold War

World War II left vast areas of Europe and Asia in ruins. Nazi Germany surrendered to the Allied powers (American, British, French, and Russian) during May 1945. Still, the Empire of Japan continued fighting until the middle of August, only surrendering after the Soviets invaded Japanese Manchuria, and US atomic bombs destroyed two Japanese cities. The war had

William J. "Wild Bill" Donovan
- The father of American intelligence and director of the Office of Strategic Services (OSS), the precursor to the Central Intelligence Agency, was an American soldier, lawyer, intelligence officer, and diplomat.

reduced the world population of 2 billion by 4 percent, with the deaths of 80 million people. The shifting balance of power resulted in new borders. During October 1945, a coalition of governments established the United Nations (UN), tasked with promoting international cooperation in the hope of avoiding future global conflicts.

Winning peace proved very evasive to achieve as Europe quickly divided into regions of Western and Soviet domination. Germany split in two, with East Germany going to the Communists. The German capital, Berlin, divided into four parts with the Americans taking control of the Tempelhof sector, the British taking the Gatow, the French the Tegel, and the Soviets the Karlshorst sector of the occupation zone. To complicate matters, Berlin lay deep within East Germany, and thus it became a centerpiece in the new Cold War, where strained relations existed even before the end of World War II.

Truman succeeded President Roosevelt, who died of a stroke during April 1945 and immediately alienated Soviet foreign minister Vyacheslav Molotov by announcing his intent to take a "tougher" stance with the Soviets. After the war, the Soviet Union annexed the Sakhalin and the Kuril Islands and sought to expand its territory further. Tensions increased as the Soviets invaded Poland and Hungary and sought revenge for millions of Russians killed during the massive German military operations of World War II.

Why and What the Fear of Russia

Geography always played a role in Russian history. For eight centuries, Russia had endured a cycle of chaos that saw it rise as an aggressive global power, only to collapse into a coalition of independent states, and then rise again. This cycle occurred more about geographic constraints than political ideology.

Geographically speaking, Russia suffered from an inherently weak position. It, the largest country in the world, covered 13 time zones (split now into four mega-zones). Some 75 percent of its landmass consisted of uninhabitable tundra that became marshland in the summer, making domestic trade difficult. Russia found maritime trade difficult given that its rivals, including Turkey, blocked its only warm-water port on the Black Sea. Thus, the nation struggled to develop economically.

Furthermore, Russia's "heartland" stretched from St. Petersburg south through Moscow and into the Volga region. It lay on a series of open plains that made it vulnerable from all sides, forcing Russia to expand its borders and influence outward to create a buffer zone between its heartland and rival regional powers. As Catherine the Great famously put it: "I have no way to defend my borders except to extend them." The most prolonged example of this expansion occurred during the Soviet era, which continued from 1922 to 1991. During this time, the cold wastes of Siberia to the east shielded the Russian heartland, 14 other Soviet republics, and seven Eastern Bloc countries.

Thus, Russia existed in a perpetual cycle divided into three parts: collapse, resurrection, and fragility. It started with a catalyst that caused governance to break down and disrupt the social order, leading to collapse. Historically, this catalyst took many shapes. The 13th century brought the Mongol invasion; in the seventeenth century, the Time of Troubles; and in the twentieth century, the Russian Revolution. From each collapse came the next stage of Russia's cycle: resurrection.

Typically, the predominant system of government during each crisis transformed into something new—usually with a strong personality at the fore. Such transformations tended to create a stable system in which Russia could consolidate itself and its borderlands, fostering a sense of national identity, uniting Russians and allied populations under a common patriotic philosophy.

Ivan III represents an excellent example. He threw off the Mongol yoke and united Russia. Another, the first Romanov Tsar (or Czar), Mikhail I, led Russia out of the civil wars of the sixteenth century. The "Greats," Peter and Catherine, transformed Russia into a global empire, and communist revolutionary Lenin turned Russia into what the world knew as the Soviet Union.

None, however, overcame Russia's geographic challenges. The problematic pattern of trying to consolidate the heartland while expanding Russian influence practically ensured an eventual collapse. As inevitable stress points emerged—political, social, national security, and economic—Moscow tightened its grip and acted more aggressively within and along its border.

Those leaders who survived the turmoil, once seen as the saviors of Russia, evolved into more authoritarian and often ruthless dictators who quashed all dissent while assertively defending the country's outlying areas and borders. A fragile situation developed when the Russian rulers found they lacked the stability their predecessors enjoyed and that they had less time to devote to consolidation and nation-building. Brutal leadership often emerged from these crumbling systems.

These authoritarian despots used extreme measures in times of crisis. Droughts and famines followed Ivan III's and his successor's successful tenures. Ivan IV—aka The Terrible—severely restricted freedom of movement and lashed out in a series of wars against the Polish-Lithuanian Commonwealth.

The most famous, or perhaps infamous, autocrat, Joseph Stalin, led the Soviet Union from 1929 to 1953, transforming the Soviet Union from a peasant society into an industrial and military superpower. He ruled by terror, unfortunately, and millions of his citizens died during his brutal reign. Stalin sent any potential enemies to forced labor or death.

Stalin's policy of forced industrialization led to

one of the most significant human-made famines in history. Stalin suppressed all domestic dissent and anything that smacked of Western influence. He sought to promote Soviet control across the globe by helping establish communist governments throughout Eastern Europe. He directed Soviet scientists to develop nuclear weapons and long-range rockets, so the US would not have a monopoly on the super-weapons and the means to deliver them, plus he oversaw an aggressive campaign of foreign espionage. Stalin's leadership set the stage for the Cold War.

What and Who Ended World War II

During the Potsdam Conference, held at Cecilienhof, occupied Germany, during July 1945, President Truman informed Stalin that the US possessed a new weapon, one more powerful than any other. Though this development shifted the balance of power and posed a potential threat to the Soviet Union, Stalin professed to be unimpressed. The Soviet leader already knew about it because he had spies in the US atomic program.

Stalin nonetheless honored his agreement to declare war on Japan. Russian forces turned around from the European Theater and headed for the Far East to invade Manchuria, also swiftly taking possession of the Southern Sakhalin and the Kuril Islands. Russian influence and expansion continued with the installation of a Communist government in Poland and the annexation of several occupied nations as Soviet Socialist Republics and others in Eastern Europe converting into Soviet-controlled satellite states.

Japan had ruled Korea since 1910. An agreement between the US and Soviets on the surrender of Japanese forces in Korea left the peninsula split in two at the 30th parallel. The North found itself under Soviet occupation while US forces occupied the South. The subsequent inability of the two superpowers to agree on terms for Korean independence led to Korea's division into two political entities. North Korea (formally the Democratic People's Republic of Korea) received backing from the Soviet Union in opposition to the pro-Western government of South Korea (formally the Republic of Korea). Tensions between the two resulted in the outbreak of the Korean war and had repercussions that continue well into the twenty-first century.

Russian intervention in Manchuria, coupled with Truman's decision to authorize the use of atomic weapons, brought a swift Japanese unconditional surrender that ended World War II.

Vietnam a Supporting Ally of the Office of Strategic Services

On July 16, 1945, a seven-man OSS advance team parachuted into a jungle camp called Tan Trao near Hanoi, the capital of Vietnam. They met with Ho Chi Minh, leader of the Viet Minh independence movement, for organizing a guerrilla force of 50 to 100 men to attack the railroad from Hanoi to Lang Son to prevent Japanese troops from entering China. They also received the tasks of locating Japanese military bases and depots and sending intelligence reports to OSS agents in China.

On 22 August 1945, OSS agents Archimedes Patti and Carleton B. Swift, Jr., were in Hanoi, Vietnam, on a mercy mission to liberate allied POWs. The Japanese detained upon arrival, Jean Sainteny, a French government official who accompanied the team. Although formally surrendered, the Japanese forces in Vietnam remained the only local authority capable of maintaining law and order. Thus, even with the war ended, the Imperial Japanese military remained in power to some degree. They detained French troops and Sainteny until the Viet Minh took control. In what became known as the "August Revolution," the Viet Minh toppled the French colonial government.

Within two weeks, the Viet Minh seized control of most villages and cities throughout the country, including Hanoi. On September 2, Ho announced the formation of the Provisional Democratic Republic of Vietnam. In the more politically diverse southern part of the country, various groups vied for control. This divisiveness allowed the French to push back and regain control in South Vietnam. By March 1946, with the country primarily split in half, the French signed an accord they never intended to honor. Relations began to worsen almost immediately, and Vietnam succumbed to a full-scale guerrilla war against the French colonialists. The resulting

First Indochina War lasted until July 1954, with the United States supporting France and paying 80% of their cost of the war. One should note that before the end of World War II, President Roosevelt opposed the French returning to resume the colonial rule of Indochina. President Truman, however, did not resist the French.

Special Operations Team # 13, code-named "Deer" consisted of Maj Allison Thomas (team leader), 1st Sgt William Zielski (radio operator), PFC Henry Prunier (interpreter), Lt. Rene Defourneaux (assistant leader, a.k.a. Raymond Douglas), SSgt Lawrence R. Vogt (weapons instructor), Sgt Aaron Squire (photographer), and PFC Paul Hoagland (medic). – Via Wikipedia

With the surrender of Japan, Vietnam, for the first time in 2,000 years, became free of occupation, if only briefly. Following Roosevelt's death, Ho Chi Minh sought military aid from President Truman to stop the French from returning to rule his country. Truman ignored this request, sending Ho only a set of dueling pistols and a small number of Colt .45 semi-automatic pistols instead to make a symbolic show of US support.

An Opportunity Lost to Prevent the Vietnam War

The withdrawal of OSS teams from Vietnam presaged a lost opportunity for the United States. Ho Chi Minh viewed this and Truman's pistol gift as an insult and proof that America had no intention of helping his homeland. Desperate to keep Vietnam free of occupation, Ho declared the independent Democratic Republic of Vietnam and sought the assistance of Communist Russia to keep the French out. Thus, North Vietnam emerged from World War II as a communist and nationalist liberation movement under the now Marxist–Leninist leader Ho Chi Minh.

Demobilization of the Office of Strategic Services

For most Americans, the end of World War II offered hope that the United States would return to a climate of normalcy and peace. For the Truman administration, this included demobilizing wartime agencies such as the OSS. William Donovan's political and military rivals feared the man they called "Wild Bill," and worried about the possible consequences of creating a peacetime intelligence service modeled on the OSS.

But the US and its allied leaders had much more to fear than Donovan or to protect their turf. They quickly learned that the Soviet Union had international ambitions that would likely plunge the world into more wars.

Before the surrender of Japan, Soviet amphibious forces landed in Korea. Hours before the Nagasaki bombing, the Soviets invaded China. Soon they captured Changchun, the capital of Manchukuo (Manchuria), and only five days later won the Sakhalin Island's capital, thus bringing Communism to the Far East.

The Cold War Began with No US Intelligence Capability

Shortly after accepting the Japanese surrender, the US became embroiled in the conflict between France and Vietnam as the government of the French Republic's Far East Expeditionary Corps attempted to resume its pre-war colonialism in Indochina.

Unfortunately for the US and its allies, President Truman had signed an Executive Order terminating the OSS, effective 1 October 1945. Consequently, the US government chose to support the French in Vietnam while once again weakening American intelligence-gathering capabilities. Truman had repeated the same error that encouraged the attack on Pearl Harbor. The president had, in effect, eliminated the one intelligence service that could best advise him and the nation of an impending enemy attack. Why?

Was it to prevent the OSS from producing intelligence to show that Japan had surrendered because of the Russians entering the war and not because of the US dropping the atomic bombs?

Was it to prevent the OSS gathering

information counter to the stated reasons for the US propping up the French at the beginning of the war in Vietnam?

America's lack of intelligence information on activities in North Korea encouraged the invasion that sparked the Korean War. An American president choosing to abandon his best intelligence asset at such a crucial time seems unthinkable now, but this is what happened. This same lack of intelligence-gathering led the US to support France in its war in Vietnam.

Historically, organized intelligence, both overt and clandestine—at best considered a tolerable wartime necessity—became a thing viewed as unsavory in peacetime. Truman thought the OSS a provision without permanent status in the American governmental system. An organized intelligence agency would likely have advised Truman to stay out of Vietnam (as the CIA did later for President Lyndon Johnson, who hated the Agency and others, including the press, for not supporting his course of escalating war and anti-Communism in Southeast Asia).

Despite the Soviet Union continuing its conquest of other nations, Truman abolished all OSS analysis and collection. He transferred the OSS counterintelligence services to the Department of State and the War Department (later renamed the Department of Defense). Worse yet, Truman reduced the scale of intelligence collection, as if he thought intelligence-gathering caused wars.

While the previously stated reasons for disbanding the OSS are speculation, disbanding that office is known to have eliminated the central sources of Cold War intelligence needed against the powerful communist Soviet Union. Unfortunately, the disabling of the nation's intelligence agency did not stop the Soviet threat to the US or its allies. If anything, Truman's shutdown of the country's intelligence service showed weakness that emboldened the Soviets.

Truman soon realized that ending intelligence collection activities did not eliminate the spread of communism. He reluctantly realized that the nation needed a peacetime centralized intelligence harvesting system to keep the peace. This time, he and other national leaders appreciated the need to have an agency independent of any of the policy-making branches of government.

By 1947, two years after the end of World War II, many nations were making significant changes throughout the world. In the Republic of Korea, Syngman Rhee became the president following a left-wing boycott, and in Malaya, a new communist insurgency began against the British Commonwealth.

The Americans and British united their zones of control in Germany to form the Bizone, also known as Bizonia. Rigged elections in the United Nations gave the Communist bloc 80 percent of the vote. At the same time, Truman fueled the political and military tension between the US and the Soviets by announcing the Truman Doctrine that explicitly opposed the Soviet Union's expansion of communism. The Doctrine provided aid to Greece and Turkey to prevent them from falling into the Soviet sphere of influence and further pledged the support of all nations threatened by Soviet expansionism. This difference in policy brought the Cold War to a head as the Soviets consolidated control over the Eastern Bloc states. The US countered this with a strategy of global containment to challenge Soviet power by extending military and financial aid to the countries of Western Europe.

The creation of the Atlantic Pact—the North Atlantic Treaty Organization (NATO) —led by supreme allied commander Eisenhower, replaced previous temporary wartime alliances against Nazi Germany. Based in Brussels, Belgium, NATO became an intergovernmental military alliance based on a treaty signed on 4 April 1949. The coalition left the two superpowers with profound economic and political differences.

Neither side faltered. The US continued extending its military and financial aid to anti-communist governments, and the Soviet Union kept spreading its aggressive Communism.

The world continued to experience radical change as Europe slowly recovered from World War II, and the Cold War escalated. The two sides became locked in a stalemate as the Communist Party took control of Czechoslovakia. At this point, the Cold War became more of a contest of wills than an actual conflict between the two superpowers. However, the threat of nuclear weapons continued to underpin the mutual suspicions and reactionary foreign policy of both sides.

This threat prompted Secretary of State George C. Marshall to outline plans for a comprehensive program of economic assistance to the war-ravaged countries of Western Europe. Truman signed the Marshall Plan into effect, which eventually cost the US government $12.4 billion. The Soviets refused an offer to participate in the Plan and barred Eastern Bloc countries such as East Germany and Poland from reaping the benefits.

The Cold War heated up the following month with the defection of Igor Gouzenko, a clerk working at the Soviet embassy in Ottawa, Canada, who provided proof to the Royal Canadian Mounted Police of a Soviet spy ring operating in Canada and other western countries. If any still doubted that the Soviets were more foe than an ally to the West, the Gouzenko affair changed their perceptions. Germany, the city of Berlin, still suffered from the enormous damage sustained during the war. It found its prewar population of 4.3 million people reduced to just 2.8 million.

"How to Close the Gap?"

Adding to their woe, Soviet Premier Stalin ordered a blockade of all land routes from West Germany to Berlin to starve out the French, British, and Americans in the city. Stalin wanted it all.

The Soviets launched a massive propaganda campaign condemning Britain, America, and France by radio, newspaper, and loudspeaker.

HOW TO CLOSE THE GAP

They severed land and water connections between non-Soviet zones and Berlin. At the same time, Russia halted all rail and barge traffic to and from the city, stopping the supply of food to the civilian population in sectors other than the Soviet sector. The German people were beginning to starve, and the three Western powers responded by launching a massive resupply effort by air known as the Berlin airlift.

During the Berlin airlift, an Allied supply plane took off or landed in West Berlin every 30 seconds. At the beginning of the operation, the air fleet delivered about 5,000 tons of supplies to West Berlin every day. The planes made nearly 300,000 flights in all. For more than a year, hundreds of American, British, and French cargo planes flew in provisions from Western Europe. By the end, those daily loads had increased to about 8,000 tons. The Allies carried about 2.3 million tons of cargo in all.

The eyes of the world watched throughout the tense blockade. The West finally celebrated as the airlift defeated the Soviet's attempt to starve out West Berlin. The Soviet blockade ended with the re-opening of access routes as of 12 May 1949. However, the US continued the airlift until September for fear of the Soviets re-establishing the embargo. Germany remained split with the Bizone merging with the French zone of control to form the Federal Republic of Germany with Bonn as its capital.

While the US announced new occupation policies in Germany, India, and Pakistan gained independence from the United Kingdom. A month later, the Soviet Union formed the Communist Information Bureau to dictate the actions of leaders and communist parties across its spheres of influence. Also, Communist Russia remained entrenched in the Far East.

The UN responded to the Communist Information Bureau's actions with a resolution calling for the withdrawal of foreign soldiers from Korea, for free elections, and unification of the peninsula. The United States and its coalition removed their troops, but the Soviet Union did not. Why?

Chapter 2 – Why the CIA Established

With World War II over and the OSS disbanded, America found its knowledge of Soviet intentions severely hampered. The world saw Russia engaged in brutal military conquests against their neighboring countries and Communism spreading across the globe like cancer.

The American people feared Soviet Russia, expecting any day to see parachutes floating earthward carrying armed troops waving the red hammer-and-sickle flag of the Soviet Union. American communities conducted air defense drills where everyone ran for the nearest bomb shelter. School children hid beneath their desks. Released in 1950, school children from 1951-52 onward viewed the adult-oriented civil defense film *Survival Under Atomic Attack* and *Duck and Cover.* The Korean War was on, and the Soviet Union was engaged in atomic testing and weapons production.

Paranoia erupted when a private pilot named Kenneth Arnold reported seeing a fleet of nine unidentified flying objects, or UFOs, flying past Mt. Rainier, Washington, at speeds he estimated to be more than 1,200 miles per hour. Military officials and the FBI took seriously the numerous UFO sightings occurring across the country over the next several weeks. Many people wondered if these objects represented a new Soviet threat. Debris found on a ranch near Roswell, New Mexico, caused brief excitement when the Army Air Corps initially identified it as being from one of these mysterious objects. Over time, in popular culture, UFOs came associated with possible extraterrestrial visitors. Many science fiction films of the Cold War era were thinly veiled allegories for Soviet infiltration and invasion fears.

Truman sought advice from Bill Donovan and Admiral William D. Leahy, who helped hammer out details for the creation of a new intelligence agency. The president also asked for advice from the military services, the State Department, and the FBI. In January 1946, he established the Central Intelligence Group (CIG) at a critical time when full-scale civil war broke out in Asia between the Kuomintang-led Republic of China of the Communist Party of China.

The formation of the CIG provided strategic warnings to US defense analysts and conducted significant clandestine collection activities. Unlike the OSS, the CIG enjoyed access to all-source intelligence as it functioned under the direction of a National Intelligence Authority composed of a presidential representative and the secretaries of State, War, and the Navy. The president appointed the first deputy of Naval Intelligence, RAdm (Rear Admiral) Sidney W. Souers, USNR (United States Navy Reserve), as the first Director of Central Intelligence.

Why the CIA's Importance

Less than two years after its formation, the CIG struggled to keep up with the world's turmoil. The Soviets were racing to build their atomic bomb, citizens around the world claimed to have spotted flying saucers in the sky, and Truman feared the growing threat of a Soviet nuclear Pearl Harbor. On September 18, 1947, he signed the National Security Act of 1947, which established the US Air Force as a separate military service on equal footing with the Army and Navy and established the National Security Council. The Act also allowed Truman to create a civilian foreign intelligence service tasked with gathering, processing, and analyzing national security information from around the world, using both human and technical intelligence.

He disbanded the National Intelligence Authority and the CIG and created the Central Intelligence Agency (CIA). In 1949, the Central Intelligence Agency Act authorized the agency to use confidential fiscal and administrative procedures and exempted it from most limitations on the use of Federal funds. At the outset of the Korean conflict, the CIA had only a few thousand employees and lacked its intelligence-gathering capabilities. Instead, it relied heavily upon the Office of Reports and Estimates, which drew upon State Department telegrams, military dispatches, and other public documents.

The Truman Doctrine

Truman also created an American foreign policy doctrine to counter the Soviet geopolitical expansion. He initially announced the Truman

Doctrine to Congress on 12 March 1947. He developed it further during July 1948 when he pledged to contain Soviet threats to Greece and Turkey and to support other nations threatened by Soviet communism. The Truman Doctrine helped spur and promote the formation of NATO in 1949.

President Truman shouldered the CIA with the task of watching all threats worldwide, but with a focus on Russia. He tasked the CIA with determining what kinds of strategic weapons, and how many of them, the Soviet Union, had. He also wanted to know if and how the Soviets intended to use them.

The Communist Party ruled the Soviet Union, a Marxist–Leninist State with enforcement by the secret police, the dreaded KGB. Power lay in the hands of a dictator (such as Stalin) or a small committee called the Politburo. The party controlled the press, the military, the economy, and numerous other organizations. It also managed the client states in the Eastern Bloc. The Supreme Soviet gave funding to Communist parties around the world, sometimes in direct competition with similar efforts by the People's Republic of China.

In an election speech, Joseph Stalin stated that capitalism and imperialism made future wars inevitable. He accused the United States of "striving for world supremacy." The West, predominantly democratic and capitalist with a free press and independent organizations, stood in opposition to the Communists. A small neutral bloc of nations, treading a fine line in between, arose with the Nonaligned Movement that sought good relations with both sides. As the Korean proverb, "When the whales fight, the shrimp get squashed" (In Africa: "When the elephants fight, the mice suffer.")

The CIA Becomes Technical

Those in the intelligence business recognized the Soviet Union as a strict target for traditional espionage operations. Accordingly, the CIA turned to the technical means of collection to peer beyond what had come to be known as the "Iron Curtain."

During the first 15 years of its existence, the CIA primarily deployed collection and analytic assets to detect and preempt a Soviet nuclear first strike.

In the United States, the Red Scare ran rabid with Sen. Joseph McCarthy accuses numerous American celebrities of being members of the Communist Party. McCarthy also targeted alcoholics and those who he saw as sexual deviants.

He eventually became the chairman of the Government Committee on Operations of the Senate, widening his scope to "investigate" accused dissenters. He continued for over two years, relentlessly questioning numerous government personnel in various departments. This rampant witch-hunt dubbed "McCarthyism" caused panic and increased public fear of communism.

On 29 August 1949, the Soviet Union tested its first atomic bomb known to Americans as "Joe 1," named after Joseph Stalin. Suddenly, the Soviets ranked as the world's second nuclear power.

The Soviet representative to the United Nations followed this demonstration of power by vetoing UN membership for Ceylon, Finland, Iceland, Italy, Jordan, and Portugal.

The Cold War escalated on October 1, 1949, when Communist revolutionary Mao Zedong proclaimed the foundation of the People's Republic of China (PRC). The establishment of the populous single-party state added a quarter of the world to the communist camp. Additionally, Mao's forces had been holding the American consul in Mukden and his staff as virtual hostages for nearly a year during the revolution against the Nationalist Chinese government. The Chinese communists accused the Americans of being spies. By the time the crisis ended, US relations had suffered severe damage with the new Chinese Communist government to the extent the disastrous prospect of war had included conflict with the PRC as well as with the Soviet Union.

Elsewhere in the world, the Greek Civil War, which had been raging since 1946, finally ended. In what might legitimately be the first proxy war of the Cold War era, the Greek government (backed by the US and England through the Truman Doctrine and the Marshall Plan) fought the Democratic Army of Greece (supported by the Communist governments of Albania, Yugoslavia, and Bulgaria). Government forces ultimately defeated the communist insurgents, a major Cold War victory for the West.

Indonesia, a Southeast Asian nation made up of thousands of volcanic islands and home to hundreds of ethnic groups, had been a Dutch colony until it succumbed to the Japanese occupation during World War II. After the war, an armed and diplomatic struggle ensued when the Netherlands attempted to re-establish Dutch rule. During December 1949, international pressure forced the Dutch to recognize Indonesian independence under the control of Sukarno, the founding father and first president of the island nation. During the 1950s, however, Sukarno moved Indonesia away from democracy towards authoritarianism, maintaining his power base by balancing the opposing forces of his army and the Communist Party of Indonesia. All of these forced US intelligence agencies devote more resources to watching the region for any signs of Soviet influence.

The National Security Council first became secretly involved in the Italian elections. At the same time, the US extended $400 million of military aid to Greece and Turkey, signaling its intent to contain communism in the Mediterranean.

Cold War tensions also increased in the Middle East, where the British and Soviet troops had occupied Iran to secure oil fields during World War II. The British withdrew after the war, but the Soviets remained as Stalin expanded Soviet political influence in Azerbaijan and Kurdistan. Additionally, the Soviets were instrumental in establishing the communist Tudeh Party of Iran. Soviet-backed uprisings led to the establishment of the People's Republic of Azerbaijan and the Kurdish People's Republic.

In Southeast Asia, the Soviet Union had declared the Democratic People's Republic of Korea the legitimate government of the entire Korean peninsula and backed up Kim Il-sung as the rightful leader.

European borders and allegiances continued to shift. The Republic of Austria divided into four zones of control: American, British, French, and Soviet. Enver Hoxha established the People's Republic of Albania, with himself as Prime Minister.

In the West, the global spread of Soviet-style communism engendered fears like those inspired by the German Nazi movement of the late 1930s. Any notions of world peace faded as wartime alliances crumbled, giving way to suspicion and dread. The fallout with Russia came as no surprise to those who paid attention to world events. As early as 1946, former British Prime Minister Winston Churchill condemned the Soviet Union's policies in Europe and declared that "From Stettin in the Baltic to Trieste in the Adriatic, an iron curtain has descended across the continent." From that day forward, the term "Iron Curtain" became associated with the Communist Bloc; the West became known as the "Free World."

A decade after leaving office, Harry Truman wrote a letter to the *Washington Post* decrying the CIA's clandestine "cloak and dagger" activities. He claimed the Agency "was intended merely as a center for keeping the President informed on the world at large and the United States and its dependencies in particular." He added, "It should not be an agency to initiate policy or to act as a spy organization." But that's what it did precisely, and throughout the Cold War, the CIA undertook numerous covert intelligence operations in support of foreign policy objectives on an ever-broadening scale.

Clouds of World War III

For centuries, soldiers in wartime sought the highest ground or structure to get a better view of the enemy, using at first, tall trees, then church steeples and bell towers. By the time of the American Civil War and the Franco-Prussian War of 1870-71, observers used hot-air balloons to get up in the sky for a better view of the "other side of the hill."

With the advent of dry film, it became possible to carry cameras into the sky to record the disposition of enemy troops and emplacements. Indeed, photoreconnaissance proved so valuable during World War I that in 1938 Gen Werner von Fritsch, commander in chief of the German Army, predicted: "The nation with the best aerial reconnaissance would win the next war."

By World War II, lenses, films, and cameras had undergone many improvements. So had the airplane, which could fly higher and faster than the primitive craft of World War I. Now armed forces used photoreconnaissance to obtain information about potential targets before a bombing raid, then afterward to assess the effectiveness of the attack.

Peacetime applications of high-altitude photography at first included only photo mapping and surveying for transcontinental highways and mineral and oil exploration. Nations gave little thought to using photography for peacetime espionage until after World War II when the Iron Curtain rang down and cut off most forms of communication between the Soviet Bloc of nations and the rest of the world.

The year 1949 found the Soviet military engaged in planning, production, and deployment activities with the utmost secrecy. The Soviet Union concealed its strategic capabilities—bomber forces, ballistic missiles, submarine forces, and nuclear weapons plants. The Soviet air defense system, a prime concern in determining US retaliatory policies, was also an unknown subject.

Tight security along the Soviet border severely curtailed the movement of human intelligence sources. Also, the Soviet Union made its conventional means of communication— telephone, telegraph, and radiotelephone—much more secure, thereby significantly reducing the information available from these sources.

The stringent security measures imposed by the Communist Bloc nations blunted traditional methods for gathering information. Secret agents used covert means to communicate the intelligence they gathered through wiretaps, postal intercepts, traveler's reports, and other eavesdropping methods. The entire array of intelligence tradecraft seemed ineffective against the Soviet Bloc, and no other means were available.

Early Postwar Aerial Reconnaissance

At the end of World War II, the United States captured vast quantities of German reconnaissance photos and documentation on the Soviet Union. This material rapidly became outdated with the interrogation of prisoners of war returning from Soviet captivity, the primary source of current intelligence on Soviet military installations. The intelligence community obtained information about Soviet scientific progress. It established several programs to debrief German scientists taken to the Soviet Union after the end of the war and now allowed to leave.

Interrogation of returning Germans offered only incomplete fragments of information, and this source would not last much longer. Thus, in the late 1940s, the US Air Force and Navy began trying to obtain aerial photography of the Soviet Union. A recent study of this subject by Robert Hopkins III is *Spyflights and Overflights: US Strategic Aerial Reconnaissance 1945-1960*.

When President Truman authorized sensitive intelligence (SENSINT) flights, the Air Force's initial effort involved Boeing RB-47 aircraft. The reconnaissance version of the B-47 jet-propelled medium bomber) equipped with cameras and electronic "ferret" equipment that enabled aircrews to detect tracking by Soviet radars, aerial refueling enabled long-range missions.

At that time, the Soviet Union had not yet wholly ringed its borders with long-range tracking radars, and much of the interior also lacked radar coverage. Thus, when the RB-47 crews found a gap in the air defense warning network, they would dart inland to take photographs of some of the critical open targets.

These "penetration photography" SENSINT flight missions occurred along the northern and Pacific coasts of Russia. In 1952 an RB-47 aircraft carried out a photographic mission inland across the north USSR (See "A Daytime Overflight of Soviet Siberia," - Bibliography, Declassified NRO Documents.) Such intrusions sometimes brought protests from Moscow.

Around this time, Soviet policy began undergoing notable change, with air defense units showing much more aggression in defending their airspace, attacking all aircraft that came near the border of the Soviet Union. On 8 April 1950, Soviet fighters shot down a US Navy Privateer patrol aircraft over the Baltic Sea.

Following the outbreak of the Korean War, the Soviet Union extended its "severe air defense policy" to the Far East, where, in the autumn of 1951, Soviet aircraft downed a twin-engine US Navy Neptune bomber near Vladivostok. The US suspected. Soviet fighters' involvement in bringing down an Air Force RB-29 lost in the Sea of Japan on 13 June 1952.

Not only the United States felt the effects of the new aggressive Soviet air defense policy, Britain and Turkey also reported attacks on their planes.

The US flew an estimated 'hundreds of mostly peripheral reconnaissance missions per year during peak activity in the 1950s. If Presidents Truman and Eisenhower curtailed flights, the British often took up the slack by flying missions using RB-45s and Canberras. Before the U-2 deployed, the USAF's Project Homerun used nearly 50 RB-47s to survey the northern USSR. One of those missions flew 450 miles inland to photograph the city of Igarka, Siberia. According to "Project Homerun Operations," in early May 1956, the final mission of six RB-47s, flying parallel, penetrated Soviet airspace, photographing a wide swath. (This report released by NRO in 2018 - see Bibliography.)

"We were surrounded by American air bases," Khrushchev remembered. "Our country was literally a great big target range for American bombers operating from airfields in Norway, Germany, Italy, South Korea, and Japan. The Americans. kept sending planes deep into our territory, sometimes as far as Kyiv."

- William Taubman, *Khrushchev—The Man and His Era.* 2003, pp. 243, 244)

"Every time the invisible line of the border was crossed, there would be another TASS protest reporting that, after the necessary measures were taken, 'the plane withdrew in the direction of the sea.' Everyone understood this ambiguous expression as meaning that it was shot down. This dangerous game went on for years."

- Sergei Khrushchev, *Nikita Khrushchev and the Creation of a Superpower.* 2000, p.153

Spy flights "provoked the Russians' abiding rage." Implicit in the flight program: any fliers shot down and captured had to 'be written off.' Families—due to the needs of national security or the requirements of diplomacy—were routinely told their loved ones had gone missing on training or weather assignments.

- William E. Burroughs, *By Any Means Necessary: America's Secret Air War in the Cold War.* 2001, pp. 39, 136, 141

"We were engaged in this titanic struggle with a potential enemy and adversary that lived in secrecy, and about which we did not know, and therefore we were triggering paranoia in them as they were triggering it in us.' [– David Halberstam, author of *The Fifties*]. There were ongoing attempts to see behind the Iron Curtain. The Air Force was using bombers rigged with cameras to photograph military targets. Russian fighters attacked. In this secret air war, nearly 100 Americans disappeared."

- *The American Experience: Spy in the Sky.* WGBH Video, 1996 - see Bibliography

The Arms Race

The United States realized the arms race when the Soviet Union detonated its first atomic bomb in 1949. US experts never expected the Soviets to catch up with nuclear weapons technology so soon, but they had embedded spies in the American nuclear program from the very beginning. The US Congress responded to the detonation of "Joe 1" (as nicknamed in the American press) by the passing of the Central Intelligence Act to supplement the earlier Act of 1947. This new legislative action permitted the CIA to use confidential fiscal and administrative procedures. The Act exempted the CIA from many of the general limitations on expenditures and authorized placing CIA funds in the budgets of other departments and transferring the funds to the agency without restrictions. This authorization ensured the secrecy of the CIA's budget, an essential consideration in covert operations.

Regarding the Cold War, the United States faced two primary threats, the Soviet Union in Europe and Asia and China in the Far East along the Pacific Rim. Gathering useful intelligence on these threats proved difficult because the closed society of the Soviet Union and the Communist states of Eastern Europe essentially curtained off the outside world. The same applied to the People's Republic of China. The one window into those hidden lands came from above.

When the CIA came into existence in 1947, no one foresaw that in less than a decade, it would undertake a significant program of overhead reconnaissance. Traditionally, the military services were responsible for such a role, and previous flights deep into unfriendly territory had taken place only during wartime. By the early 1950s, however, US military leaders had the urgent and growing need for strategic intelligence on the more significant Communist powers and their satellite states.

Taking on a serious risk, Air Force and Navy aircraft conducted peripheral reconnaissance and shallow-penetration overflights. Still, these missions paid a high price in lives lost and increased international tension. Furthermore, many critical areas of the Soviet Union and China lay well beyond the operating range of existing planes.

The Air Force had therefore begun the development of a high-altitude reconnaissance aircraft able to conduct deep-penetration reconnaissance missions over the Soviet Union. President Dwight D. Eisenhower and his civilian scientific advisers feared that the loss of such an aircraft deep inside the hostile territory in peacetime could spark a war. They, therefore, authorized the development of a new nonmilitary plane flown by civilian pilots and operated only in the highest secrecy. The president assigned the primary responsibility for this new reconnaissance program to the CIA with the Air Force providing vital support. This effort spawned first the U-2 and later the A-12 reconnaissance platforms, delivering unprecedented performance characteristics.

With the sound of war drums and saber-rattling, the clouds of war began swirling over the CIA, only three years old. Would it be able to develop resources to deter seemingly inevitable war around the globe? World War III with Russia and China appeared more than just 'possible.' The minute hand of the symbolic Doomsday Clock inched closer to midnight in its countdown to a global catastrophe as the world moved closer to a nuclear war. The People's Republic of China severed diplomatic relations with the United Kingdom (UK). The last Kuomintang soldiers surrendered at the same time the Soviet Union and the People's Republic of China signed a pact of mutual defense. Kuomintang leader Chiang Kai-shek moved his capital to Taipei, Taiwan Island, in a standoff with the People's Republic of China.

During January 1950, the People's Republic of China and the Soviet Union recognized the Viet Minh's Democratic Republic of Vietnam, based in Hanoi, as the legitimate government of Vietnam.

The following month the US and UK recognized the French-backed State of Vietnam in Saigon, led by the mercurial former Emperor Bao Dai, as the only legitimate Vietnamese government. The Cold War between the Soviet Union and the United States also split Korea into two regions, with separate governments. Then, both parties claimed legitimacy in governing Korea. Neither side accepted the border as permanent.

The Soviets flaunted their developing strategic capabilities every May Day ("International Worker's Day," celebrated annually by the Communists on the first day of May). This ritual became an exercise in perception vs. reality. The world watched as the Soviets created the impression that it vastly overshadowed the US regarding military might. The annual May Day parade in Moscow featured a seemingly endless stream of troops, tanks, fighters, bombers, and missiles. Astute observers noted that some of the aircraft flying overhead were merely the same ones shown repeatedly, to give the impression of more numbers.

As the Soviets produced and deployed bomber forces, ballistic missiles, nuclear weapons, and submarine forces, they made every effort to conceal any weaknesses while making a show of strength. The overall situation made it impossible for the US to develop an effective strategy for retaliatory measures in the event of war. At almost all levels, the US government realized the need to seek information on Soviet strategic forces and their capabilities but found this a difficult challenge.

The CIA now found its entire array of intelligence tradecraft ineffective at a time when rumors concerning the Soviet Union's large bomber forces shocked Washington analysts. The strength of Soviet border security and internal counterintelligence efforts curtailed US covert human intelligence (HUMINT) activities. Communications intelligence (COMINT), the collection of telephone, telegraph, and radiotelephone transmission signals similarly hampered. Russia classified the Moscow telephone book itself and made it unavailable for inspection.

The actions of Soviet counterintelligence affected how the CIA collected and analyzed the vast amount of information gathered by other agencies and the various branches of the military. Out of necessity, as the Cold War escalated, the CIA performed more operational functions and conducted intelligence-gathering on its own.

The Air Force had expanded its aerial reconnaissance fleet to include the English Electric Canberra, later produced in the US as the Martin B-57. The jet-powered bomber made an excellent long-range aerial reconnaissance platform, particularly after modification (with extended wings) to fly higher than Soviet fighters. However, these shallow penetration reconnaissance flights only covered the border areas over Soviet airspace and were vulnerable to Soviet countermeasures.

The US could only fly peripheral missions using oblique photography to penetrate the veil of secrecy around military activities in the Soviet interior. Consequently, these missions covered only the Soviet Union west of the Urals, west of the Volga River, and only a few of the strategically important regions. The rest of the Soviet Union was effectively unknown and out of reach.

Electronic intelligence (ELINT) reconnaissance missions provided signal data on the existence of Soviet radar sites. By using this information to avoid detection, it made possible actual intrusions into Soviet territory to photograph targets impossible to image from the periphery.

While the existing Navy and Air Force aircraft were flying their risky reconnaissance missions over the Soviet Union, the CIA began planning for a more systematic and less dangerous approach using innovative technology.

Richard S. Leghorn, with Kodak, firmly believed in the need for what he called pre-D-day reconnaissance observation of a potential enemy before the outbreak of actual hostilities. He preferred this in contrast to combat reconnaissance in wartime.

In papers presented in 1946 and 1948, Leghorn argued that the United States needed to develop such a capability. Doing so would require high-altitude aircraft and high-resolution cameras. The outbreak of the Korean War provided Leghorn an opportunity to put his ideas into effect. Recalled to active duty with the Air Force, Lt. Col Leghorn became the head of the Reconnaissance Systems Branch of the Wright Air Development Command at Dayton, Ohio, during April 1951.

Altitude remained the key to success and survivability for overhead reconnaissance missions. Since the best Soviet interceptor at that time—the MiG-17—struggled to reach 45,000 feet. Leghorn reasoned an aircraft exceeding 60,000 feet as being safe from Soviet fighters for the time being. Recognizing that the fastest way to produce a high-altitude reconnaissance platform was to modify an existing airframe, he began looking for the highest-flying aircraft available in the Free World. This search soon led him to a British twin-engine medium bomber—the Canberra—built by the English Electric Company.

The Canberra had made its first flight during May 1949. Its maximum speed of 469 knots (870 kilometers per hour) and its service ceiling of 48,000 feet made the Canberra a natural choice for high-altitude reconnaissance work. The British Royal Air Force quickly developed a reconnaissance version of the Canberra, the PR3 (the PR stood for photoreconnaissance), which began flying during March 1950. At the higher altitudes without oxygen and much atmospheric air pressure, the Canberra aircrewmen, for the first time, would need a pressure suit and oxygen pre-breathing.

At Leghorn's insistence, the Wright Air Development Command invited English Electric representatives to Dayton, Ohio, in the summer of 1951 to help find ways to make the Canberra fly even higher. The Air Force adopted the bomber version of the Canberra produced by the Glenn L. Martin Aircraft Company under license as the B-57 medium bomber.

Leghorn and his English Electric colleagues designed a new Canberra configuration with very long high-lift wings, new Rolls-Royce Avon 109 engines, a single pilot, and an airframe stressed to less than standard military specifications.

They calculated that a Canberra so equipped might reach 63,000 feet early in an extended mission and as high as 67,000 feet as the declining fuel supply lightened the aircraft. Leghorn believed that such a modified Canberra could penetrate deep into Soviet and Chinese airspace and photograph up to 85 percent of the most desired intelligence targets in those countries.

Leghorn persuaded his superiors to submit his suggestion to the Pentagon for funding. He had not, however, cleared his idea with the Air Research and Development Command. By then, Soviet air defenses were aggressively attacking all aircraft, making border incursions, sometimes even those over Japanese airspace. Existing reconnaissance aircraft, such as the RB-45 and RB-47, were increasingly vulnerable to anti-aircraft artillery, missiles, and fighters. The US Air Force and Navy lost numerous planes and airmen to hostile fire during Cold War reconnaissance missions.

The two superpowers never engaged directly in full-scale armed combat. However, they heavily armed their military services in preparation for a possible all-out nuclear world war. Each side now maintained a nuclear arsenal to deter an attack by the other side. Both realized that such an attack would lead to the destruction of the attacker. This offensive/defensive doctrine became known as "mutually assured destruction" (MAD).

The struggle for dominance between the East and West not only focused on conventional and nuclear forces. It also included proxy wars around the globe, psychological warfare, massive propaganda campaigns and espionage, rivalry at sports events, and technological competitions such as the Space Race. The victory of the Communist side in the Chinese Civil War and the outbreak of the Korean War in 1950 expanded the conflict.

The Proxy War in Korea

When World War II ended, Soviet forces remained entrenched in North Korea while Stalin set his sights on invading the southern end of the Korean peninsula. Meanwhile, President Truman announced his intent to withdraw all remaining American forces by June 1949. Gen William Lynn Roberts, the US commander of the Korean Military Advisory Group, voiced utmost confidence in the Republic of Korea (ROK) Army, going so far as to boast that defending against a North Korean invasion would merely be "target practice." Stalin saw the situation differently. Once the Americans left, he could take South Korea without getting the Soviet Union embroiled in a war with the United States.

The United States signaled a lack of interest in South Korea. In spring 1950, Secretary of State Dean Acheson failed to include Korea in his outline of the strategic Asian Defense Perimeter even though Soviet troops remained in North Korea and indications that Stalin intended to unify the North and South under Communist rule. American soldiers had entirely withdrawn from Korea because military strategists were more concerned with the security of Europe against the Soviet Union than with the situation in East Asia.

The Soviets had cracked America's secret codes used to communicate with the US embassy in Moscow. Reading these dispatches convinced Stalin that Korea lacked enough importance to the US to warrant a nuclear confrontation. The fact that the Americans had not intervened to stop the communist victory in neighboring China bolstered this impression. Thus, Stalin calculated that the US government lacked the will to fight another war in the Far East so soon after World War II.

Stalin approached the Korean situation using a repertoire of deception known as Maskirovka. Soviet doctrine concealed true intentions and denied responsibility for any questionable activities by employing all means, political and military, to further their geopolitical ambitions. Its Maskirovka objective kept the Americans and others guessing.

Stalin had carefully engineered the Berlin incident in 1948 to show the West that it could not bully the USSR. Gromyko believed that Stalin embarked on closing off East Germany, knowing that the conflict would not lead to nuclear war. Both the USSR and the USA initiated intimidating reactions not to appear intimidated. In a kind of reverse nuclear brinksmanship, the People's Republic of China leader Mao Zedong also showed himself to be a master in the 1950s at calibrating how far the PRC could push its foreign policy objectives without triggering general war or incurring a nuclear strike from the United States.

For the objective of moving against South Korea, Soviet military support would remain hidden in the shadows. North Korean soldiers would take to the battlefield in a "war of unification," but if the US and its allies intervened, the Soviets would provide covert military support to the Communists.

Timing and the Maskirovka strategy paid off

for the Soviets, finding the South Koreans at the time entirely unprepared for repelling an attack. With them having no tanks, anti-tank weapons, or heavy artillery, South Korea could only commit its forces to battle in a piecemeal fashion.

On Sunday, 25 June 1950, the Soviet Union gained their opportunity. Communist North Korean troops with Soviet-supplied tanks, heavy artillery, and aircraft crossed the 38th parallel and invaded the ROK. The war was on.

For the second time in recent history, a surprise attack caught the United States unprepared. In two days, the North Korean forces had sacked the South Korean City of Seoul. In five days, South Korean forces that once numbered 95,000 men were down to less than 22,000.

Dean Acheson immediately notified Truman of the invasion. The president acted quickly and decisively, instructing Acheson to contact the UN to seek a resolution condemning the attack and to offer aid to assist South Korea.

Joseph Stalin
Ruled the USSR from the mid-1920s to
his death in 1953

Chapter 3 - War in Korea and Related Developments

Why the Korean War Intelligence Controversy?

By all appearances, the North Korean attack surprised the Truman administration, the US Army's Far East Command under Gen Douglas A. MacArthur, and the fledgling CIA. Within days, administration and congressional critics leveled charges of the agency with not fulfilling its primary mission of providing a warning to the president. The critics rated it an intelligence failure of the highest magnitude. Truman feared the regional conflict would quickly escalate into another world war should the Chinese or Soviets become directly involved. It didn't come to that, but it did push Cold War tensions to the brink.

The crisis began when approximately 75,000 soldiers of the North Korean People's Army Forces poured across the 38th parallel to invade the South. Soviet and Chinese support of North Korean aggression ignited a proxy war with the United States, which of course, supported the South.

Thus, less than three years after its creation, the CIA became involved in its first "hot war." Suddenly, the newly formed agency found itself conducting an array of espionage and covert operations unilaterally and in support of US armed forces taking part in a UN coalition effort to stem the Communist invasion.

The surprise attack raised fundamental questions about the CIA, not providing a warning to US policymakers that North Korea would attack its southern neighbor. Not so, according to some. In any case, the CIA took responsibility for the lack of intelligence, though a few critics mentioned the absence of warning from any intelligence services of other allied nations who were now mobilizing for war in the Far East as well.

What about the lack of forewarning from other countries operating under the UN aegis? This group would include Australia, Belgium, Canada, Colombia, Ethiopia, France, Greece, Luxembourg, the Netherlands, New Zealand, the Philippines, South Africa, Thailand, Turkey, and the United Kingdom.

Few asked that question. Instead, critics of the CIA before and after the Korean invasion focused on the relatively few references to Korea in the agency's intelligence reporting. They noted the lack of any predictive estimates or other "actionable" warning information that might have allowed US policymakers to anticipate Korean events before they reached the crisis stage.

Before the invasion, most analysts and pundits focused on the Soviet threat to the security of Europe. Critics seemed to forget how only six months earlier, Dean Acheson failed to include Korea in his outline of the strategic Asian Defense Perimeter. Nor did they fault President Truman for withdrawing all American soldiers from the Korean peninsula, an action that all but waved a flag to signal the Soviet bear to come and get South Korea.

The CIA had only 5,000 employees worldwide at the time of the invasion. Only 1,000 of them were analysts, and only three employed as operations officers in Korea.

Contrary to what some historians claimed in later years, CIA analysts frequently reported on Korea during the prewar years. Admittedly, the CIA geared its intelligence harvesting perspective, primarily toward highlighting the Soviet Union's involvement in Asia rather than local Korean politics.

At the time, most of America viewed communist movements around the world as Kremlin-controlled, and therefore, saw North Korean activity closely interrelated with other Soviet-induced crises, but not of any greater or lesser importance. Ultimately, Korea became just one of many fronts in the Cold War.

Tensions between the East and West further heightened when Soviet air defenses downed a twin-engine US Navy P2V Neptune near Vladivostok in 1951 and an Air Force RB-29 over the Sea of Japan during June 1952. Two months later, Soviet aircraft began violating Japanese airspace, and it wasn't long before a Soviet fighter stalked and shot down another RB-29 flying over the island of Hokkaido. Aerial reconnaissance of the Soviet Union and surrounding areas had become a deadly business.

Despite these growing risks, senior US

officials firmly believed that such missions were necessary to overcome the lack of information about the Soviet Union. Material support of the North Korean invasion significantly strengthened the US perception of the Soviet Union as an aggressive nation determined to expand its borders and influence. US military officials were more determined than ever to avoid being surprised by a Soviet attack. Nonetheless, such an attack occurred in Korea, where unexpected.

Before the attack, the CIA had noted only signs of Soviet mischief-making and proxy-sponsored "tests" of American resolve. The CIA Office of Research and Reports had already indicated the possibility that some regional crises might occur during the period from 1947 through 1950. Understandably, with so much other Communist activity throughout the world, CIA reporting on Korea failed to stand out either in intelligence publications or the minds of policymakers.

In all fairness to the agency, the CIA reported explicitly in 1949 on the potential for war in Korea. When President Truman sought military proposals for withdrawing US forces from the peninsula, the CIA warned of removing them, most likely leading to war.

In the buildup to the Korean invasion, the Soviets again successfully employed a Maskirovka strategy. Almost from the moment of the last US military departure, North Korea began deploying its People's Army southward. CIA agents reported the acquisition of heavy equipment and armor. While recognizing the present program of propaganda, infiltration, sabotage, subversion, and guerrilla operations against southern Korea, the Agency failed to realize an invasion imminent. During the next two years, the Agency reorganized and hired additional personnel, resulting in a more extensive CIA and a new Directorate of Intelligence (1952).

China joins the Proxy War Against the USA

Three months after the Korean invasion, the People's Republic of China mobilized its troops along the Yalu River, where China bordered North Korea. The PRC subsequently sent around 300,000 soldiers into North Korea as reinforcements. The Chinese intervention in Korea escalated an offensive that pushed UN coalition forces back toward South Korea.

Following the capture of Seoul by Chinese troops, the Korean War became stalled along a defensive line at the 38th Parallel—in effect, cutting the Korean peninsula in half.

The Politics of Why the CIA

CIA analysts knew the Soviets were exerting a strong influence on events in North Korea but failed to anticipate the invasion. The attack—the first open military action of the Cold War—raised the stakes considerably. Unless the Soviets were seriously considering an all-out global conflict, such a move did not make sense.

To Americans, it appeared that the Soviet-orchestrated global Communist push would come through Eastern Europe. CIA analysts never expected a Soviet proxy attack through Asia.

Meanwhile, the Korean conflict took a terrible toll in just three years, taking the lives of some five million soldiers and civilians. Almost 40,000 Americans died, and more than 100,000 others wounded.

Despite these overwhelming numbers, the conflict failed to receive the same media attention in the US as did World War II. Officially, the United States never formally declared it a war. Truman described the conflict as a "police action."

It became known among veterans as "The Forgotten War" or "The Unknown War" from the lack of public attention it received both during the conflict and afterward. Ironically, the 1970s television show M*A*S*H, a comedy set in a field hospital in South Korea, became the most enduring representation of the war. It aired in 1972 and provided audiences with half-hour episodes weekly for a little more than ten years, more than three times the duration of the war itself. The final episode during February 1983 became the most-watched in television history.

To the CIA leadership, Korea represented the second US war sparked by the lack of prior intelligence to foresee a sneak attack. They vowed the agency would never again let such a thing happen.

To this end, the CIA took a step to do something that would typically have been the exclusive province of the Air Force. To understand why a civilian intelligence agency and not the US military air arm took the initiative to develop and field a high-flying reconnaissance jet, one needs to know a little bit about Gen Curtis LeMay, chief of the Strategic Air Command (SAC) from 1948-1957 and later Air Force Chief of Staff, known to some as "Old Iron Pants."

Why the CIA and not the Air Force

After the Korean conflict ended with an armistice and the creation of a demilitarized zone (DMZ) between the North and South, US military leaders like LeMay turned their thoughts to potential future wars.

Three things warranted concern. First and foremost, the US needed intelligence to determine Soviet capabilities and intentions. Secondly, the US lacked accurate geographical locations for Soviet targets, and thirdly the leaders felt concerned about the Air Force's dependence on the use of radar-guidance bombing technology.

At this point, the US had produced only the first generation of simple, crude atomic bombs with relatively low yield (equivalent to approximately 20,000 tons of TNT, or 20 kilotons). To effectively destroy a target, bombardiers relied on radar guidance systems. However, LeMay distrusted radar because the technology had performed poorly thus far.

The SAC chief had recently directed an exercise in which his crews made simulated bomb runs against various targets in American cities, including Wright-Patterson Air Force Base near Dayton, Ohio. Evaluators noted that the bombers had performed miserably and gave LeMay a "scathing" review.

Afterward, LeMay called a meeting of his lead navigators and radar observers at SAC headquarters. He blamed the radar navigation system as the primary cause and assigned Maj Harry Cordes, the same Harry Cordes, who first soloed in the U-2 during May 1956, to rewrite the AN/APQ-13 radar manual for the B-29.

LeMay also selected Col Edward Perry, an experienced lead navigator, to head a study group of radar observers from the 309th Bomb Wing. This assignment sent Perry to the Pentagon in Washington to work with the Air Force Intelligence Staff in analyzing all available maps and charts of potential targets within the Soviet Union.

This analysis provided LeMay with its best judgment of the expected results from the coordinated radar-guided bombardment. Perry and Cordes worked well together, and LeMay came to rely on their opinions. Additionally, Cordes proved himself by serving as the radar observer on a B-29 crew that won the annual SAC bombing competition in which radar bomb scoring evaluated aircrews for targeting effectiveness.

Competitors simulated unguided bomb drops on targets including SAC and Navy radar stations and Army Nike missile sites located near Amarillo and San Antonio, Texas; Denver, Colorado; Salt Lake City, Utah; and Kansas City, Missouri. Air Force radar bomb scoring evaluators endeavored to make the exercise as realistic as possible by using various ground radar, computers, and other electronic equipment such as jammers to disrupt the bombers' radar navigation systems. Results of the SAC bombing evaluation highlighted US and NATO concerns about the nuclear threat posed by the Soviet Union and how to counter it effectively.

Western military commanders continued to wonder the extent of the Soviets developing their nuclear arsenal and the magnitude of the Soviet military buildup in Eastern Europe. Aerial reconnaissance still mostly consisted of flying camera-carrying bombers over or near denied territory, exposing aircrews to extreme danger. Another method involved the use of specially instrumented balloons such as projects Mogul, GENETRIX, and Moby Dick. The loss of a Mogul balloon in the summer of 1947 had inadvertently triggered the infamous "Roswell Incident." Human intelligence, or HUMINT, required inserting agents into Soviet territory, an activity nearly impossible under the prevailing circumstances. Without reliable knowledge of Soviet military capabilities, international fears and tensions continued to increase.

Beginning in 1954, President Eisenhower adopted strategic overflight reconnaissance in peacetime as a national policy. He approved and introduced to the international community the terms "Open Skies" and "Freedom of Space" to account for its practice overhead. Eisenhower authorized development and organization of all innovations—the high-performance cameras, aircraft, and satellites—needed to execute the policy that included Project AQUATONE.

Dwight Eisenhower eventually assigned the direction of American strategic overflight reconnaissance projects to a civilian agency reporting to the secretary of defense, later named the National Reconnaissance Office, NRO. Although the US Air Force retained a prominent role in the NRO, many of its members debated the wisdom of assigning the mission to a civilian organization. The service, after all, had nurtured the concept of strategic overflight reconnaissance and developed the technology. The president's actions posed a problem with the Strategic Air Command, who operated the long-range reconnaissance aircraft and remained firmly committed to its JCS wartime mission.

If SAC served the nation as its intercontinental striking arm, its commander, General LeMay, believed that it would distract his wartime force, dividing it into operating special "boutique" flying units in peacetime. As MGen Foster Lee Smith remembered: "The command needed, wanted, and got strategic intelligence. It didn't need that sexpot mission in the SAC nunnery." Later, when SAC began operating the U-2 and SR-71 aircraft in the 1960s, it had to deal with lower morale that attends elite units in a wartime force just as General LeMay envisioned. Air Force leaders could not imbue the idea of their mission as avoiding a war.

Understanding of the Soviet Union's technical capabilities became the prime objective for the CIA at this time. Agency analysts viewed the Air Force's brute force approach as too dangerous and no match for Soviet air defense systems. The solution, they felt, lay in innovative technology and improved scientific methods for gathering covert intelligence. This led to the formation of an organization that evolved into the CIA Directorate of Science and Technology, an in-house think tank and operations group that developed advanced spy gear to breach the Iron Curtain.

To this end, the CIA recruited the best scientific and engineering minds in the fields of radar, electronic warfare, and aeronautics. Innovative approaches included methods for spoofing Soviet air defenses to trick radar operators into revealing their frequencies and capabilities and developing aerial reconnaissance platforms that could successfully evade Soviet surface-to-air missiles. Some of the most innovative solutions came from the civilian sector.

Civilian Support for an Aerial Reconnaissance Aircraft

Richard Leghorn, with a degree in physics from MIT and a reserve commission as an Army second lieutenant, became the first to articulate a vision of how to meet the intellectual demands of the Cold War. He began working with Eastman Kodak, but in the early 1940s accepted active duty assignment to the Aeronautical Photographic Laboratory at Wright Field. There he worked with such top-notch optical scientists and engineers as James G. Baker, Amrom Katz, Richard Philbrick, and Duncan Macdonald. Leghorn remained with the Aeronautical Photographic Laboratory until late 1942 when he received orders to report for pilot training.

In April 1943, Leghorn became the commander of the 30th Photographic Reconnaissance Squadron. The squadron arrived in England during January 1944 and began flying missions over northern France, photographing German forces and transporting equipment in preparation for the D-day invasion. After the landings on 6 June 1944 at Normandy, Leghorn's unit flew in support of the US First Army as it advanced through France during the Battle of the Bulge, and finally, during its drive into Germany in the spring of 1945.

Following his 1951 recall to active duty with the Air Force during the Korean War, Leghorn became the head of the Reconnaissance Systems Branch of the Wright Air Development Command in Dayton, Ohio. There, he strived to design a surveillance aircraft capable of climbing above 60,000 feet, an altitude not approachable by any Soviet plane. Thus, Leghorn saw the future of high-altitude aerial reconnaissance.

He envisioned a lightweight aircraft that operated from one or more overseas bases. SAC planners had already become adept at such deployments while flying long-range bombers on nuclear alert missions from remote locations such as Nouasseur Air Base in French Morocco. Still, persuading Gen LeMay to back an unarmed aircraft proved challenging, and adding armament to satisfy him would only burden the airframe with too much weight, making high-altitude flying impossible.

Leghorn transferred to the Pentagon in early 1952, where he helped plan the Air Force reconnaissance requirements for the next decade. There, Leghorn worked for Col Bernard A. Schriever, assistant for Development Planning to the Air Force Deputy Chief of Staff for Development. In this position, he collaborated with Charles F. "Bud" Weinberg, a colleague from Wright Field, and Eugene P. Kiefer, a Notre Dame graduate in aeronautical engineering who had designed reconnaissance aircraft at the Wright Air Development Center during World War II. Leghorn felt these three reconnaissance experts working together might promote high-altitude photoreconnaissance, and eventually garner high-level Air Force support.

Richard Leghorn

Project LINCOLN - The BEACON HILL Report

In 1951, Air Force deputy chief of staff MGen Gordon P. Saville had added 15 reconnaissance experts to Project LINCOLN, an existing air defense study underway at MIT. This group purposely determined the best ways to conduct surveillance of the Soviet Bloc countries. Members of note included James Baker and Edward Purcell from Harvard University, Allen Donovan from Cornell Aeronautical Laboratory, Edwin "Din" Land of the Polaroid Corporation, Richard S. Perkin of the Perkin Elmer Company, Stewart Miller of Bell Laboratories, and Lt. Col

Richard Leghorn as the Wright Air Development Center liaison officer. Leghorn, further, promoted the idea of 'open skies' that was later developed by President Eisenhower.

The LINCOLN experts assembled at a secretarial school on Beacon Hill in Boston, where Kodak physicist Carl Average chaired the effort under the adopted code name, "The BEACON HILL Study Group."

Polaroid founder Edwin Land often saw the solution to a problem in his head, but often failed to write it down. He studied chemistry at Harvard but left after his freshman year. After developing a revolutionary new polarizing film, Land returned to Harvard, though he never finished his studies or received a degree.

Early on, Land established his R&D lab, known as the Color Lab in an old, abandoned Kaplan's Furniture Warehouse located in Cambridge. From his oversized office, located on the second floor and accessible by dusty, wooden stairs, he developed the instant film innovations that played a significant role in the development of photographic reconnaissance and intelligence-gathering efforts. Some say that while out for a walk, Land invented the color film that developed and printed in 60 seconds.

In early 1952, the BEACON HILL Study Group spent weekends conducting briefings on the latest technology and projects focusing on aerial reconnaissance. They discussed innovative approaches using high-flying aircraft, camera-carrying balloons, and even an "invisible" dirigible, a lighter-than-air giant, flattened airship with a blue-tinted, non-reflective coating that cruised at an altitude of 90,000 feet at slow speeds with a large camera lens photographing targets of interest.

The LINCOLN group returned to MIT at the end of February 1952 and spent three months writing a classified report. The introduction of the 15 June 1952 BEACON HILL report said that since the Soviets now had nuclear weapons, "intelligence and reconnaissance are more important to the United States by several orders of magnitude than ever before."

The report went on to advocate a radical approach to obtain the information needed for national intelligence estimates. The report mainly covered radar, radio, and photographic surveillance. It examined the use of overhead passive infrared and microwave reconnaissance accompanied by the development of advanced vehicles to carry such sensors. Mainly, the report called for rapid progress towards the production of high-altitude reconnaissance aircraft.

The Iron Curtain

The characteristic closed society within the Communist Bloc nations provided the driving influence behind the need for advanced aerial reconnaissance in the West. Allies in the Free World needed to see beyond the veil of the Iron Curtain.

The Soviet Union blocked itself and its satellite states from open contact with the West and non-Soviet-controlled areas. The eastern side of the Iron Curtain contained the countries connected to or influenced by the Soviet Union. Consequently, the western countries had no idea of occurrences on the Soviet side of this imaginary wall that symbolized the Cold War.

Within the Eastern Bloc, imaginary walls soon became real boundaries of concrete and steel with watchtowers overlooking a barrier marked with warning signs. Nowhere else did this become more prominent than the city of Berlin, where both East and West German border guards used dog patrols along the strip of land between the actual borderline and the barrier. Barbed wire fences soon appeared, and rumors flowed that the Soviet Union intended to erect a real wall in Berlin to separate the city.

Assumptions About Soviet Radar Capability

In the early 1950s period, many analysts felt confident of a reconnaissance aircraft flying at 70,000 feet or higher as being beyond the reach of Soviet fighters and anti-aircraft missiles. One such expert was John Seaberg, an aeronautical engineer for the Chance Vought Corporation, and an Air Force reservist recalled to active duty during the Korean War. Following the armistice, the Air Force teamed him with German aeronautical experts Woldemar Voigt and Richard Vogt to develop a new aircraft that combined the latest turbojet engines and new high-efficiency wings to

reach ultra-high-altitudes previously unattained.

Together, they sought to develop a plane capable of reconnaissance missions at high subsonic speeds while cruising at altitudes at or above 70,000 feet. The airplane would carry a single crewmember and up to 700 pounds of observation equipment. Seaberg hoped for an operational mission radius of around 1,500 nautical miles with new cameras developed to take photos with extraordinary detail from high altitude.

The MiG-17, at the time the best Soviet pursuit fighter, could barely reach 45,000 feet. Seaberg and others believed that contemporary Soviet air defense radar would have difficulty tracking aircraft flying above 65,000 feet. Soviet air defenses at that time relied heavily on American-built SCR-584 radar equipment obtained through the WW II Lend-Lease program. The US Army used the SCR-584 system, a microwave-type radar designed and built by the MIT Radiation Laboratory, to control antiaircraft guns. The Soviets also acquired some SCR-270 long-range radars

Seaberg and his team believed that if a plane ascended to 65,000 feet before entering the range of the SCR-584 (or a Soviet equivalent), it could penetrate the airspace largely undetected. While a valid assessment at the time, it did not remain so for long. The Soviet Union, unlike Britain and the United States, continued to improve radar technology after the end of World War II.

The early SCR-584 detected bomber-size targets at a maximum range of only 40 miles and easily tracked them at 18 miles, even up to 90,000 feet. One flaw inherent in the SCR-584 involved its high-power usage burning its cavity magnetron tube (an electron tube for amplifying or generating microwaves). With the flow of electrons controlled by an external magnetic field, the magnetron quickly burned out and required frequent replacement.

The SCR-270 early warning radar could remain on for much more extended periods and had a greater horizontal range (approximately 120 miles), but the curvature of the earth limited it to a maximum altitude of 40,000 feet.

Even after evidence of improved Soviet radar capabilities became available, however, many advocates of high-altitude overflight continued to believe that any aircraft flying above 65,000 feet were safe from detection by Soviet radars.

The author's radar at Beatty was an SCR-584 (short for Signal Corps Radio #-584) microwave radar developed by the MIT Radiation Laboratory during World War II. It replaced the earlier and much more sophisticated SCR-268 as the US Army's primary antiaircraft gun-laying system¶

The author's radar at Beatty was an SCR-584

Development of Post-World War II Soviet Defense

The massive surprise invasion of June 22, 1941, by Nazi Germany, became the Pearl Harbor for the Soviet Union. As one Russian documentary states: "Postwar threw all strength and efforts into the development of the Soviet A-bomb and its means of delivery. The first postwar generation of the American reconnaissance aircraft constantly flew along the USSR border, trying to 'peep' into its territory. The enemy was, most of all, interested in the regions where the development of the missiles and atomic weapons were on the way. In the second half of the 50s, the NATO spy aircraft registered several hundred episodes of the USSR border crossing annually." (***Wings of Russia***, Episode 18, minutes 14-23. See Bibliography)

What were some specific steps in the development of Soviet defense from 1945 in response to the conditions at that time? A 1975 US Army study ***History of Strategic Air and Ballistic Missile Defense, Vol. 1: 1945 – 1955*** (see Bibliography) identifies many such Soviet efforts and accomplishments. Here are some highlights and chronology mentioned and analyzed in the study.

Budget pressures plus need of the MiG-15 for the Korean War and for arming Warsaw Pact forces slowed Soviet fighter development. But, 1953 still marked the appearance of the supersonic MiG-17. The 1955 May Day parade saw the introduction of the YAK-25 and MiG-19 interceptors, in addition to long-range Bear bombers. But increasingly capable fighters were not useful for defense without integrated early warning:

"The Soviets decided that their wartime WW2 approach to an early warning was inadequate. Indeed, it was necessary to greatly expand the use of radar of various kinds." (p. 181)

In 1954, the Soviet air defense system became more organized and, in a position, to establish an integrated radar system. PVO Stravny ("Air Defense of the Nation") became co-equal with other Soviet armed forces. Also, important and decisive was the integration of German know-how in efforts to produce jet engines, aircraft, rockets, and radars:

"By the end of the war, the Soviets had captured a considerable one group of German scientists which had been working on surface-to-air missiles was put to work at Scientific Research Institute 88, directed to design a surface-to-air missile effective up to 98,000 feet and was to carry a 500 Kilogram warhead." (p. 180)

1950: German scientists tasked to study guidance problems of the SA-1 (p. 237)

"By 1945, the Soviets had enough knowledge to manufacture copies of Western radars, through the assistance of German scientists and engineers. From 1945 – 1946 and later, we find that Germans were forcibly evacuated and taken from East Germany." (p. 182) -some tens of thousands of German scientists, engineers, and technicians took to the USSR under 5-year contracts.'

"Forced work on radar systems took place at the Scientific Research Institute 160, about 22 miles from Moscow." (p. 182)

1950: SCAN ODD radar developed with German help (p. 183)

1953: Patty Cake a uniquely Soviet design, not

a copy (p. 184)

1954: Token radars become more widespread, developed since 1951 based on US AN-CPS-6 V-beam radar: "Soviet technicians were more successful at maintaining them at an operational level than the US had initially anticipated based on US experiences. It found the basic design of the Token was considerably simpler." (p. 184)

The Soviets improved radar power further by synthesizing at first two separate radars and next four separate radars into an integrated system housed in a single facility.

Why the Increased Fear of Surprise Soviet Attack?

When Dwight D. Eisenhower became the US president during January 1953, he endorsed Seaberg, Leghorn, Weinberg, and Kiefer's belief that the US needed a high-flying reconnaissance plane. He quickly made known his dissatisfaction with the quality of the intelligence estimates of Soviet strategic capabilities. The paucity of reconnaissance on the Soviet Bloc dismayed him.

During March 1953, William E. Lamar, chief of the New Developments Office, Bombardment Aircraft Branch, at the Wright Air Development Center, drew up a proposal calling for high-altitude reconnaissance aircraft. Shortly afterward, the Air Force implemented the ideas of the previously mentioned study called the BEACON HILL Report.

The BEACON HILL Report addressed the president's concerns and those of the US political and military leaders regarding the Soviet progress toward achieving military parity with the United States. This concern came from the Soviet Union's alarming advances in nuclear weapons, demonstrated when it detonated a hydrogen bomb manufactured using dry lithium deuteride fuel.

In an incident two months later, Soviet troops aggressively crushed an uprising in East Berlin. US officials such as Secretary of State John Foster Dulles viewed the Soviet Union as a threat to future peace. A top-secret RAND study amplified their concerns and fears by pointing out the vulnerability of the SAC's US bases to a surprise attack by Soviet bombers. (RAND, Research ANd Development, the post-WWII non-profit 'think tank' with government contracts for research on issues of intercontinental warfare.) Even without the RAND study, Air Force leaders feared the Soviets might already possess a larger bomber force than the US, the fear of a "bomber gap" based on experience from Korea.

When the Cold War dawned following World War II, the Russians cloaked entire cities in secrecy, and any maps gave no hint of their true location, or in some cases, even their existence. In 1949, the Soviet Union detonated an atomic bomb, and by October 1951, the US saw signs that a Soviet Tu-4 bomber had dropped a nuclear weapon in an air-burst test. The USSR followed this with the detonation of the thermonuclear weapon in 1953. Early intelligence estimates projected the Kremlin as having as many as 600 Tu-4 bombers in service and up to 100 atomic bombs in the stockpile, raising fears of a Soviet strike. Newly inaugurated President Eisenhower had a "Pearl Harbor complex" that deeply convinced him of the need for accurate reconnaissance.

The US entered the Korean War with propeller-engine bombers and fighters virtually unopposed in the beginning. When the Soviets provided North Korea with the agile MiG-15, suddenly US pilots found themselves engaged in uneven conflict with Soviet pilots in fast jets. In response, US P-51 squadrons rapidly converted to jet fighters, including the F-80 and F-86. Soon, large-scale jet-versus-jet air battles took place along the disputed Korean border in what became known as "MiG Alley."

No one wanted another "MiG" surprise. Memories of the Soviet escalation of the war in the Far East, combined with the RAND study, prompted the Air Force to establish the Intelligence Systems Panel (ISP). This new advisory group proposed ways for implementing the construction of high-flying aircraft and high-resolution cameras. Experts looked at the BEACON HILL Study Group report for recommendations. The CIA contributed Edward L. Allen of the Office of Research and Reports (ORR) and Philip Strong of the Office of Scientific Intelligence (OSI) to the group.

When the Intelligence Systems Panel first met at Boston University on 3 August 1953, Strong emphasized the poor state of US knowledge of the

Soviet Union. He informed the others that the best intelligence available on the Soviet Union's interior dated back to the German Luftwaffe imagery gained during World War II. Worse yet, the German photography covered only the Soviet Union west of the Urals, west of the Volga River, and only a few of the important regions. Picture postcards made up some of the "intelligence."

The English Electric Canberra, the highest-flying aircraft available to America and its allies at the time, reached 48,000 feet. The British already produced a photo-reconnaissance variant, but the Air Force instead asked English Electric's help for their further modification of the Martin B-57 version with enlarged wings and other changes.

This aircraft, the American-licensed production of the Canberra, with enlarged and computer-designed wings, new engines, and a lighter-than-normal airframe, reached 67,000 feet. However, ARDC bandaged onto it a design change that resulted in the RB-57D plane being able only to reach 64,000 feet.

Additionally, the Air Force, recognizing that the Soviets vastly improved their radar capabilities, had briefed the Intelligence Systems Panel members on the need to seek ways to provide updated photography of all the Soviet Union. The Air Force had discussed with them the multiple sensors used in existing aircraft such as the RB-47, RB-52, and RB-57?

In 1952, the Air Force also had begun testing Project FICON (an acronym for "fighter conveyor"), under which engineers adapted a giant, 10-engine B-36 bomber to launch and retrieve a Republic RF-84F Thunderstreak reconnaissance aircraft. Other concepts discussed included reconnaissance versions of the Navajo and Snark missiles, a high-altitude balloon program, and a new high-altitude reconnaissance aircraft.

Some of the Air Force's schemes to collect intelligence on the Communist Bloc exceeded existing levels of technology. Others, though technically feasible, involved dangerous political consequences such as what occurred to a British overflight of the Kapustin Yar area in early 1953. The British had overflown and photographed the Soviet missile test range there flying a high-altitude Canberra containing modifications

inspired by Richard Leghorn's collaboration with English Electric Company designers in 1951. Despite the capabilities of this variant equipped with Rolls-Royce Avon-109 engines, long, fuel-filled wings, a range of 4,300 miles and an altitude of 65,880 feet, Soviet air defenses detected the airplane and nearly shot it down.

Lt Col Joseph J. Pellegrini, assigned to head all US aerial espionage activities and responsible for approving all new reconnaissance aircraft designs, reviewed Leghorn's design and ordered extensive modifications.

According to Leghorn, Pellegrini showed no interest in a special-purpose aircraft suitable only for covert peacetime reconnaissance missions. He believed that all Air Force reconnaissance aircraft should have the capability of operating under wartime conditions, and therefore, insisted that Leghorn's design meet the standard specifications for combat aircraft. These requirements included heavily stressed airframes, armor plates, and other apparatus that made a plane too heavy to reach the higher altitudes necessary for safe overflight of the Soviet Bloc.

Leghorn, frustrated by the rejection of his original concept, transferred to the Pentagon in early 1952 to work for Col Bernard A. Schriever, Assistant for Development Planning to the Air Force's deputy chief of staff for Development.

By 1955, after its alteration by Pellegrini's staff, Leghorn's concept, the RB-57D achieved an altitude of only 64,000 feet. The USAF deployed this model in 1956 as "Black Knight."

Formative Years of Air Research and Development Command

When Dwight D. Eisenhower became the US president during January 1953, he endorsed Seaberg, Leghorn, Weinberg, and Kiefer's belief that the US needed a high-flying reconnaissance plane. He quickly made known his dissatisfaction with the quality of the intelligence estimates of Soviet strategic capabilities. The paucity of reconnaissance on the Soviet Bloc dismayed him.

During March 1953, William E. Lamar, chief of the New Developments Office, Bombardment Aircraft Branch, at the Wright Air Development Center, drew up a proposal calling for high-altitude reconnaissance aircraft. Shortly afterward, the Air Force implemented the ideas of the previously mentioned study called the BEACON HILL Report.

The BEACON HILL Report addressed the president's concerns and those of the US political and military leaders regarding the Soviet progress toward achieving military parity with the United States. However, this concern came from the Soviet Union's alarming advances in nuclear weapons, demonstrated when it detonated a hydrogen bomb manufactured using dry lithium deuteride fuel.

In an incident two months later, Soviet troops aggressively crushed an uprising in East Berlin. US officials such as Secretary of State John Foster Dulles viewed the Soviet Union as a threat to future peace. A top-secret RAND study amplified their concerns and fears by pointing out the vulnerability of the SAC's US bases to a surprise attack by Soviet bombers. (RAND, Research ANd Development, the post-WWII non-profit 'think tank' with government contracts for research on issues of intercontinental warfare.) Even without the RAND study, Air Force leaders feared the Soviets might already possess a more significant bomber force than the US, the fear of a "bomber gap" based on experience from Korea.

When the Cold War dawned following World War II, the Russians cloaked entire cities in secrecy, and any maps gave no hint of their actual location, or in some cases, even their existence. In 1949, the Soviet Union detonated an atomic bomb, and by October 1951, the US saw signs that a Soviet Tu-4 bomber had dropped a nuclear weapon in an air-burst test. The USSR followed this with the detonation of the thermonuclear weapon in 1953. Early intelligence estimates projected the Kremlin as having as many as 600 Tu-4 bombers in service and up to 100 atomic bombs in the stockpile, raising fears of a Soviet strike. Newly inaugurated President Eisenhower had a "Pearl Harbor complex" that genuinely convinced him of the need for accurate reconnaissance.

The US entered the Korean War with propeller-engine bombers and fighters virtually unopposed in the beginning. When the Soviets provided North Korea with the agile MiG-15, suddenly US pilots found themselves engaged in uneven conflict with Soviet pilots in fast jets. In response, US P-51 squadrons rapidly converted to jet fighters, including the F-80 and F-86. Soon, large-scale jet-versus-jet air battles took place along the disputed Korean border in what became known as "MiG Alley."

No one wanted another "MiG" surprise. Memories of the Soviet escalation of the war in the Far East, combined with the RAND study, prompted the Air Force to establish the Intelligence Systems Panel (ISP). However, this new advisory group proposed ways for implementing the construction of high-flying aircraft and high-resolution cameras. Experts looked at the BEACON HILL Study Group report for recommendations. The CIA contributed Edward L. Allen of the Office of Research and Reports (ORR) and Philip Strong of the Office of Scientific Intelligence (OSI) to the group.

When the Intelligence Systems Panel first met at Boston University on 3 August 1953, Strong emphasized the poor state of US knowledge of the Soviet Union. He informed the others that the best intelligence available on the Soviet Union's interior dated back to the German Luftwaffe imagery gained during World War II. Worse yet, the German photography covered only the Soviet Union west of the Urals, west of the Volga River, and only a few of the essential regions. Picture postcards made up some of the "intelligence."

The English Electric Canberra, the highest-flying aircraft available to America and its allies at the time, reached 48,000 feet. The British already produced a photo-reconnaissance variant, but the Air Force instead asked English Electric's help for their further modification of the Martin B-57 version with enlarged wings and other changes.

However, this aircraft, the American-licensed production of the Canberra, with enlarged and computer-designed wings, new engines, and a lighter-than-normal airframe, reached 67,000 feet. However, ARDC bandaged onto it a design change that resulted in the RB-57D plane being able only to reach 64,000 feet.

Additionally, the Air Force, recognizing that the Soviets vastly improved their radar capabilities, had briefed the Intelligence Systems Panel members on the need to seek ways to provide updated photography of all the Soviet Union. The Air Force had discussed with them the multiple sensors used in existing aircraft such as the RB-47, RB-52, and RB-57?

In 1952, the Air Force also started testing Project FICON (an acronym for "fighter conveyor"), under which engineers adapted a giant, 10-engine B-36 bomber to launch and retrieve a Republic RF-84F Thunderstreak reconnaissance aircraft. Other concepts discussed included reconnaissance versions of the Navajo and Snark missiles, a high-altitude balloon program, and a new high-altitude reconnaissance aircraft.

Some of the Air Force's schemes to collect intelligence on the Communist Bloc exceeded current levels of technology. Others, though technically feasible, involved dangerous political consequences such as what occurred to a British overflight of the Kapustin Yar area in early 1953. The British had overflown and photographed the Soviet missile test range there flying a high-altitude Canberra containing modifications inspired by Richard Leghorn's collaboration with English Electric Company designers in 1951. Despite the capabilities of this variant equipped with Rolls-Royce Avon-109 engines, long, fuel-filled wings, a range of 4,300 miles and an altitude of 65,880 feet, Soviet air defenses detected the airplane and nearly shot it down.

Lt Col Joseph J. Pellegrini, assigned to head all US aerial espionage activities and responsible for approving all new reconnaissance aircraft designs, reviewed Leghorn's plan and ordered extensive modifications.

According to Leghorn, Pellegrini showed no interest in a special-purpose aircraft suitable only for covert peacetime reconnaissance missions. He believed that all Air Force reconnaissance aircraft should have the capability of operating under wartime conditions, and, therefore, insisted that Leghorn's design meet the standard specifications for combat aircraft. These requirements included heavily stressed airframes, armor plates, and other apparatus that made a plane too heavy to reach the higher altitudes necessary for safe overflight of the Soviet Bloc.

Leghorn, frustrated by the rejection of his original concept, transferred to the Pentagon in early 1952 to work for Col Bernard A. Schriever, Assistant for Development Planning to the Air Force's deputy chief of staff for Development.

By 1955, after its alteration by Pellegrini's staff, Leghorn's concept, the RB-57D achieved an altitude of only 64,000 feet. The USAF deployed this model in 1956 as "Black Knight."

FORMATIVE YEARS OF AIR RESEARCH AND DEVELOPMENT COMMAND

Before tasking the CIA to produce a plane capable of reconnaissance missions overflying the Soviet Union, the ARDC, the Air Research and Development Command, United States Air Force (Andrews Air Force Base), shared the concerns for this need as did the civilian committees. Before becoming general officers and involved with ARDC, most personnel came as highly educated engineers who had previously participated in the modifications to the B-29s used to drop the first atomic bomb, under Project Silverplate at Wright Field. Unlike General LeMay's Strategic Air Command, the Air Technical Services Command (ATSC) saw the need to advance aviation technology and did contribute to the reconnaissance needs of the United States.

Most of these generals received an assignment to the ARDC and its successor organizations as a step in their becoming flag officers. Consequently, each officer assignment usually limited the officer to a maximum of two years, followed by either reassignment, promotion, and reassignment, or retirement. (The military services' personnel promotion and reassignment protocol later became a significant factor in the president choosing the CIA over the Air Force to run the secret flight-test facility that would become known as Area 51.)

ATSC organized the Air Research and Development Command (ARDC) in the fall of 1950 as an outgrowth of the Air Materiel Command's (AMC) engineering capability that the ATSC had developed over the years. The ATSC intended for the Air Research and Development Command, ARDC, to be an organization that emphasized the growing importance of aviation technology as a separate organization from the AMC. Reporting to ARDC would be such famous organizations as The Wright Air Development Center (WADC), which had earlier names of Air Development Center

(ADC) and Air Development Force (ADF).

WADC attempted to bring all the labs and flight tests under one central management structure. The key organizations in WADC were Plans and Operations, Division for Aeronautics (containing such vital labs as the equipment lab, aircraft lab, and the power plant lab), flight test, research, weapons components, and weapons systems. Many of the key personnel from the AMC's engineering arm and Headquarters USAF became the initial cadre for ARDC, and it later became the Air Force Systems Command (AFSC). Its lead organization in the new management structure would be WADC, and it later became The Aeronautical Systems Division (ASD).

From the cadre of outstanding ARDC engineers came two officers who would play such a key role in our nation's quest for national reconnaissance assets. These were Col Osmond Ritland and Col Jack A. Gibbs. Both officers received a promotion to general officer rank for their performance in these critical airborne and space-borne projects. Ritland helped CIA's Dick Bissell lead the development phase of the U-2, and he took a lead role in the Corona reconnaissance satellite, again working with Dick Bissell. Gibbs helped Bissell manage the operational aspect of the U-2 and the search for the CIA's successor aircraft until the summer of 1958 and as an engineering manager for the very first attempt to purposely reduce the radar cross-section (RCS) of a government aircraft, the U-2, under project Rainbow starting in 1956. Both officers were natural selections for these assignments because Ritland had been chief of the Aircraft Lab, and Gibbs had been chief of the Wind Tunnel Branch and chief of the Aircraft Lab in the AMC's organizational structure.

While ARDC initially located in Washington, DC, MGen David Schlatter, a senior member of the USAF Liaison Committee, Atomic Energy Commission, Washington, DC, and assistant deputy chief of staff operations for atomic energy played a significant role in setting up the management structure. As ARDC's first commander, General Schlatter, West Point graduate in the Class of 1923, later a lieutenant general, held this position until June 1951 when then Lt Gen Earle Partridge replaced him. General

Partridge held this position until June 1953. Early on and until October 1951, the ARDC Vice Commander BGen Ralph Swofford, West Point graduate of Class of 1930, managed vital engineering development in the AMC.

During October 1950, ARDC moved to Wright-Patterson AFB (WPAFB) in Dayton, Ohio, where it remained until the spring of 1951 when it relocated to Baltimore, Maryland, on 11 May 1951. During November 1950, BGen Fred Dent, Schlatter's head of the engineering division of AMC, held the additional duty as commanding general of the Air Development Center (ADC). In April 1951, the ADC changed its name to the Air Development Force (ADF), and Dent continued to command this organization until ADF changed to WADC. Brigadier General Dent controlled WADC through all these name changes. Upon the reorganization of AMC engineering arms into ARDC, Gen Schlatter assumed additional duties as director of research and development at the ARDC. Upon the restructuring, Gen Schlatter replaced Gen, Sam Brentnall, who had been the director of R&D of the AMC and managed such vital capabilities as the power plant lab and the aircraft lab.

MGen Sam Brentnall, West Point graduate in the class of 1928, an All-American football player from 1925-1928, and a graduate of the Air Corps Engineering School in 1938 had studied at Stanford University, graduating with a mechanical engineering-aeronautical degree. In the WWII and post-war era, Gen Brentnall served in the Pentagon as deputy chief of the Materiel Division at Hq, USAAF, and Wright Field in the AMC in such critical postings as deputy director of R&D and became the director of R&D during October 1949.

BGen Fred Dent, West Point graduate in the class of 1929, had entered military service in the Corps of Engineers in 1929 but transferred to the Air Corps in 1930. Dent, also a graduate of the Air Corps Engineer School at Wright Field in 1935, upon graduation, entered the Massachusetts Institute of Technology and graduated in July 1938 with an MS in Aeronautical Engineering. Having served early on as a test pilot at Wright Field, he had held several management positions in the 8th Air Force as commander of a bomb group and the commander of the 95th Bomb

Wing.

During April 1945, Dent became the chief of the engineering branch, material division at Hq, USAAF. In Jan 1947, he became the head of the equipment lab at Wright Field, and in Dec 1948, he became deputy chief of the engineering division at the AMC. In Nov 1950, Dent received a promotion to brigadier general and named the head of the engineering division there, replacing BGen Ralph Swofford. Dent assumed command of WADC in 1951.

MGen Donald Putt became the second commander of WADC during June 1953 with a promotion to lieutenant general. A highly skilled aviator with a BSEE from Carnegie Tech in 1928, a graduate of the USAAF engineering school in 1937, and an MS in Aeronautical Engineering in 1938 at Caltech, Putt had an outstanding background in engineering and intelligence of the pre-WWII era and WWII. He had served in the Air Technical Services Command at Wright Field. He became involved in Project Silverplate while at Wright Field in a close association with Colonel Demler and Colonel Gibbs, both later leaders in WADC.

In 1952 Putt received an assignment to Wright Field as chief of WADC until he became the Vice Commander of ARDC in Baltimore, Maryland, and then the Commander of ARDC during June 1954. Lt Gen Thomas Power replaced Putt as commander of ARDC in 1954. During the years of the quest for a national reconnaissance capability, Gen Putt served as the Commander of ARDC until he received a promotion to the rank of lieutenant general with an assignment to Hq, USAF in 1954 as the Deputy Chief of Staff for Development and the military director of the USAF Scientific Advisory Board (SAB).

Upon Lt Gen Putt's retirement after serving as DCS for R&D at Hq, USAF MGen Roscoe C. 'Bim' Wilson replaced him with a promotion to lieutenant general. Gen Wilson and MGen Sam Brentnall both graduated from West Point in the class of 1928. Wilson graduated from the Air Corps Engineering School in 1933 and then served at Wright Field in engineering and science as well as an aviator. In 1957, while assigned to WESG as the Air Force member to WESG in the Office of the Assistant Secretary of Defense for R&D, Gen

Wilson played a vital role in the late Gusto era and its transition to the CIA Project OXCART.

MGen Albert Boyd, while a brigadier general, became Vice Commander of WADC, ARDC in February 1952, and Commander of WADC during June 1952. He stayed in command of WADC as a major general until 1955. Boyd became Deputy Commander of Systems at Hq, ARDC, in Baltimore during August 1955. Boyd retired during November 1957. Like most of the other general officers in the leadership of WADC, Boyd had advanced in rank from an aviation cadet, enhanced his education in engineering, and had previously served during World War II with future WADC leadership such as BGen Jack Gibbs. Most of the senior officers of WADC knew each other from having served on Project Silverplate. These included then Colonels Jack A. Gibbs, Homer Boushey, and Norm Appold.

MGen Marvin Demler served as chief of staff and then vice commander of WADC. In 1954, Marv Demler transferred to Hq, ARDC in Baltimore, and 1956 became the ARDC Deputy for Research and Development. In 1958 he would receive an assignment to Hq, USAF, to work for Lt Gen Putt and then for Lt Gen Wilson as Director of Research and Development, Director of Aerospace Systems, and Director of Advanced Technology.

After the CIA contract award to Lockheed for the U-2, Gibbs became part of the ARDC team briefed by CIA's Dick Bissell. Gibbs went on to play critical roles on AQUATONE, RAINBOW, GUSTO, and then OXCART until the summer of 1958. Col William Burke replaced Colonel Gibbs in the summer of 1958 when Colonel Gibbs returned to the USAF.

USAF Colonel (later MGen) Osmond suggested Groom Lake in Nevada to Richard Bissell for a top-secret training facility, a hidden venue that later became the center of Area 51 flight test operation.

Ritland had served as aircraft lab chief until 1949 when he received an assignment to the Special Weapons Program at Kirtland AFB, New Mexico, during February of 1950. There, Ritland organized and commanded the 4925th Test Group (Atomic), to become responsible for developmental testing of all equipment needed in attaining an Air Force nuclear weapons capability. His test group also assisted the effects tests and led the development of new airborne sampling techniques for the detection and analysis of atomic weapons tests. Some of the CIA U-2 pilots participated in this atomic cloud sampling program.

Colonel Ritland attended the ICAF, graduating in 1954. Ritland received an assignment to Hq, USAF DCS R&D as a special assistant to the chief of staff for R&D, then Lt Gen Don Putt. In this assignment, he worked with CIA's Richard Bissell to help conceive and develop the airborne national reconnaissance capability realized in the U-2. Ritland stayed in this position until March 1956 when Col Jack A Gibbs replaced him. Ritland would continue to work with CIA's Dick Bissell on the Corona reconnaissance satellite in his appointment to the ARDC.

USAF Col (later BGen) Jack A. Gibbs graduated from Oregon State College in 1936 with a BS in mechanical engineering with an aeronautics specialty. In 1947, Gibbs earned an MS in aeronautical engineering from Caltech, specializing in supersonic aerodynamics. He received a regular commission in the US Army Corps of Engineers in 1938. He then attended pilot training at Randolph and Kelly Fields, graduating as an Air Corps pilot in the class of 39-B.

At Wright Field, Colonel Gibbs worked with crucial USAF R&D officers Colonel Ritland, Brigadier General Dent, Major General Brentnall, Major General Schlatter, Major General Putt, Brigadier General Swofford, Colonel Demler and Brigadier General Boyd, all of them also highly regarded by senior officials of the AEC and NACA.

In early 1956, then Colonel Gibbs became the CIA's Deputy Project Director for the U-2, reporting to Dick Bissell to help manage the operational phase of project AQUATONE. In the assignment to the CIA, Gibbs also managed research and development projects for RAINBOW and GUSTO, which led to OXCART in 1958.

Throughout his career in R&D, Col Norm Appold, a highly skilled aeronautical engineer with many years in R&D in AMC, WADC, ARDC, and AFSC, brought to the table his specialty of propulsion systems for aircraft and rockets. He had a degree in chemical engineering from the University of Michigan before WWII and an MS in Aeronautical Engineering from Caltech in 1947. He, also a skilled pilot, had flown B-24s in the famous Ploesti raids on Romanian oil fields in 1943. ARDC/AFSC Commanders Bernard Schriever and Thomas Power highly regarded him as a crucial player in the ARDC quest for a national reconnaissance capability in the 1952-1962 era.

AEC and the Armed Forces Special Weapons Project in nuclear weapons

Appold's involvement included the Bell X-16, RB-57D, and Suntan projects. He also served as a member of the ARDC review team for the CIA's Project OXCART, which came from the GUSTO efforts run by the CIA. He served as the ARDC program manager for Suntan, which attempted to leapfrog the CIA's U-2 with the hopes of regaining USAF control of covert overflight of the USSR.

BGen Jack Gibbs emerged from this crop of highly qualified and motivated military officers and the politics involved to create the U-2 and serve as Richard Bissell's deputy for Project AQUATONE.

We do well to recall that President Dwight D. Eisenhower authorized the U-2 and its follow-up the A-12 Blackbird plane, plus the reconnaissance satellites starting with Corona. Other books go into greater detail on the leadership role taken by President Eisenhower in coming to grips with the critical times of the Cold War. However, this book will explore some of his thinking and actions in more detail later. For now, consider how senior analyst Dino A. Brugioni assessed President Eisenhower's role of oversight as Commander-in-Chief of air research and development (Brugioni 2010, p. x-xi):

"With the advent of atomic weapons, Eisenhower recognized that global wars had become inconceivable. Probably more than anything else during his administration, Eisenhower realized the value of aircraft and satellites in acquiring intelligence. Establishing a new primary source of objective US intelligence gave Eisenhower a strategic advantage that allowed him to define US positions in world affairs with clarity and emphasis. And much of what he did occurred behind the scenes. R. Cargill Hall, an eminent historian, may have stated it best: 'In his memoirs, Eisenhower mentioned none of the technical advances or changes in intelligence operations for which he claimed responsibility. One would be hard-pressed to think of a contemporary politician able to resist claiming credit for such an intelligence revolution, or willing to carry the secrets with him to the grave.'"

Lt Gen Donald L. Putt

Gen Frank F. Everest

BGen Jack A. Gibbs - Deputy Project Director of the U-2 Program Under Richard Bissell. - Photo credit - Jim Gibbs*

Allen Welsh Dulles, director of Central Intelligence from February 26, 1953, to November 29, 1961, under Presidents Dwight Eisenhower and John F. Kennedy. *Wikipedia.*

Chapter 4 - Searching for New Aerial Reconnaissance

National Security Council Directive 5412 on Covert Operations

During the Korean War, President Truman had signed the CIA Act stipulating the classification of the CIA's activities and its budget, enabling any other government agency to transfer funds to the CIA "without regard to any provisions of law." The Act further granted the US government plausible deniability of any responsibility for exposed CIA actions or activities.

Also, the Truman administration's concern over Soviet "psychological warfare" prompted the new National Security Council (NSC) to authorize peacetime covert operations. The NSC made the Director of Central Intelligence (DCI) responsible for psychological warfare, at the same time establishing covert action as an exclusive executive branch function.

The CIA became a natural choice for this function, at least in part because the agency controlled unvouchered funds that enabled it to undertake operations with minimal risk of exposure in Washington.

Everyone Covering Their Butts

Peacetime covert activities were new to the United States. However, during the Korean conflict, the CIA's covert operations had proliferated. The Agency had adapted to the demanding situation. Wartime commitments and other missions soon came to make covert action the most expensive and bureaucratically prominent of the CIA's activities. Secrecy and CIA successes gave the Departments of State and Defense cause to become jealous of the CIA alone having this power. As a result, the military created a new rival covert action office in the Pentagon.

The NSC had directed the CIA to conduct "covert" rather than merely "psychological" operations. In these operations, the CIA hid from unauthorized persons any US government responsibility for activities planned or executed. If uncovered, the US government could plausibly deny any responsibility for them. Such operations excluded those undertaken during armed conflict.

During wartime, the military forces could provide "deception for military operations" without "engaging in espionage."

Then, President Eisenhower, on 15 March 1954, reaffirmed the CIA's responsibility for conducting covert actions abroad. However, his confirmation came with the caveat that the CIA must advise in advance the representatives of the secretary of state, the secretary of defense, and the president before initiating any major covert action programs. Eisenhower went even further by designating a Planning Coordination Group as the body responsible for coordinating covert operations. This "Special Group," emerged as the executive body to review and approve covert action programs carried out by the CIA.

The Covert Actions Oversight Group changed its name to the 303 Committee, the name coming from National Security Action Memorandum No. 303 dated 2 June 1964. McGeorge Bundy, then the National Security Advisor, became the chairman of the committee, and the successor name to the "Special Group" became the 40 Committee. (On February 18, 1976, President Gerald Ford issued Executive Order 11905 to replace the 40 Committee with the Operations Advisory Group. The new group was composed of the President's Assistant for National Security Affairs, the Secretaries of State and Defense, the Chairman of the Joint Chiefs of Staff, and the Director of Central Intelligence.)

Air Force's Proposal for a New Reconnaissance Aircraft

To stabilize the situation and reduce Cold War tensions, Eisenhower in 1955 had earlier proposed to Soviet Premier Nikolai Bulganin at the 'Big Four' summit in Geneva that both their nations along with others arrange for unfettered overflights to photograph each other's military installations. President Eisenhower promoted this concept as "Open Skies."

Bulganin at first appeared to respond favorably, but First Secretary Khrushchev made his contempt clear to Eisenhower when the two men talked together at the summit. Khrushchev, whose primacy in Moscow was not yet clear to American experts, informed Eisenhower that he did not agree with Bulganin and considered the overflight suggestion to be a transparent espionage plot. He later told his son, Sergei, that he believed Washington would use the flights to refine its targeting plans for a nuclear strike against the Soviet Union.

When Khrushchev rejected this "Open Skies" proposal, Eisenhower decided to give the green light for the development of a reconnaissance aircraft capable of overflying the Soviet Union above the range of Soviet countermeasures. While Eisenhower acknowledged the need for such missions, he feared eventual shoot-down of one of the planes. His fear proved well-founded.

The US needed an aircraft capable of evading such radar equipment as the SCR-584 used by both the East and West. The Air Force had transferred several SCR-584 radar sets to NACA for use in tracking the X-1 and X-2 planes. The Air Force and CIA saw that for this, NACA would have to be able to modify and improve the radar's tracking capabilities. If NACA could do this, then so could the Soviets. (For the X-15 flight testing, NACA's successor now named NASA successfully improved the radar and proved the need for a new reconnaissance aircraft capable of evading detection by the same model radar that Russia used.)

To meet this need, the Wright Air Development Center oversaw the design competition of the two Air Force projects (dubbed "Bald Eagle") that eventually produced the Bell X-16 concept. The development of the modified RB-57 platform was also well underway. About this time, Eisenhower became concerned about the Soviet Union shooting down such an aircraft over their territory and considering it an act of war. Instead, he believed the new reconnaissance platform should be a civilian aircraft with a civilian pilot to lessen the potential for provocation in the eyes of the Communists.

In 1952, with interest in high-altitude reconnaissance growing, several Air Force agencies had begun developing an aircraft to conduct such missions. At the suggestion of Richard Leghorn, the Air Research and Development Command (ARDC) had awarded Martin Aircraft Company a contract to modify a single B-57 with high-lift wings and improved engines.

At about the same time, John Seaberg was working with Woldemar Voigt and Richard Vogt at the Wright Air Development Center (WADC) on the development of a new aircraft whose design combined the high-altitude performance of the latest turbojet engines with high-efficiency wings to reach ultra-high altitudes. Rather than modifying an existing airframe, they started with a "clean sheet" and built a set of specifications.

When, during March 1953, Seaberg submitted his request for proposals, Wright Air Development Center leadership decided not to approach the major airframe manufacturers, believing a smaller company would give the new project a higher priority and produce a better aircraft more quickly. During July 1953, the Bell Aircraft Corporation of Buffalo, New York, and the Fairchild Engine and Airplane Corporation of Hagerstown, Maryland, both received study contracts to develop designs for the new high-altitude reconnaissance aircraft while they continued to pursue the idea of a large winged version of the Martin B-57. With two engines, the RB-57 could carry a heavier side-looking radar and camera, but not to an extreme altitude.

The Intelligence Systems Panel, chaired by James Baker of Harvard University, showed an interest in the potential of high-altitude reconnaissance. During a May 1953 meeting of the Intelligence Systems Panel at Boston University, representatives from industry and academia discussed the problem and focused on the limitations inherent in the use of a modified B-57. Allen Donovan of Cornell Aeronautical Laboratory noted that he did not believe the aircraft would meet the reconnaissance requirements addressed in the BEACON HILL Report. He felt the airframe too heavy to achieve the desired altitudes with a useful sensor payload, making the RB-57 vulnerable to Soviet air defenses. Donovan insisted that they needed "a penetrating aircraft capable of flying above 70,000 feet for the entire mission."

Donovan met with an old Air Force acquaintance, Lockheed Vice President L. Eugene Root, and learned about the Air Force's design competition for a high-altitude reconnaissance plane. Also, Philip Strong of the CIA had recently informed him of Lockheed Aircraft Corporation already designing a lightweight, high-flying aircraft. Upon Donovan's telling the Intelligence Systems Panel, Baker urged him to evaluate the Lockheed design and asked him to gather additional ideas from other aircraft manufacturers concerning high-altitude aircraft. Donovan couldn't make the trip to California until late summer, but at Lockheed, he saw what the Intelligence Systems Panel needed.

The Lockheed Skunk Works

In the early days of World War II, the Germans had explored the latest revolution in aeronautical technology: jet-powered aircraft. Test flights of a German jet fighter began in 1941, with full-scale production starting in 1944. In the US, the War Department turned to Lockheed to build a jet fighter prototype to counter the emerging threat in the skies over Europe.

Realizing the essence of time, chief research engineer Kelly Johnson declared that he could design and build a new jet fighter aircraft in just 180 days. He proposed creating a straight-winged, single-engine plane that would come to be known as the XP-80. Once Lockheed received a contract, he assembled a handpicked team of 22 engineers and manufacturing people and got to work.

With the war effort in full swing and Lockheed amid producing numerous other fighter and bomber aircraft at the company's Burbank facilities, the company lacked available space for Johnson's modest yet critical effort. Consequently, the organization operated out of a rented circus tent.

Johnson quickly established a highly efficient and streamlined process that enforced secrecy and fostered close cooperation between the design and manufacturing teams. Living up to his motto, "Be quick, be quiet, be on time," Johnson cautioned each member of his team to carry out the design and production of the new XP-80 in strict secrecy and never to discuss the project outside the small organization. He even warned the team members to be careful about how they answered the phones.

It happened that the tent sat next to another manufacturing plant that produced a strong and foul odor. Irv Culver, a Lockheed engineer and a fan of Al Capp's comic strip, "Li'l Abner," espoused a running joke about a mysterious and malodorous moonshine still deep in the forest called the "Skonk Works." Supposedly, the Skonk Works brewed a potent beverage called Kickapoo Joy Juice from skunks, old shoes, and other ingredients.

One day, Culver's phone rang, and he answered by saying, "Skonk Works, inside man Culver, speaking." His contemporaries quickly adopted that name for their mysterious division of Lockheed. Later, to avoid copyright issues, "Skonk Works" became "Skunk Works." Thus, the legendary Lockheed Skunk Works gave birth to a name more commonly used than the official designation: Advanced Development Projects. Although Johnson himself hated the moniker, the once informal nickname eventually became a registered trademark of the company.

Johnson's people worked to complete the XP-80 project rapidly under a veil of total wartime secrecy. The company granted the Skunk Works full autonomy, and under Johnson's oversight, the team produced the XP-80 prototype in just 143 days. The first flight took place during January 1944, and production models entered service soon after. The P-80 (later F-80) Shooting Star proved highly successful. Although it saw only limited service in World War II, the Air Force used this type extensively in Korea. It spawned a family of later variants, including the RF-80, T-33, and F-94. More importantly, its development expanded the Skunk Works organization and work philosophy that played a key role in the development of several revolutionary Cold War aircraft.

About the time of Donovan's visit to Lockheed, the company also became aware of the design competition in the fall of 1953 from John H. (Jack) Carter, now the Assistant Director of Lockheed's Advanced Development Projects Division. While at the Pentagon on business, Carter sought out Eugene P. Kiefer, an old friend and colleague from the Air Force Office of Development Planning (more commonly known by its office symbol, AFDAP). Kiefer told Carter about the competition for a high-flying aircraft and expressed his opinion of the Air Force going about the search in the wrong way by requiring the new aircraft to be suitable for both strategic and tactical reconnaissance.

When Carter returned to California, he sought out Lockheed vice president L. Eugene Root, the top civilian official at AFDAP. Carter proposed to Root that Lockheed submit a design for an aircraft capable of reaching altitudes between 65,000 and 70,000 feet, and a speed near Mach 0.8. Carter emphasized the need to accelerate development to allow for operational capability as soon as possible.

Carter noted that the proposed aircraft would have to reach altitudes of as high as 70,000 feet and correctly forecast, "If a practical aircraft at speeds near Mach 0.8 could realize the extreme altitude performance, it should be capable of avoiding virtually all Russian defenses until about 1960." Carter added, "To achieve these characteristics in an aircraft which would have a reasonably useful operational life during the period before 1960 would, of course, require very strenuous efforts and extraordinary procedures, as well as a nonstandard design philosophy."

To save weight and increase altitude, Carter recommended eliminating conventional landing gear and building a lightweight airframe that did not need to meet combat load factors. Naturally, the company tasked Kelly Johnson with designing such an airplane and overseeing its construction.

Already one of the world's leading aeronautical engineers, Johnson had many successful military and civilian designs to his credit by now, including the P-38, F-80, F-104, and Constellation. He had earned a reputation for completing projects ahead of schedule while working in the Skunk Works. This reputation helped Lockheed earn admission to the competition.

Johnson started with a radical design based upon the fuselage of the F-104 jet fighter but incorporating a high-aspect-ratio sailplane wing. The proposal, designated CL-282, saved weight and increased the aircraft's altitude capabilities by stressing the airframe to only 2.5 units of gravity (g's) instead of the military specification strength of 5.33 g's. For the power plant, Johnson selected the General Electric J73/GE-3 non-afterburning turbojet engine with 9,300 pounds of thrust (the same engine chosen for the F-104, that became the basis for the CL-282 design).

The CL-282 adopted many of its design features from gliders. Thus, the wings and tail were detachable. Instead of conventional landing gear, Johnson proposed using two skids and a reinforced belly rib for landing—a standard sailplane technique—and a jettisonable, wheeled dolly for takeoff. Other features included an unpressurized cockpit and a 15-cubic-foot payload area that could carry 600 pounds of sensors. The CL-282's maximum altitude would be 70,000 feet with a 2,000-mile range.

Johnson submitted the CL-282 design to BGen Bernard Schriever's Office of Development Planning in early March 1954. On the recommendation of Kiefer and Weinberg, Schriever asked Lockheed to submit a specific proposal for development and production. Lockheed did so the following month, and Johnson followed up with a proposal for Lockheed to construct and maintain a fleet of 30 aircraft.

In early April, Johnson presented a full description of the CL-282 along with construction and maintenance estimates to a group of senior Pentagon officials that included Schriever's superior, LtGen Donald L. Putt, Deputy Chief of Staff for Development, and Trevor N. Gardner, Special Assistant for Research and Development to the Secretary of the Air Force.

Afterward, Johnson noted, the civilian officials were very much interested in his design, especially so with Gardner, but the generals were less enthusiastic. According to Weinberg, General LeMay stood up halfway through the briefing, took his cigar out of his mouth, and told the briefers, "This is a bunch of shit. I can do all that stuff by putting cameras in my B-36s." Besides saying that, LeMay added his disinterest in a single-engine plane with no wheels or guns. The general then left the room, remarking that this whole business was a waste of his time.

Meanwhile, the CL-282 design proceeded through Air Force development channels. In mid-May, Seaberg and his colleagues carefully evaluated the Lockheed submission and eventually rejected it in early June. Johnson's choice of the unproven General Electric J73 engine became one of their main reasons for doing so. Engineers at Wright Field considered the Pratt and Whitney J57 to be the most powerful engine available, and the designs from Fairchild, Martin, and Bell all incorporated this power plant. They perceived the absence of regular landing gear as a shortcoming of the Lockheed design.

Air Force reconnaissance experts who gained their practical experience during World War II in multi-engine bombers preferred multi-engine aircraft, which became another factor in the rejection of the CL-282. Also, aerial photography specialists in the late 1940s and early 1950s emphasized focal length as the primary factor in reconnaissance photography and, therefore, preferred large aircraft capable of accommodating long focal length cameras.

This preference reached an extreme in the early 1950s with the development of the cumbersome 240-inch Boston camera, a device so large that it required partial disassembly for installation in any aircraft that carried it. Senior photo analyst Dino Brugioni describes the Boston camera as weighing three tons: "The first photo Arthur Lundahl and I saw from this was of New York City. The aircraft was seventy-two miles away, and yet we could see people in Central Park." – Brugioni 2010, p. 75

Many Air Force officers felt that two engines were always better than one because besides being able to lift more weight if one failed, the spare engine could get the aircraft back to base. However, with the RB-57D: "Should one engine fail at low airspeed and full power, corrective reduction in power of the other engine could not be done quickly enough to maintain control of the aircraft." – Mikesh p. 141

Aviation history showed that single-engine planes had often proved more reliable than multi-engine aircraft. According to the black humor of some airmen, "The second engine took you to the crash site." Furthermore, a reconnaissance aircraft deep in enemy territory would have little chance of returning if one of its engines failed, forcing the plane to descend low enough for missile or gun attacks.

January 1954 found three firms with submitted proposals for a high-altitude jet. Fairchild's entry offered a single-engine plane known as the M-195, with a maximum altitude potential of 67,200 feet. Bell Aircraft offered a twin-engine craft called the Model 67 (later designated X-16) with a maximum altitude of 69,500 feet. Martin's Model 294 design

called for a big-wing version of the B-57 with a cruise altitude of 64,000 feet.

During March 1954, Seaberg and other engineers at Wright Field, having evaluated the three contenders, recommended adopting both the Martin and Bell proposals. They considered Martin's version of the B-57 an interim project that they could quickly complete and deploy while Bell developed the more advanced concept.

Air Force Headquarters had quickly approved Martin's proposal to modify the B-57 and expressed great interest in the Bell X-16. By this time, rumors of the competition for a new reconnaissance airplane had reached the Lockheed, which submitted their unsolicited proposal.

Consequently, on 7 June 1954, Lockheed received a letter from the Air Force rejecting the CL-282 proposal because of it having only one engine, being too unusual and because of the Air Force's commitment to the Martin B-57. By this time, the Air Force had also selected the Bell X-16 with a formal contract. Despite this setback, Lockheed continued to work on the CL-282 and began seeking new sources of support for their plane.

During Allen Donovan's visit to the Skunk Works in the fall of 1953 for the Intelligence Systems Panel as a follow-up to the May 1953 meeting at Boston University, Lockheed chief engineer Johnson had shown him the plans for the company's unsuccessful entry. A lifelong sailplane enthusiast, Donovan recognized the CL-282 design as being the desirable single-engine jet-propelled glider design that could attain the extreme altitudes necessary for carrying out reconnaissance of the USSR.

Upon his return from visiting the Skunk Works, Donovan contacted James Baker to suggest an urgent meeting of the Intelligence Systems Panel. Other commitments, however, forced the panel to wait until September 1954 to hear Donovan's report during a session at the Cornell Aeronautical Laboratory.

By September 1954, the Air Force had already committed to developing both the RB-57 and the Bell X-16. Despite Intelligence Systems Panel support for Lockheed's CL-282, these commitments prevented Trevor Gardner, Assistant

Secretary of the Air Force for Research and Development, from offering any funds to Lockheed to pursue the concept. Lockheed would have to look elsewhere, and so the company turned to scientists serving on some of the high-level advisory committees.

When the Intelligence Systems Panel finally met during September 1954 to hear Donovan's report on his visit to Lockheed, those in attendance were somewhat upset to learn that the Air Force had not informed them of the high-altitude aircraft design competition. Their annoyance slightly abated once Donovan began describing Kelly Johnson's rejected design.

To design a successful high-altitude reconnaissance aircraft capable of flying above 70,000 feet, Donovan insisted on three essential requirements: single-engine, a sailplane-type wing, and an airframe built with low structural load factors.

Donovan favored a single-engine configuration because it would be substantially lighter than a multi-engine aircraft. He felt a sailplane-type wing would be necessary to provide maximum lift in the rarefied atmosphere of extreme altitudes. Engineers estimated that above 70,000 feet, the power curve of a jet engine would fall to just six percent of its sea-level thrust. Most of all, Donovan emphasized the need for low structural load factors. He explained how to strengthen the wings and the wing root area closest to the fuselage. This would require not meeting standard military specifications that allow the airframe to withstand high speeds and sharp turns, but that added too much weight.

Donovan insisted that only Kelly Johnson's CL-282 met those requirements. He noted that because it would fly well above any Soviet fighters, the CL-282 did not need to meet combat aircraft specifications. He ultimately convinced the other Intelligence Systems Panel members of the merits of the CL-282 proposal, presaging a conflict with Air Force officials who had already rejected it.

Lockheed CL-282 Supporters and the CIA

Although the Air Force's uniformed hierarchy had decided for the Bell and Martin aircraft, some high-level civilian officials continued to favor the Lockheed design. Trevor Gardner, Special Assistant for Research and Development to USAF Secretary Harold E. Talbott and the most prominent proponent of the Lockheed proposal, enjoyed many contacts in West Coast aeronautical circles. Before coming to Washington, he headed the Hycon Manufacturing Company that produced aerial cameras in Pasadena, California. He attended Johnson's briefing on the CL-282 at the Pentagon in early April 1954 and believed that this design showed the most promise for reconnaissance of the Soviet Union. Gardner's assistant, Frederick Ayer, Jr., and Garrison Norton, an adviser to Secretary Talbott, shared this belief.

Gardner and others preferred the CL-282's higher potential altitude and smaller radar cross-section. They tried to win over Gen LeMay, but he remained uninterested in an unarmed aircraft. This left Gardner, Ayer, and Norton with little choice when they approached Philip G. Strong, the CIA chief of the Office of Scientific Intelligence's operations staff.

At that time, the Agency's official involvement in overhead reconnaissance limited itself to advising the Air Force on the problems of launching large camera-carrying balloons for reconnaissance flights over hostile territory. Strong, a colonel in the Marine Corps Reserve who later advanced to the rank of brigadier general, served on several Air Force advisory boards and kept well informed on developments in reconnaissance aircraft design. From Gardner's enthusiasm for the CL-282, Strong at first thought Air Force officials also supported the Lockheed design until he learned of the Martin RB-57 and Bell X-16.

Gardner, Norton, and Ayer met with Strong in the Pentagon on 12 May 1954, six days before the Wright Air Development Center began its evaluation of the Lockheed proposal. Gardner described the CL-282 and showed the drawings to Strong, who, after the meeting, summarized his impressions of the Air Force's search for a high-altitude reconnaissance aircraft.

At that time, the military services were responsible for conducting reconnaissance overflight missions while the CIA relied on more traditional collection methods. Even the Director of Central Intelligence, Allen Dulles, opposed the idea of the Agency getting involved with aerial reconnaissance, favoring human agents over technical intelligence-gathering methods. Unfortunately, camera-carrying bombers were inadequate for the task.

At about this time, the Intelligence Systems Panel became involved, advising the Air Force and CIA on the need for a new aerial reconnaissance platform. It had become apparent that the RB-57 would not meet the necessary 70,000-foot altitude requirement. Proponents within the CIA told the panel about the CL-282 and the aspects of its design advantages that the Air Force saw as flaws. The CL-282 appealed to Allen Donovan because he believed the panel sought a sailplane-type, high-altitude aircraft.

President Eisenhower's Concerns

The Air Force's reluctance to risk overflying the Soviet Union and the capability of the Soviets to launch a surprise nuclear attack on the United States concerned the Eisenhower administration. In early 1954, Trevor Gardner received a RAND Corporation study showing that a Soviet surprise attack might destroy 85 percent of the Strategic Air Command bomber force before it could get off the ground. This acted as a wake-up call.

Richard "Dick" Bissell, the GS-18-rated CIA Manager who Selected and Established Area 51 was known as "Mr. B" when he visited the Lockheed Skunk Works and the Area 51 Flight Test Facility

Gardner met with Dr. Lee DuBridge, president of the California Institute of Technology and chairman of the Office of Defense Mobilization's Science Advisory Committee. Afterward, Gardner criticized the committee for not dealing with such essential problems as the possibility of a surprise attack, a "nuclear Pearl Harbor." DuBridge invited him to speak at the Science Advisory Committee's next meeting. After hearing Gardner, the committee members decided personally to approach President Eisenhower.

President Eisenhower told the committee on 27 March 1954 of the discovery of a new model of Soviet bomber, the M-4 Bison. He expressed concern that Russia might use these new long-range aircraft to conduct a surprise attack on the United States and asked the committee to advise him on how to prevent such a possibility from occurring.

In response to the president's request, DuBridge asked MIT President James R. Killian, Jr., to meet with other Scientific Advisory Committee members. They met at MIT on 15 April 1954 and discussed the feasibility of a comprehensive scientific assessment of the nation's defenses.

As a follow-up, on 26 July 1954, Eisenhower authorized Killian to create and lead a panel of experts to study the country's technological capabilities to meet some of its current problems.

Killian set up shop in offices located in the Old Executive Office Building, where he organized 42 of the nation's leading scientists into three special project groups to investigate US offensive, defensive, and intelligence capabilities with an additional communication working group.

The Technological Capabilities Panel (TCP) first met on 13 September 1954. They eventually met on 307 separate occasions over the next 20 weeks for briefings, field trips, and conferences. Members conducted meetings with every major unit of the US defense and intelligence establishments. Afterward, they drafted a report to the National Security Council with the most up-to-date information available on the nation's defense and intelligence programs.

The TCP group chaired by Edwin Land was the one that focused on intelligence. As soon as Land saw Philip Strong's drawing of the Lockheed CL-282 proposal, he telephoned James Baker of the Intelligence Systems Panel to say, "Jim, I think I have the plane you are after."

Chapter 5 - The CIA U-2 Project AQUATONE

Even if the Air Force had gone along with the Lockheed concept, Edwin Land, the brilliant scientist who developed instant photography, believed the military operating the CL-282 during peacetime might provoke a war. Therefore, another member of the panel and he proposed to Dulles through Dulles' aide, Richard M. Bissell, Jr., that the CIA fund and operate this aircraft.

Richard Bissell, a leader, and innovator became a CIA legend because he dared to take risks during one of the nation's darkest periods, the Cold War. As Deputy Director/Plans, a purposefully innocuous title, Bissell was chief of the Agency's clandestine service, a web of spies spread around the world. He, as he later put it, "had taken to covert action like a duck to water."

Bissell seemed an unlikely spymaster. He was bookish, slightly stooped, and a bit clumsy. He was the type who couldn't step on a playing field without breaking a leg, yet he enjoyed the dangerous sport of scaling vertical cliffs in the hills outside New Haven. At Yale, Bissell and his roommates often donned sneakers, smeared their faces with black greasepaint to not reflect the streetlights while climbing and clambering about the steep slate rooftops of the college halls, before dropping through the open window of a startled friend—and demanding cocktails.

Richard M. "Dick" Bissell, Jr., joined the CIA after already having an active and varied career. He essentially had written the Marshall Plan and was the real brains behind the complex task of designing a program for European recovery following World War II. No one doubted his sharp intellect and personality.

Born in Hartford, Connecticut, during September 1909 to a wealthy and influential New England family, he attended the prestigious Groton School in Massachusetts for his primary education. Bissell then went on to study at Yale University and the London School of Economics. In 1934, Bissell returned to Yale, earning his doctorate. He became a full professor at MIT in 1948.

During World War II, he served as the executive officer of the Combined Shipping Adjustment Board, managing American shipping. Following the war, he became the deputy director of the Marshall Plan from 1948 until the end of 1951, when he joined the Ford Foundation staff.

Two years later, he contracted with the CIA to study possible responses by the United States against the Soviet Bloc in the event of a repeat uprising like the East Berlin riots the previous year during June 1953.

Between 60 and 80 construction workers went on strike in East Berlin in response to a 10 percent increase in work quotas. Their numbers quickly swelled for a general strike and protest the next day. One hundred thousand protesters gathered at dawn. They demanded the reinstatement of old work quotas and, later, the resignation of the East German government. At noon, German police trapped many of the demonstrators in an open square. Soviet tanks fired on the crowd, killing hundreds, and ending the protest. From this, Bissell held little hope for clandestine operations against the Soviet Bloc nations.

After that, Bissell joined the CIA in late January 1954, where he would become a legend because he dared to take risks during one of the nation's darkest periods, the Cold War. Under Bissell's leadership, the Agency would develop the U-2 and debunk the myth of the Soviet Union being ahead of America in producing missiles and weapons. "The U-2 program would, in the view of many agency hands, making him the second most powerful man in government."

Bissell coordinated the operation aimed at overthrowing Guatemalan President Jacobo Arbenz. While preoccupied with this project, Bissell met with Strong in mid-May 1954 concerning the concept of Lockheed's proposed spy plane. Still preoccupied with the Guatemalan operation, Bissell forgot about the CL-282.

Although Dulles remained reluctant to have the CIA conducting overflights, Land and James Killian of MIT had told President Eisenhower about the aircraft, and Eisenhower had agreed to have the CIA as the operator. Though Dulles finally agreed, some USAF officers opposed the project in fear of it endangering the RB-57D and X-16. Nonetheless, the USAF's Seaberg helped persuade his agency to support the CL-282, albeit with the higher-performance J57 engine.

Final approval for a joint USAF-CIA project came during November 1954, allowing the CIA to deal with sophisticated technology for the first time, marking the beginning of the Science and Technology Directorate station at Area 51, Nevada.

Too Secret to Explain

On 26 November 1954, the day after Thanksgiving, Richard Bissell, GS-18 special assistant to Allen Dulles, learned about President Eisenhower approving a secret program that his boss, Dulles, wanted him to take charge. Dulles described the project as too secret for him to explain. Instead, he gave Bissell a packet of documents to acquaint himself with the project. This became a project more highly classified than the Manhattan Project that developed the atomic bomb.

Bissell knew all along of the proposal to build a high-altitude reconnaissance plane, however, only in the most general terms. Now, he knew the details concerning the proposed project sending planes over the Soviet Union.

Late on the morning of 2 December 1954, Dulles sent Bissell to the Pentagon to represent the CIA at an organizational meeting for the U-2 project. Herbert I. Miller, chief of the Office of Scientific Intelligence's Nuclear Energy Division, and soon to become the executive officer of the overflight project accompanied Bissell to the conference.

Bissell and Miller arrived at the Pentagon the following afternoon to meet with a group of key Air Force officials that included Trevor Gardner and Lt Gen Donald L. Putt. Bissell attended the meeting with neither Dulles nor him exactly knowing the agency already tasked with running the project. Once the DCI learned this, he instructed Bissell to "work it out."

The participants spent little time delineating Air Force and Agency responsibilities in the project, taking for granted the CIA's handling the security. Much of the discussion centered on methods to divert Air Force materiel to the program. It mostly concerned the Pratt & Whitney J57 engines. The CIA feared Pratt & Whitney having a separate contract for the engines might jeopardize project security.

The Air Force promised to turn over several J57 engines in production for B-52s, KC-135s, F-100s, and RB-57s. When Bissell asked about paying for the airframes built by Lockheed, the others greeted his query with silence. Everyone expected the CIA to come up with the funds. The meeting adjourned with Bissell volunteering to consider it.

After the meeting, Bissell advised Dulles, the DCI of the CIA, of his having to use money from the Contingency Reserve Fund for covert activities to launch the project.

Through this fund, the DCI, with the president and the Director of the Budget's approval, paid the costs of the Central Intelligent Agency's covert activities. Nonetheless, Dulles told Bissell to draft a memorandum for the president for the funding of the over-flight program and to start putting together a staff for Project AQUATONE, the project's new code-name. The USAF used the name "OILSTONE" for its support to the CIA.

Philip Strong of the Office of Scientific Intelligence (OSI) did not stop with Bissell in his effort to drum up support for high-altitude overflight. He persuaded DCI Allen W. Dulles to speak to the Air Force concerning taking the initiative for gaining approval to overfly the Soviet guided missile test range at Kapustin Yar.

Allen Dulles favored the classical agent form of espionage rather than technology. Consequently, he had always supported the idea of the Air Force overflying the Soviet Union, with the CIA only in a supporting capacity. Thus, Dulles lacked enthusiasm concerning the CIA's taking a "military" role. Besides that, he disliked the Lockheed CL-282 design for the Air Force. He had never mentioned the CL-282 or any of the other proposed high-altitude aircraft during the numerous meetings between the Air Force and the CIA while exploring the overflight proposal.

Now in the summer of 1954, civilian scientist and engineer advisory boards and committees still supported the Lockheed CL-282 proposal. Unfortunately, the proposal lacked the official support of the key decision-makers of both the Air Force and the CIA.

The lack of support included the Scientific Advisory Committee formed in 1948 by the Air Force's Scientific Advisory Board. The

Committee had provided scientific advice and assistance after disbandment at the end of the war. However, for Lockheed's needs today, the Truman administration had made little use of the new advisory committee.

Lockheed had received a break in 1951 when the Air Force sought more assistance from scientists to meet the SAC request for information concerning targets behind the Iron Curtain. MGen Gordon P. Saville, the Air Force's deputy chief of staff for Development, had sought new ways of conducting reconnaissance against the Soviet Bloc by adding 15 reconnaissance experts to Project LINCOLN, an existing undertaking at the Massachusetts Institute of Technology on air defense.

The concern for a Soviet attack on the CONUS had grown from the 1953 discovery of a new Soviet intercontinental bomber at the Ramenskoye airfield, south of Moscow. Jet engines rather than turboprop-powered the new Myasishchev-4 (Bison), making the Bison the Soviet equivalent of the US B-52 just going into production. It had surprised US intelligence to see the Soviets produce such an aircraft without stealing the technology.

At first, the new Project Staff, (renamed the Development Projects Staff during April 1958) had consisted of Bissell, Miller, and the small existing staff in Bissell's Office of the Special Assistant to the DCI, Director Central Intelligence.

During the months following the establishment of the project, its administrative workload had increased. The project handled the growth by adding a finance officer and a contracting officer early in 1955, then during May 1955, an administrative officer, James A. Cunningham, Jr., a former Marine Corps pilot working in the Directorate of Support. Cunningham stayed with the U-2 project for the next ten years.

As Cunningham recounted, "There was then established in the CIA a cell concerned with the entire U-2 project. It was this cell, then, which I joined during May of 1955 at a time when it had a total of 5 people. Our sole job was to get on with the work of creating an airplane with a limited number of people from the Agency and from the Air Force, who dropped all other concerns. At its peak in Headquarters, this small cell took in around 123 people in all."

In 1955, only four people in the White House knew about the U-2 reconnaissance proposal. The project had to achieve maximum security, which meant making the project staff self-sufficient with its own administrative, financial, logistical, communications, and security personnel. Bissell kept funding for Project AQUATONE separate from other Agency components. The deployment of the units overseas followed this compartmentation, paying its personnel and operating costs outside of regular Agency accounts. Bissell did not know it at the time, but he fathered a new directorate within the CIA that would soon become the Science and Technology Directorate.

Meanwhile, at Lockheed, Kelly Johnson and twenty-five engineers redesigned the airplane to provide for new landing gear, different engine, different camera bay, and a means of further improving performance. Eighty-one people, including all shop personnel, worked on the plane known as the U-2.

During December 1954, the CIA ordered twenty aircraft with Kelly promising delivery of the first one in eight months. He froze the design on 10 December 1954 and presented the first status of the plane along with a cost letter for $20M to Washington by the middle of the month, mid-December 1954. The government made the first check out to Kelly Johnson personally and sent it to his home to maintain secrecy on the program. Lockheed completed the initial wind tunnel tests before Christmas and began tooling on 27 December 1954.

The production aircraft differed considerably from the original CL282. Lockheed and the CIA were now calling on Pratt & Whitney to produce an engine for a 70,000-foot altitude. Lockheed had changed the tail configuration to a low mounted horizontal stabilizer and incorporated a bicycle landing gear arrangement instead of the original takeoff dolly/landing skid.

Pressurizing the cockpit to some extent enabled pilots to operate for periods of up to 10 hours without full pressure suits. (Partial pressure suits provided pilot protection in the event of aircraft pressure loss.)

The innovative design included weight-saving

measures such as un-boosted flight controls, no ejection seat (changed after the loss of several pilots), and a bicycle landing gear system with jettisonable outrigger supports (called "pogos") to hold the wings level for taxi and takeoff.

The aircraft became a masterful blend of innovative technologies, the successful marriage of a multitude of component airframe sensors adapted for optimum use at high-altitudes, modern pilot physiological support equipment, a finely tuned engine, and a special, non-freezing fuel called JPTS.

Additional weight savings came in the form of a new camera and film design. The Killian panel, when considering state-of-the-art reconnaissance photography, identified potential candidates for use. All were too heavy, exceeding the 750-pound Q-bay capacity of the aircraft, prompting the CIA to finance a development effort that yielded an

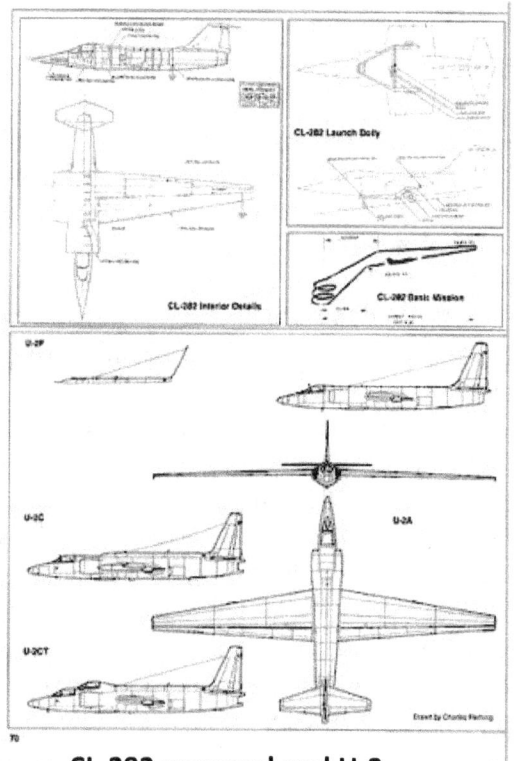

CL-282 proposal and U-2

empty camera weighing 400 pounds, 500-pound loaded for an 8-hour mission. Best of all, picture resolution captured items of 2 ½ feet (75 cm) across in size. The new camera arrived at Watertown for installation into the U-2 before it advanced into the flight test phase.

The first fuselage came out of the tooling on

21 May 1955; one week after the Soviet Union completed the Warsaw Pact. Problems with the wings kept everyone working frantically to finish the first U-2 on July 15, 1955, and at the cost of $19 million.

Under the authority of the president and the Air Force, the CIA controlled the missions of the U-2 plane that the CIA codenamed AQUATONE, and the Lockheed Skunk Works called the Angel because it flew so high.

The CIA dubbed the U-2 as the "Utility 2," disguising it as a civilian weather plane to mislead the Communist Intelligence Service. The USAF inventory had a Utility category of planes for support, transport, courier duty—such as Cessna U-3, Helio U-5 Twin Courier, Piper U-7 (previously L-21).

The U-2 earned the name "Dragon Lady," the term going back to the 1940s to associate the word "Dragon" with British projects to gain information concerning German rocket programs. Eventually, in the reconnaissance world, the term "Dragon" began referring to individuals processing scientific or technical information. Because of the Air Force's interest in obtaining the U-2 aircraft with the CIA the executive agent of the program, "Project Dragon Lady" became the official name given to the process through which the Air Force acquired the U-2 with the contractor go-ahead during January 1956, First Article received during September 1956, and deployment in 1957.

Despite the various names, Lt Col Leo P. Geary, an Air Force project officer at OILSTONE and later the Air Force's assistant Director of Operations, informed Culbertson that Johnson wanted the Lockheed CL-282 known as the U-2. (Colonel Geary was a key person throughout the program as the Hq, USAF project officer while never actually assigned directly to the CIA. This is an important but critical distinction in the CIA's organization of the early U-2 Air Force colonels).

Norm Nelson - CIA

Interestingly, Lockheed had developed the CL-282 on its own and devoted considerable effort to promote it. However, Lockheed required persuasion to undertake the project during November 1954 because of the company's substantial commitments to several other civilian and military projects. The situation concerned the CIA enough to send Norm Nelson to monitor the Lockheed Advanced Development Products Company, the Skunk Works, on the U-2.

At first, Kelly Johnson refused to have a CIA monitor looking over his shoulder. Nonetheless, the CIA insisted on having a representative there to ensure security. Johnson gave in, however, with conditions. Kelly Johnson refused to allow whomever the CIA assigned to the project to have a desk or phone within the Lockheed complex. Thus, the CIA's representative, Mr. Norm Nelson, had to locate a public payphone to make his progress reports.

Norman E. Nelson had studied aeronautical engineering at the University of Cincinnati during the Great Depression. Afterward, he began a toy company after WWII in Dayton, Ohio, and taught weather courses at the University of Dayton during a brief stint in the 1940s.

During the second world war, he had worked at Wright Field again in Dayton and followed that with active duty during the Korean War. Interestingly, while stationed in Hollywood, he spent a significant part of his time taking the train up and down the coast of California, traveling both to UC Berkeley and their war programs and Boeing in Seattle.

Following the Korean War, he left the Air Force a captain and went to the Doak Aircraft Company in Torrance, California, where he became vice-president in charge of engineering. This innovative company produced an early VTOL (Vertical Take-Off and Landing) aircraft that required no runway and could hover and fly forward. One of those now resides at the Smithsonian.

When Douglas Aircraft purchased Doak in 1961, the CIA recruited Nelson to work at the Lockheed Advanced Development Products Company, the Skunk Works, on the follow-up A-12/YF-12/SR-71 Blackbird program with Kelly Johnson.

Nelson later became a regular Lockheed employee and eventually the chief of Advanced Design at the rotary-wing branch. After Lockheed, he worked at McCullough Aircraft Company and Hughes Tool Company, where he ran some exciting projects, including a small single-person aircraft, a gyrocopter for McCullough, and the Glomar Explorer project for Hughes. He spent much of this time at various locations throughout the US.

From Hughes, Nelson returned to the Skunk Works in 1976 and became the program manager and chief engineer of the stealth fighter programs, working with Ben Rich. Out of this came the F-117 Stealth Fighter at Area 51, which served so successfully in the first Gulf War. In 1984, he became vice president and general manager of the Skunk Works and held that position until he retired in 1988.

Kelly Johnson received a call from Trevor Gardner on 17 November 1954, asking him to come to Washington for conversations on the project. Lockheed's senior management instructed him not to commit to any program during the visit. His instructions were to obtain the information and return.

When he returned to California, Johnson noted in his project log how the secrecy aspect surprised him. Gardner, when drafted for the project, had informed Johnson that he might have to take a leave of absence from his job at Lockheed. Gardner feared the secrecy of this unique project, and the need not to talk about it would make it impossible for Kelly Johnson to handle Project AQUATONE in addition to managing his other projects.

Kelly Johnson did not need persuasion to undertake such a bold step forward in aircraft design. He used the Gardner statement to persuade Lockheed's senior management to approve the project, which they did after meeting with Johnson when he returned to California on the evening of 19 November 1954.

Four days later, on 23 November, the Intelligence Advisory Committee (IAC) approved DCI Dulles' request to undertake the Lockheed CL-282 project. The members of the IAC believed aerial reconnaissance and photography had an improved capability for filling the "Intelligence Gap" on the Soviet Bloc. The following day, Deputy Director of the Central Intelligence, Lt Gen Charles Cabell drafted, and Dulles signed a three-page memorandum asking President Eisenhower to approve the overhead reconnaissance project.

Later in the afternoon, Dulles attended a meeting with the secretaries of State and Defense and senior Air Force officials. Following the meeting, Dulles and Cabell presented the document to the president and received verbal authorization to proceed. The CIA found it now in the surveillance plane design and flying business. The question "where" would soon arise.

Norm Nelson   Kelly Johnson

Other CIA Aviation Activities

The US Air Force and Navy shared a desire not to want the CIA involved in aerial reconnaissance. However, the generals and admirals felt flying spy missions could spark a war. The president realized this and relied on the CIA to do things in a way that provided him plausible non-attribution and deniability. Project AQUATONE was the first accomplishment of the Science and Technology branch of the CIA to meet this need, followed by Project OXCART in 1962.

The U-2 Project AQUATONE became just one of the CIA's covert air activities that the Air Force would not or could not do. Mostly, the CIA's aviation companies would conduct military operations while posing as a civilian air carrier, covertly flying into areas denied the US armed forces by the 1954 and later, the 1962 Geneva Accords treaty restraints.

Thus, the U-2 reconnaissance program underway did not qualify as a first and only in CIA aviation. On the operations side, the CIA had first entered the aircraft business during August 1950. At the direction of the National Security Council, the CIA had formed a Delaware corporation named Airedale. Airedale created a subsidiary company called CAT, Inc. CAT purchased 40% of the assets of Civil Air Transport (CAT), an airline started in China in 1946 by Gen Claire Lee

Chennault (of Flying Tigers fame) and Whiting Willauer. The remaining sixty percent of the company remained with Chinese investors. CAT Inc. also formed Asiatic Aeronautical Company Ltd, a Republic of China company.

Even during the Korean War, the CIA's air proprietary, CAT, dropped agents and supplies into Kirin Province as part of a project known to the pilots as Operation Tropic and experimented with ways to retrieve them from the air via harness system—part of an effort to start a resistance movement in Manchuria, China.

In 1957, Airedale changed its name to Pacific Corporation. CAT, Inc. changed its name to Air America, Inc. In 1959 after settling objections from Air France, Asiatic Aeronautical Company, Ltd changed its name to Air Asia Company, Ltd the same year. CAT, Civil Air Transport remained in existence throughout the tenure of Air America from 1950 through 1976, and for several years the flag carrier of the Republic of China.

The CIA's CAT, "the most shot-at airline in the world," changed its name to Air America with its slogan "Anything, Anywhere, Anytime, Professionally." Air America Airline operations in China did not have enough work to keep the asset afloat. Consequently, the National Security Council farmed the airline out to various government entities that included the USAF, US Army, US Navy, and for a brief time the French Republic.

The CIA's aviation activities would change when the Vietnam War brought a burst of covert air transport business to the CIA. The CIA would realize a need to increase its covert aviation activities to meet the CIA's needs around the world. To do so, Airedale would change its name to Pacific Corporation.

Pacific Corporation would control several aviation front companies. Its affiliates would include Actus Technology, the Civil Air Transport that would become Air America, Air Asia Co., Intermountain Airline, Seaboard World Services, Southern Air Transport, Thai Pacific Services, and Bird & Sons to name a few.

The CIA's civilian Air America Airline would operate from the Marana Army Airfield near Tucson, Arizona. However, the airline would fly all over Southeast Asia. Mostly, it would be carrying civilians, diplomats, spies, refugees, commandos, and sabotage teams. It would also carry doctors, war casualties, drug enforcement officers, and even visiting VIPs. This all occurred on the operations side of the CIA. Following this pattern, the CIA's new Science and Technology Directorate would eventually station at Area 51, known by the CIA as Station D, while flying reconnaissance missions from remote deployments around the world.

The CIA would also identify its front company, Intermountain Airline, as Intermountain Aviation or Intermountain Airway. While the DS&T set up shop in Nevada, the CIA's covert airline, Intermountain, performed covert operations for the CIA in Southeast Asia and elsewhere. Intermountain Airlines eventually became the Evergreen International Airline.

The CIA based its Evergreen International cargo airline in McMinnville, Oregon, to operate contract freight services. It chartered and scheduled flights, as well as wet lease services. With it, the CIA conducted services for the US military and the US Postal Service, as well as ad hoc charter flights.

When Agent Orange disrupted local food production in Southeast Asia, Air America delivered "rice drops," meaning thousands of tons of food, live chickens, pigs, water buffalo, and cattle. In addition to the food drops were the logistical demands of the Vietnam War, with Air America pilots flying thousands of flights transporting and air-dropping ammunition and weapons (or what the CIA would refer to as "hard rice") to friendly forces.

The CIA's shell corporations would conceal the agency's involvement by shuffling aircraft continuously between them, altering their aircraft registration numbers, a tactic the agency used when a Russian missile shot Gary Powers' U-2 down over Russia, and while flying A-12 reconnaissance missions out of Kadena.

Evergreen would acquire from the CIA's Air America fleet a significant aircraft maintenance and storage facility at the Pinal Air Park in Marana, Arizona. At its peak, the CIA's aviation companies employed almost 20,000 and flew about 200 airplanes.

The CIA would base its Southern Air Transport (SAT) in Miami, Florida. Southern Air Transport was a cargo airline best known for its mid-1980s role in the Iran-Contra affair. In October 1986, an SAT plane shot down in Nicaragua carried four loads of US weapons bound for Iran from the US through Israel. On the return flights, it carried weapons destined for the Nicaraguan Contras from Portugal. Air Asia Company Limited, headquartered in Taiwan, would provide the aircraft maintenance, repair, and overhaul (MRO) service for the CIA flights.

Forging a CIA-Air Force Partnership

The CIA's U-2 project headquarters concurrently moved forward with procuring the aircraft and equipment. It recruited personnel and planned for the testing and operational phases. Richard "Dick" Bissell began what he later described as "a rather civilized and amicable battle" with the Air Force to hammer out a charter for joint USAF/CIA project participation. For the Air Force to passively mow the grass, fuel the planes, and wash the windshields did not come easy, especially when some civilian gloriously flew off into the wild blue yonder on a mission that the flyboys thought they should be doing. These civilians would also be making several times more money than the AF recon pilots.

At the initial interagency meetings to establish the U-2 program during December 1954, the participants found it difficult to work out a clear delineation of duties between the CIA and the Air Force. They could agree only with the Air Force supplying the engines, and the CIA paid for the airframes and cameras.

A myriad of details remained unsettled. The CIA and Air Force representatives worked on an agreement to assign specific responsibilities for the program. These negotiations were difficult.

Dick Bissell experienced his first significant encounter with USAF Gen Nathan F. Twining, Chairman of the Joint Chiefs of Staff, on 7 March 1955 while preparing for a meeting. On 25 February, Bissell had prepared a briefing paper where he summarized project developments to date and recommended giving urgent attention to advance preparations for acquiring Air Force support in the operational phase. The project needed to complete research and planning in the fields of aeromedicine and after-intelligence requirements and flight planning, meteorology, and logistics. The project required selecting and completing an organizational structure to recruit and train test pilots and Air Force personnel holding important positions.

General Curtis LeMay   Colonel Douglas T. Nelson

The briefing paper passed to General Twining in advance of the meeting, wherein recommending the designation of a single officer responsible for all the activities of the Air Force in support of and as a participant in the project. Clothing the officer with this authority and responsibility placed him in a better position to arrange, in the most secure manner possible, for access to the variant resources of the Air Force upon which he hoped to draw. The plan called for him to join the CIA Project Officer in developing organizational plans for approval by appropriate authorities in the CIA and the USAF. The plan positioned him to secure other Air Force project personnel at an early date.

In further preparation for the 7 March 1955 meeting with the Air Force chief of Staff Nathan Twining, Dick Bissell prepared a background paper for the Director and General Cabell. He first warned them of General Twining wanting the Air Force to have operational command, and then recommended the Director take a general line with the chiefs of staff.

DCI Allen Dulles took up discussions with Twining on this subject following Bissell's meeting earlier that month. Twining wanted the Strategic Air Command, headed by Gen Curtis E. LeMay, to run the project once the planes and pilots became ready to fly; Director Dulles opposed such an arrangement. His opposition dragged the CIA-USAF talks on for several months with Twining determined to have the

Strategic Air Command in full control once the aircraft deployed.

Even with General LeMay wanting nothing to do with the U-2, he sent Col Douglas T. "Doug" Nelson on temporary duty from the Strategic Air Command Headquarters to monitor the program and report to him.

Nelson, who later retired as a major general, had earned his student pilot's license before old enough to get a permit to drive a car. He flew his first plane in 1930 at the age of nine. For $5.00, Nelson flew an old Waco, taking off on the beach in Seaside, Oregon. He soloed on his 16th birthday in an E-2 Taylor Cub on a rainy day over a grass strip, also in Oregon. His early military service took him to the Middle Eastern desert, Alaska, and the China-Burma Theater of Operations. During the war, he flew 590 combat hours, primarily in C-46s.

In 1946, after a tour of duty as a fighter aircraft instructor pilot, Nelson separated from the US Army Air Force. He became a commercial airline pilot, flying DC-3s with West Coast Airlines until November 1947 when he received a regular commission and returned to active duty. He received an assignment to the Strategic Air Command, where he served in the 33rd Fighter Group before attending Air Tactical School in Florida.

Nelson spent October 1948 to March 1949 in Palestine as part of the United Nations Truce Force. While there, he sometimes relied on camels as his mode of transportation with the Syrian Camel Corps. He then returned to the States and became a B-29 lead crew aircraft commander at Walker Air Force Base, New Mexico.

Nelson served at the Strategic Air Command headquarters, Offutt Air Force Base, Nebraska with the Tactics Branch and Recon Division, Directorate of Operations when, in 1956, General LeMay appointed him the Strategic Air Command Project Officer to support the CIA U-2 Program.

General LeMay's sent Colonel Nelson to watch the CIA. Having Nelson there did not bother Bissell, who felt it none of the CIA's business how the Air Force organized its activities if it did not interfere with the agency's role given it by the president. However, LeMay's sending a senior officer challenged the character of the project, forcing Bissell to impose certain requirements having a bearing on the organization.

General LeMay could not accept the thought of his Air Force stepping aside while the CIA flew planes against the nation's enemies. Having LeMay's Air Force colonel present caused many to conceive the CIA's U-2 project as a clandestine, intelligence-gathering operation based on military pilots flying the missions.

For clandestine intelligence-gathering, Bissell wanted the project to have the least amount of military aura and the greatest amount of "civilian footprint" possible. He required a rigorously secure, secret CIA operating facility/flight testing range and not an Air Force Base. He required close and continuous policy control by the senior policymakers of this government and felt that only the CIA's project headquarters in Washington could maintain that control.

The CIA and Bissell thought of also having non-Americans to fly reconnaissance missions. Bissell made the initial policy decision to proceed based on it conforming to this concept. This way, the CIA avoided describing the project as a military operation conducted by any offensive arm of a regular military establishment. The CIA now thought, "science and technology," and did not want another CAT, Nationalist Chinese Civil Air Transport airline owned by the CIA for covert operations per the Mutual Defense Assistance Act of 1949. Ironically, the non-Americans who flew the U-2 the most would be Nationalist Chinese pilots. CIA Detachment H supported them, as described in Chapter 16.

The CIA saw Project AQUATONE as a power projection operation with a US-based facility operating as the nucleus of global aerial spy operations, doing what the Air Force refused to do, building a fleet of pure reconnaissance planes to overfly the USSR.

Bissell saw the CIA power projection as having the capacity to deploy and sustain its aerial reconnaissance rapidly and efficiently. Having power projection meant its US base must remotely operate from multiple dispersed locations outside the limited bounds of its territory. Bissell wanted the CIA capable of responding on a global scale. At this stage, this meant deploying anywhere in the world that the Soviet Union or others posed a threat to the United States and its allies.

It was globally flying brought with it the logistical difficulties inherent in projecting its U-2 detachments great distances away. It meant the U-2 project must proceed with utmost secrecy, more than that of the development of the atomic bomb.

The vital necessity for security brought with it two implications for the organization. First, they must limit knowledge of the project to the narrowest possible circle of those with a need to know. This category included only those individuals working on some aspect of AQUATONE and a few top policymakers. Second, it organized the project to give it the best possible cover.

Standard operating procedure (SOP) for the Air Force involved first a chain of command and second a policy of routinely rotating its personnel assignments. This CIA project required the Air Force to serve outside its chain of command protocols and focus its responsibility for maintaining security and ensuring close control. The Air Force would have to use unique channels rather than the standard chain of command. Both Air Force SOPs were reasons for choosing the CIA to manage the U-2 program. The CIA did not follow either of these policies.

Bissell knew that doing this the CIA way would alienate General LeMay. Nonetheless, he felt the project's character required the Air Force participants to station its leadership in Washington and give its representative the authority to deal with the CIA and with other components in the project's business. Any Air Force personnel assigned to the project must work for the CIA.

Bissell also wanted a direct channel from the Washington project headquarters to overseas units. The Air Force balked at the idea of the CIA playing down any connection to the Air Force's operational command. Bissell insisted on avoiding any identifying the project with the military. He took these requirements up with the Air Force chief of Staff Nathan Twining. However, no substantial agreement came from the meeting. A month later, Dick Bissell fired his second shot. He gave to Generals Everest and Putt for discussion a proposed memorandum addressed to the deputy chief of staff for Operations. The opening paragraph began,

"It was understood the view of the air staff that Air Force support for Project AQUATONE in its operational phase should be the responsibility of the Strategic Air Command. Assistance and support during research, development, and procurement will, however, continue as the responsibility of the deputy chief of Staff, Development."

Using this premise, Dick Bissell continued to explain the original concept of the project being a clandestine intelligence-gathering operation. Using the CIA would minimize the risk of detection and offer plausible deniability for the US government. He indicated the CIA's aims for the character of the operations.

The CIA would determine the number of aircraft, the equipment, and operating facilities. As to specific functions performed by the CIA, it would recruit and administer the civilian pilots. The CIA would furnish maintenance personnel for primary mission aircraft and equipment. It would maintain project security control, project communications, and the collection and coordination of requirements and intelligence.

Bissell also made specific recommendations as to the most efficient and most secure manner, from the CIA viewpoint, for channeling Air Force support.

In doing so, he exposed a certain lack of support and differences of opinion on the part of Generals Everest and Putt. Neither accepted the CIA's proposals. Nor did either of them present or put forth an agreed counterproposal of their own.

The project staff met on 8 June 1955. At the meeting, Col George McCafferty reported to Dick Bissell concerning Generals Twining, White, and Everest's engaging in controversy. None could agree on the Air Force's role in the project. Bissell learned that the office of the Deputy Chief of Staff, Personnel had received instructions to take no further action on the project's personnel requirements pending a settlement of the issue.

Dick Bissell sought the assistance of Mr. Trevor Gardner in trying to agree. His doing so resulted in a letter signed by the Secretary of the Air Force on 27 June 1955. Addressed to General Twining, it urged the chief of staff and his deputies to agree with the CIA as much as possible.

At the same time, Gen LeMay made it clear at a meeting with Bissell that as soon as the CIA paid for the U-2, he planned to take it over 'not too far ahead in the future.'

Considering the resistance of higher levels of the USAF to interagency cooperation, the dedicated service of Air Force project officers such as Colonel Ritland and Col Jack A. Gibbs proved crucial to the success of the U-2. A 1958 OER written by Dick Bissel on Colonel Gibbs l stated, "Colonel Gibbs has become the model for technical excellence in the program office."

CIA power projection concept – designed to be nimble and responsive.
Preparing to load at Watertown for deployment

"It's no accident that the complete Angel and all its intricate cargo can be disassembled and packed quickly, ready for airborne transport. Everything about the Angel can go aboard a cargo plane: cameras in their 'dog houses,' engines, lab equipment and supplies, ground support equipment, and of course, the Angel. Men and belongings are air transportable too. Thus, you have the plane, the pilots, the equipment, and the ground support to put it into the air, all in one neat package to send out on wings. The result of foresight and planning, engineering, precise, and rapid manufacturing. From the desert wastelands of Watertown, it's but a matter of hours to anywhere in the world where reconnaissance might be desired." - *The Inquisitive Angel*

Chapter 6 - Funding AQUATONE

Ten years earlier, in 1945, Lockheed had established the secure area called the Skunk Works to develop the nation's first jet aircraft, the P-80 Shooting Star. Kelly Johnson naturally selected this highly secured facility as the venue for the new U-2 project.

As he did ten years earlier with the P-80, Kelly Johnson invoked his motto, "Be quick, be quiet, be on time." He located his engineers and drafters not more than 50 feet from the aircraft assembly line. His philosophy was that this made the engineers instantly aware of any difficulties in construction. This way, they fixed problems of design in a matter of hours, not days or weeks.

Another thing notable about Johnson's work tactics, he eliminated the paperwork that plagued most corporations. His engineers did not waste time producing neatly typed memoranda. They made pencil notes on drawings to keep the project moving. A week after Johnson received authorization for the U-2 project, he wrote a 23-page report detailing his most recent ideas on the proposal.

Johnson's design called for a 2.5g load factor, the full limit of a transport aircraft rather than a combat plane. The design gave the U-2 the sought speed of Mach 0.8 or 460 knots at the desired altitude. The light load enabled the aircraft to obtain an initial maximum altitude of 70,600 feet that Johnson hoped to increase to an ultimate maximum altitude of 73,100 feet.

These early December 1954 specifications gave the new plane a takeoff speed of 90 knots and a landing speed of 76 knots. Its wings gave it the ability to glide 244 nautical miles from an altitude of 70,000 feet. This glide factor made the U-2 difficult to land. The excellent glide ratio, in addition to the ground effect, made the aircraft float over the runway for a very long distance. For every foot it descended, the aircraft glided another thousand feet down the runway.

Johnson and James Baker worked out various equipment combinations not to exceed a weight limit of 450 pounds in the reconnaissance bay. (The "B" camera design would be able to pan horizon to horizon and take stereo pictures straight down.) In his report, Johnson promised the first test flight on or before 2 August 1955, and completion of four aircraft by the 1st of December.

Although the final product resembled a typical jet aircraft, its construction differed from any flown by the US military. One unusual design feature, the tail assembly, saved weight by attaching to the main body with three tension bolts, much like a sailplane design.

The unique wings differed from conventional aircraft whose main wing spar passed through the fuselage to give the wings continuity and strength. The U-2 utilized two separate wing panels attached to the fuselage sides with tension bolts, again, as in sailplanes. The design without a wing spar passing through the fuselage enabled the location of the camera bay behind the pilot and ahead of the engine, improving the center of gravity.

The exclusively long, wide, and thin wings produced a combination of high-aspect-ratio and low-drag uncommon in jet aircraft design. The wings provided the most challenging design features of the entire airplane by serving as integral fuel tanks carrying almost all the U-2's fuel supply.

With the wings and tail bolted on in a fragile construction, Kelly Johnson had to look for a way to protect the aircraft from gusts of wind at altitudes below 35,000 feet. Being only a 2g plane, the delicate U-2 compared to the conventional military plane as a butterfly did to a bird. Thus, the comparatively delicate aircraft required high pilot skill and concentration. Otherwise, any wind turbulence might cause the plane to disintegrate.

Johnson again borrowed from sailplane design to devise a "gust control" mechanism. He developed the ailerons and horizontal stabilizers to position the aircraft in a nose-up attitude to enable the plane to avoid the sudden stress caused by the wind. With the wind control problem solved, the lightweight, bicycle-type landing gear became the final major design feature.

The entire structure consisted of a single oleo strut. The strut had two lightweight wheels to the front of the aircraft and two small, solid-mount wheels under the tail. Combined, they weighed only 208 pounds, yet withstood the force of touchdown for this 7-ton aircraft.

Locating both sets of wheels underneath the fuselage required equipping the U-2 with detachable "pogos" (long, curved spring-steel supports with two small wheels on them) on each wing to keep the wings level during takeoff. The pilot dropped the pogos after takeoff for their recovery and reuse. The aircraft landed on its front and back landing gear. It slowed to a near stop and tilted over onto one of the wingtips equipped with landing skids.

December 1954 arrived with Kelly Johnson working on drawings for the U-2's airframe, while Pratt & Whitney built the J57 jet engine. At this point, no final designs existed for the all-important cameras needed. Somehow, he and Baker had to replace existing cameras too bulky or with an insufficient resolution for use in high-altitude reconnaissance.

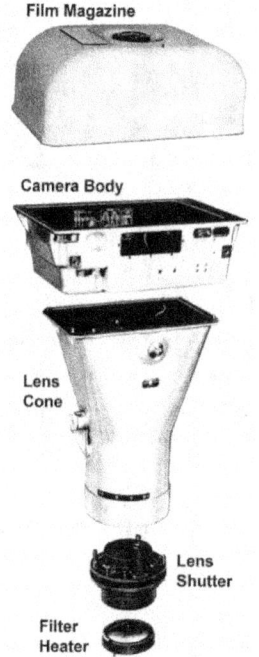

Removing the camera

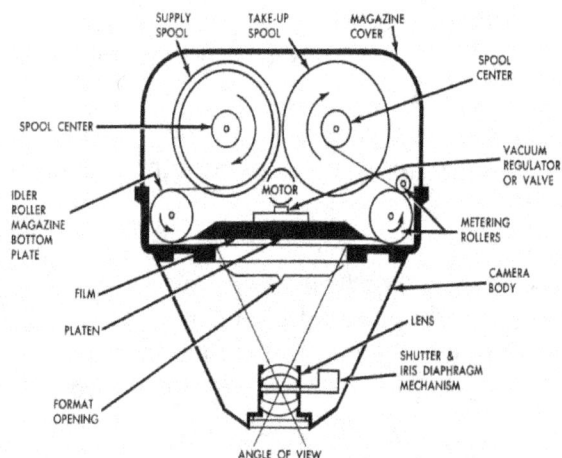

Hycon A-2 camera schematic and mounting (below)

The Fairchild K-19 and K-21 framing cameras were the workhorses of World War II aerial photography. They both used the lenses of varying focal lengths from 24 to 40 inches.

Late in the war, the trimetrogon K-17 mapping camera system, consisting of three separate cameras, became available. The K-17 system took three photographs simultaneously: one vertical image, one oblique to the left, and one oblique to the right. Unfortunately for the U-2, the trimetrogon system required too large an amount of film and lacked the sharp definition needed on the oblique.

Dr. James G. Baker, a Harvard College Observatory research associate, first became involved in aerial reconnaissance in 1940 when he established the full-scale Harvard University Optical Research Laboratory to produce high-quality lenses. After the war, he continued designing photoreconnaissance equipment for the

The standard aerial cameras available in the early 1950s could only achieve resolutions of 20 to 25 feet (7 to 8 meters) on a side from an altitude of 33,000 feet (10,000 meters), thus equating to 25 lines per millimeter.

The military accepted this level of achievement as adequate for aerial photography, finding it enough for choosing strategic bombing targets, assessing bomb damage after air raids, and making maps and charts. However, photos taken at twice that altitude and for accurate intelligence purposes required a resolution of fewer than 10 feet to discern smaller targets in detail.

The resolution factor meant altitudes above 68,000 feet required cameras, four times better than existing aerial cameras to achieve a resolution of fewer than 10 feet. Thus, some scientists doubted the CIA would obtain useful photography from altitudes higher than 40,000 feet. However, James G. Baker of Harvard and Richard S. Perkin of the Perkin-Elmer (P-E) Company of Norwalk, Connecticut, did not share this belief.

They knew better from having developed a 48-inch scanning camera that mounted in a modified B-36 bomber. Flying over Ft. Worth, Texas at 34,000 feet, this camera produced photos that distinguished two golf balls on a putting green. This photo demonstrated the high acuity of the Baker lens. However, the camera weighed more than a ton, much too large for carrying aloft in an aircraft as small as the U-2.

James Baker realized the size and weight as being major restraining factors in developing a camera for the U-2 and began working on a radically new system during October 1954, even before the CIA adopted the Lockheed proposal.

Baker estimated he needed a year to produce a working model of such a complex camera. He also knew of Kelly Johnson's promise to have a U-2 in the air within eight months. Thus, Baker needed to find an existing camera to use until the new camera became ready. He consulted with his friend and colleague Richard Perkin and decided to adapt to t the Air Force K-38 24-inch aerial framing camera (built by the Hycon Manufacturing Company of Pasadena, California) for the U-2.

Perkin suggested modifying several standard K-38 cameras to reduce their weight to the U-2's 450-pound payload limit. At the same time, Baker made critical adjustments to existing K-38 lenses to improve their acuity. He accomplished this in a few weeks and had several modified K-38s, now known as A-1 cameras ready when the first "Angel" aircraft took to the air in mid-1955. The timing earned Hycon a CIA contract for the modified K-38 cameras. Hycon, in turn, subcontracted to Perkin-Elmer to provide new lenses and to make other modifications to the cameras to make them less bulky. In return, Perkin-Elmer subcontracted Baker to rework the existing K-38 lenses and later to design an improved lens system.

In the mid-1940s, the two had collaborated to design an experimental camera with high-acuity lenses for the Army Air Corps. Baker kept his lens-designing efforts separate from his research associate duties at Harvard and his service on government advisory bodies by establishing a small firm known as Spica, Inc., on 31 January 1955.

The A-1 camera system consisted of two 24-inch K-38 framing cameras. One mounted vertically and photographed an 11.2-mile swath beneath the aircraft onto a roll of 9.5-inch film. The second K-38 operated in a rocking mount to alternately photograph the left oblique and right oblique out to 36.5 inches onto separate rolls of the same sized film.

The film supplies unwound in opposite

directions to minimize their effect on the balance of the aircraft. Both cameras used standard Air Force 24-inch focal length lenses adjusted for maximum acuity by Baker. The development of the special rocking mount by Perkin-Elmer's Dr. Rideric M. Fairbanks proved a significant factor in reducing the size and weight of the A-1. He accomplished this by the camera providing panoramic side-to-side coverage with a single lens, ending the need for two cameras.

The U-2s equipped with the A-1 camera system also carried a Perkin-Elmer tracking camera using 2.75-inch film and a 3-inch lens that enabled the camera to produce continuous horizon-to-horizon photographs of the terrain passing beneath the aircraft.

The A-2 camera system used Baker's new lenses, which returned to a trimetrogon arrangement because of problems with the A-1 system's rocking mount. The A-2 consisted of three separate K-38 framing cameras and 9.5-inch film magazines. One K-38 shot the right oblique while another filmed the vertical, and a third the left oblique. The A-2 system included a 3-inch tracking camera. All A-2 cameras came equipped with the new 24-inch f/8 Baker-designed lenses. The lens was the first large photographic compound lens to employ several aspheric surfaces. James Baker ground these surfaces and made the final optical bench tests on each lens before releasing it to the CIA.

The new A-1 system included a backup camera system, a K-17 6-inch three-camera trimetrogon unit using 9-inch film. While still developing the A-1 system, James Baker had begun working on the next generation of lenses for high-altitude reconnaissance.

In doing so, he pioneered the use of computers to synthesize optical design. His software algorithms made it possible to model lens performance and determine in advance the effect of variations in lens curvatures, glass types, and lens spacing on rays of light passing through the lenses. These ray-tracing programs required extensive computations, and for this, he turned to the most modern computer available, IBM's card program calculator (CPC) installation in nearby Boston.

The new lenses resolved 60 lines per millimeter, a 240% improvement. Baker and Fairbanks redesigned the 24-inch lens and turned their attention to Baker's new camera design, known as the B model, a new high-resolution panoramic-type framing camera with a much longer 36-inch f/10 aspheric lens system. The B camera, housed in a vibration-compensation mount, weighed 447 lbs.

The complex B camera used a single compound lens to obtain photography from one horizon to the other, thereby reducing weight by having two fewer lens-and-shutter assemblies than the standard trimetrogon configuration. Besides having a longer focal length than those used in the A camera, the B camera achieved even higher resolution—100 lines per millimeter.

The B camera used an 18-by 18-inch format obtained by focusing the image onto two counter-rotating 9.5-inch wide strips of film that overlapped. Baker designed this camera to locate one film supply forward, the other aft. Thus, as the film supplies unwound, they counterbalanced each other and did not disturb the aircraft's center of gravity.

Hycon's chief designer, William McFadden, engineered the complex B cameras to operate in two modes. In mode 1, the camera used a single lens to make seven different exposures from 73.5 degrees on the far right and far left oblique to vertical photos beneath the aircraft, covering from one horizon to the other. Mode II narrowed the lateral coverage to 21.5 degrees on either side of vertical, increasing the available number of exposures and doubled the camera operating time. Three of the seven B-camera frames provided stereo coverage.

James Baker preferred the C model as the ultimate high-altitude camera because of its 240-inch focal length. During December 1954, he made a preliminary design for folding the optical path using three mirrors, a prism, and an f/20 lens system. Before working out the details of this design, however, Baker flew to California in early January 1955 to consult with Kelly Johnson concerning the weight and space limitations of the U-2's payload compartment. Despite every effort to reduce the dimensions of the C camera, Baker needed an additional six inches of payload space to accommodate the lens. When he broached this subject to Johnson, Johnson replied. "Six inches? I'd sell my grandmother for that!"

Baker realized the 240-inch lens as both too large and too heavy for the camera bay. He scaled the lens down to a 200-inch f/11 system to meet both the weight and space limitations. Realizing this still too large, he further reduced the size, resulting in a 120-inch f/11 lens system by July 1955.

Later in the year, Baker switched to making the mirrors for the system out of a new, lightweight, foam silica material developed by Pittsburgh-Corning Glass Company. The new material design reduced the weight, enabling him to scale up the lens with a 180-inch f/14 system for a 13-by 13-inch format.

In the past, calculations for such a complex lens took years to complete. Thanks to Baker's ray-tracing computer program, he accomplished the task in 16 days.

In addition to the camera systems, the U-2 carried one other important item of optical equipment, a periscope. Designed by James Baker and built by Walter Baird of Baird Associates, the optical periscope helped pilots recognize targets beneath the aircraft and provided a valuable navigational aid.

Note here the early U-2s bore the **NACA identification on the tail** and wings as part of the CIA's cover story of designing the planes for weather research. "NACA" or the National Advisory Committee for Aeronautics became NASA in 1958. Also shown in the above image: the spring-steel "pogo" unit that kept the wings level for taxi and takeoff. Pogos released and fell off after the wings generated lift.

Unvouchered Funding

When Allen Dulles approved the concept of covert financing of the reconnaissance project, many unsettled financial details remained, including the contract with Lockheed. Nevertheless, work on the U-2 began once the project received authorization. Between 29 November and 3 December 1954, Kelly Johnson formed a team of 25 engineers, an arduous task as it meant taking them off other Lockheed projects without explaining why to their former supervisors.

A pilot entering U-2 at Area 51 for takeoff. Note the National Advisory Committee for Aeronautics markings supporting the CIA's cover story. *CIA via TD Barnes Collection.*

The engineers began working a 45-hour week schedule on the project. The project staff expanded to 81 personnel with the workweek increasing to 65 hours.

Kelly Johnson's willingness to begin work on the aircraft without a contract illustrated one of the most important aspects of this program: the use of non-vouchered funds for covert procurement.

Lockheed knew the covert procurement process from having modified several aircraft for covert uses by the CIA. It modified their P-2 Neptune this way, for example. Secret funding for such sensitive projects simplified both procurement and security procedures. It also made the funds non-attributable to the Federal Government with no public accountability for their use.

The process went back to Public Law 110, approved by the 81st Congress on 20 June 1949, designating the Director of Central Intelligence as the only government employee allowed obligating Federal money without the use of vouchers.

Using non-vouchered funds made it possible to eliminate competitive bidding and thereby limit the number of parties knowing of a given project. The use of non-vouchered funds sped up the Federal procurement cycle. It enabled a general contractor such as Lockheed to purchase much if not all, of the needed supplies for a project without resorting at each step to the mandatory procedures involving public, competitive bidding,

In mid-December 1954, President Eisenhower authorized DCI Dulles to use $35 million from the CIA's Contingency Reserve Fund to finance the U-2 project. On 22 December 1954, the CIA signed a letter contract with Lockheed, using the code-name Project OARFISH.

To shave costs, the CIA proposed giving Lockheed "performance specifications" rather than the more rigid and demanding standard Air Force "technical specifications." Kelly Johnson agreed to such a move, saving a considerable amount of money off Lockheed's original proposal to the Air Force during May 1954 that amounted to $28 million for 20 U-2s equipped with GE J73 engines. During negotiations with the CIA General Counsel Lawrence R. Houston, Lockheed had changed its proposal to $26 million for 20 airframes plus a two-seat trainer model, with the Air Force furnishing the engines.

Houston insisted on the CIA budgeting $22.5 million for the airframes because it needed the balance of the available $35 million for cameras and life-support gear. The two sides agreed on a fixed-price contract with a provision for a review three-fourths of the way.

On 9 July 1955, Director of Central Intelligence Dulles attended a conference at Air Defense Command Headquarters in Colorado, with the U-2 project the number one agenda item. Dick Bissell wrote still another briefing paper to prepare the director for the task of getting from the Air Force the decisions so urgently need to move the project forward.

He outlined the proposals advanced to date and recommended that the task force responsible for the project should have a clear responsibility for both operational planning and the actual conduct of operations. The members should have a direct line of command from headquarters to the field. Within that premise, he saw three feasible alternatives: a CIA-controlled task force drawing upon Air Force personnel and support, an Air-Force-controlled task force drawing upon the CIA for support, or the task force staff drawing on both agencies for support and control.

Herbert I. Miller, the chief of the Office of Scientific Intelligence's Nuclear Energy Division, followed up with a Memorandum for the Record dated 12 July 1955 on the Soviet's improved capabilities to detect and intercept the U-2, and the need to incorporate drone control equipment and a long-range guidance system. He emphasized the need to accomplish this by mid-1957.

As an interim measure, he proposed studying drone and guidance equipment for the U-2 that would, in effect, make it a long-range guided missile. He considered the basic elements being the electronic takeoff and landing equipment controllable by radio signal from both a ground station and a mother plane, a flight programmer, and a star-seeking guidance system.

Early 1955 - the U-2 Project Outline

The internal Agency charter for the project codenamed AQUATONE went through twelve drafts during the first month of planning before submission to the director and approval by him on 10 January 1955. In his refining process, Bissell's comprehensive six-page document expected to remain valid for three months would remain unaltered for seven years of its duration.

The President authorized the Director of Central Intelligence to obligate in the fiscal year 1955 an amount not to exceed $35 million from the reserve for aircraft procurement. The project outline estimated the cost of the airframes, photographic and electronic equipment, and some field maintenance equipment at $31.5 million with a margin of error of $2 million, within the $35 million limit.

These estimates assumed the Air Force would furnish technical assistance, supervision, and all government-furnished equipment (GFE). The assessment included forty jet engines and transportation of materiel and personnel to a yet unspecified testing site. The estimate placed pilot recruitment and training costs at close to $600,000. If the Air Force underwrote flight training, it reduced to $100,000 the amount charged to the CIA for the initial period.

The estimate in the project outline contained no allowance for the testing program since it fell within the fiscal year of 1956. Nor did it include benefits for acquisition or preparation of bases, operational costs, or costs to process the photographic and electronic products obtained from overflights.

Bissell's heading the project using covert funding under the Central Intelligence Agency Act of 1949 made the CIA's director the only federal government employee who could spend "unvouchered" government money. The project outline designated Dick Bissell as the officer in charge of the project and as the approving officer, subject to the guidance of the director and deputy director. He held an authorization to obligate funds in amounts up to $100,000 with any items in excess requiring the director's approval. The comptroller could expend funds in the manner and to the extent approved by the approving officer within the limitations as to quantity and procedures outlined in the project outline. All contractual documents required the approval of the general counsel.

Dick Bissell's appointment as the approving officer authorized his arranging for the collection of intelligence requirements and flight planning in cooperation with the Air Force as appropriate. The simple system envisioned by Dick Bissell for the establishment of demand priorities later grew into a bureaucratic committee with representation from every intelligence agency of the government.

The initial charter required Dick Bissell to maintain the closest possible security control over all phases of AQUATONE. The requirement turned into one of the most challenging, and yet unbelievably successful tasked projects for several years.

The formal contract SP-1913 signed on 2 March 1955 called for the delivery of the first U-2 during July 1955 and the last during November 1956. Lockheed had received a $22.5 million contract during March 1955 for the first 20 aircraft, and the CIA had delivered by mail the first $1.26 million to Johnson's home the following month to keep work going during negotiations. As it turned out, Lockheed made the three-fourths point contract review unnecessary by delivering the aircraft on time and under budget. Under flight test engineer Joseph F. Ware, Jr., Lockheed delivered the final aircraft at $3.5 million below budget.

Selecting the Venue for Testing the U-2

The work continued in California on the airframe, in Connecticut on the engines, and Boston on the camera system. With the construction of the first few U-2 aircraft nearing completion and flight tests needing to begin, the top officials of the Development Projects Staff now felt it time to find a venue offering safety and secrecy to flight test the U-2.

On 12 April 1955, Johnson sent project pilot Tony LeVier and Lockheed Skunk Works chief foreman Dorsey Kammerer in an unmarked, red and white-colored Beech Twin Bonanza on a two-week survey mission to scout for a new flight test location for the CIA. A few days later, Richard M. Bissell, Jr., with his Air Force liaison, Col Osmond J. "Ozzie" Ritland departed on a two-day survey. Piloted by LeVier, they reviewed fifty potential sites. They told everyone they were going on a hunting trip in Mexico. They dressed and packed appropriately to keep their hunting trip cover realistic. During their two-week mission, they photographed and explored desert areas having potential as a test site in southern California, Nevada, and Arizona.

None of the sites met the stringent security requirements of the program. Bissell rejected Johnson's proposed Site 1 (Mud Lake) because of its closeness to populated areas near Tonopah, Nevada. Ritland, however, recalled, "a little X-shaped field" located at the eastern side of Groom Dry Lake, north of Las Vegas, Nevada, outside the AEC's nuclear proving ground at Yucca Flat. The airstrip called Nellis Auxiliary Field No. 1 lay on the eastern side of Groom Dry Lake, and just outside the Atomic Energy Commission's (AEC) nuclear proving ground at Yucca Flat.

LeVier, Johnson, Bissell, and Ritland again flew out to Nevada on a two-day survey of the most promising lakebeds, including Groom Lake. They found the abandoned airfield that Ritland had remembered, but it was now sandy, overgrown, and unusable.

They debated whether to land on the airstrip, deciding that attempting to land there ran the risk of nosing the plane when the wheels sank into the loose soil, possibly killing or injuring all the key figures in the U-2 project. Instead, LeVier set the plane down on the lakebed and walked to inspect the strip covered ankle-deep with dust after more than a decade of disuse. Shell casings from gunnery practiced during World War II littered the dry lakebed. In this area of flat, dried-up land—a desert basin—water evaporated quickly, except on the clay lakebed of impermeable sediment with particles smaller than silt.

Groom Lake proved to be an ideal location for the test site, with its excellent flying weather and unparalleled remoteness. Bissell later described the playa as a perfect natural landing field as smooth as a billiard table. Thus, Groom Lake in Nevada came in as the top choice from all possible sites. At that time, neither highway 375 nor the current-day town of Rachel existed.

Kelly Johnson rated Groom Lake as his second choice for the test location. He had already designed a base around his primary lakebed, dubbed Site I, a small, temporary camp with only the most rudimentary accommodations. Johnson estimated the construction costs of such a facility at $200,000 to $225,000. Additionally, Groom Lake lay farther from Burbank than he would have liked. Its proximity to the Nevada Proving Ground (later renamed the Nevada Test Site) understandably concerned him about conducting a flight test program adjacent to an active nuclear test site. Groom Lake lay in the primary downwind path of radioactive fallout from atomic blasts.

Johnson ultimately accepted Ritland's recommendation because AEC restrictions for the atomic testing would help shield the operation from public view. Bissell and his colleagues agreed that all things considered, Groom Lake provided the ideal site for testing the U-2 and training its pilots.

Bissell returned to Washington, where he first discovered that Groom Lake lay outside the Atomic Energy Commission proving ground. Bissell undertook to secure presidential action adding the Groom Lake area to the Atomic Energy Commission proving ground. He wrote three memos, sending them to the Air Force, Atomic Energy Commission, and the Training Command in charge of administering the gunnery range. Signed by Assistant Air Secretary for Research and Development Trevor Gardner, they ensured no other range activities cropping up, which assured the needed security for Project AQUATONE.

In 1955, the Atomic Energy Commission's Federal Services, Inc. (FSI) operated the check stations on the Nevada Proving Grounds. The road split a few miles past Camp 12 on the test site with the road to the right branching off the Atomic Energy Commission property and onto the Department of Defense (DOD) restricted land. Here the CIA set up security officers recruited out of Pennsylvania to operate a security check station on the road leading into Watertown. Once the CIA moved in, no one advanced past this check station without the proper security badge and their name recorded on the authorized personnel roster.

The original rectangular, six by 10 miles (9.7 by 16.1 km) restricted airspace became part of the so-called "Groom Box," a rectangular area measuring 23 by 25 miles (37 by 40 km). The restricted area, later known as Area 51, connected to what is known today as the internal Nevada National Security Site (NNSS) where the road network of paved roads that from Groom Lake led west and south to Yucca Flat, and then further south to Mercury.

The road formerly leading to the mines in the Groom basin now wound past a security checkpoint and further east into the restricted area around the facility. The Groom Lake Road left the restricted area and descended eastward to the floor of the Tikaboo Valley, where it passed the dirt-road entrances to several small ranches before converging with what would become State Route 375, the "Extraterrestrial Highway," south of Rachel.

The Watertown no-fly restriction zone shared a border with the Yucca Flat region, the location of 739 of the 928 nuclear tests conducted by what is now the United States Department of Energy at the Nevada Test Site. (Atomic Energy Commission and the Nevada Atomic Proving Grounds at the time.)

During May 1955, LeVier, Kammerer, and Johnson returned to Groom Lake in Lockheed's Bonanza. Using a compass and surveying equipment, they laid out a place for a 5,000-foot, north-south runway at the southwest corner of the lakebed and staked out the facility's general layout.

Groom Lake, once used by Nellis for a landing strip and target practice during World War II now lay dormant, unpopulated, and unused. Radioactive fallout from atomic tests at the Nevada Atomic Proving Grounds starting from 1951 blanketed the Groom Lake facility area and a sporadically worked lead mine operated by the Sheahan family in the mountains overlooking the dry lake.

One could see why humans had avoided the

area. If the atomic fallout wasn't bad enough, Groom Lake lay in an isolated and silty valley of a desert, vegetated with scattered scrub plants. It received an average of eight inches of rain per year. The usual temperature ranged from 104 degrees F in the summer to -21 F in winter. At Groom Lake, everything chapped, baked, froze, bit, or punctured you.

The mountains around the valley contained abundant sagebrush and semiarid pinion-juniper. In the blistering heat and desolate valley surrounded by mountains, nothing moved during the day—and at night, predators, spiny reptiles, poisonous rattlesnakes, and stinging insects came out of hiding from the blazing sun amid sparse cactus or otherwise thorny plant life. The workers joked about the heat so bad; they saw a Bobcat chasing a rabbit at walking speed.

Groom Lake offered what the CIA needed—a salt flat, an ancient dry pluvial lakebed of parched clay and alkaline. Located in the Emigrant Valley at a 4,462-foot elevation, Groom Lake provided a perfect landing strip for any size or type of plane anywhere on the lakebed except when it rained.

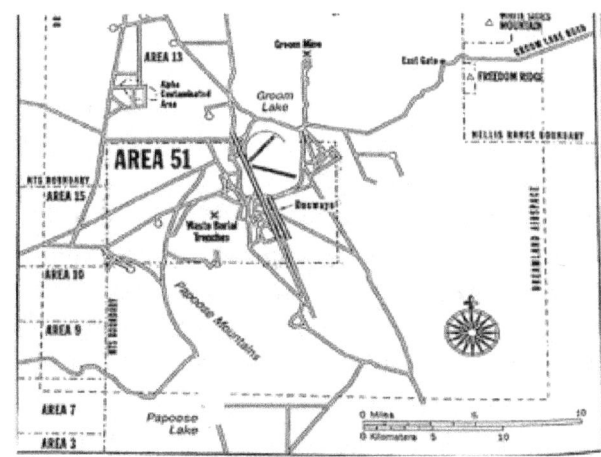

AEC Area layout from Darlington, Area 51: The Dreamland Chronicles

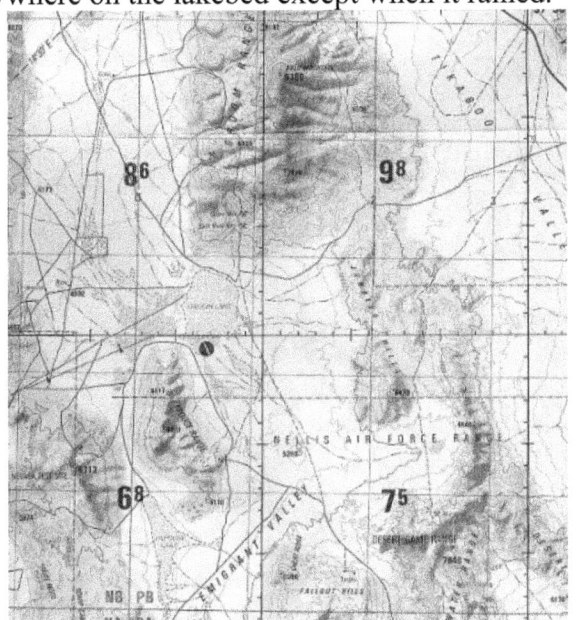

Groom Lake (at left center)

Defense Mapping Agency 1992

Receiving such little precipitation, only occasionally did water stand on the dry lakebed more noted for hosting dusty whirlwinds. No tree or bush grew here. The parched clay and alkaline surface had smoothed through the centuries to glass-like flatness from desert winds sweeping water from winter rains across the lakebed in a timeless cycle. The lakebed's surface provided enough hardness for aircraft and marked airstrips. For all practical purposes, the CIA treated the entire surface as an active runway under the authority of the Watertown air traffic control tower.

The Groom Mountains surrounding the desert valley measured at their widest point 3.7 miles from north to south and 3 miles east to west, providing visual security. Achieving unauthorized access required passing down a lonely, dusty road winding into the Atomic Proving Grounds full of atomic bomb craters. The occasional appearance of the owners of the Groom Lead Mine patented in 1876, mined off and on until the early 1950s, was the only sign of intelligent life other than the lakebed activity.

Twenty-five miles north of the Groom Mountains lay a small farming and ranching community known as the Templute Village before changing its name to Sand Springs, and then to Rachel in 1977, named after the first baby born in the valley.

Of all the remote sites to choose from, LeVier preferred Groom Lake for all these reasons—the mountainous barrier surrounding the lakebed to provide visual security, the glass-like flatness providing the ultimate runway, and its proximity to Nellis Air Force Base explaining any incidental sightings of the U-2.

LeVier and fellow Lockheed test pilot Bob Matye spent nearly a month at Groom Lake removing surface debris from the gunnery practiced during World War II. While at it, LeVier drew up a proposal for marking four three-mile-long runways on the hard-pack clay. Johnson, however, refused to approve the $450.00 expense, citing a lack of funds.

U-2 approaching Groom Lake for landing

Chapter 7 – Why the CIA Chose Nevada

The state of Nevada provided the perfect venue for the CIA to test a top-secret reconnaissance plane. Nevada, born into battle, remained a battle state to that day, ready to give "All for Our Country."

Trappers and traders, including Jedediah Smith and Peter Skene Ogden, entered the Nevada area in the 1820s. In 1843–1845, John C. Frémont and Kit Carson explored the Great Basin and the Sierra Nevada. In 1843, John C. Frémont led the first of his many expeditions through the State.

The first permanent European settlement was a Mormon trading post founded in 1850. Nevada officially became part of the United States in 1848 with the signing of the Treaty of Guadalupe Hidalgo. The 1859 discovery of the Comstock Lode, the richest known US silver deposit, made Nevada famous. Its mines produced enormous quantities of gold, silver, copper, lead, zinc, mercury, barite, and tungsten—a great strike that created boom towns and fabulous fortunes.

The United States continued to administer the new areas acquired as territories. As part of the Mexican Cession in 1848 and the subsequent California gold rush that used emigrant trails through the area, the State's area evolved first in the context of the Utah Territory before becoming the Nevada Territory; a name taken from the nearby Sierra Nevada Mountains.

The Nevada 1861 territory boundary changed three times. Each change took away portions of the Utah Territory, bringing with each change more Mormon settlers escaping US Army troops during the 1858 Mormon War. At the same time, the Mormon settlers and the gold seekers were fighting the Paiute, Shoshone, Washoe, and Hualapai during the Indian Wars.

The Comstock mining boom of 1859 in Virginia City raised the State's population to 6,857, enough to create a provisional territorial government of the Nevada Territory in 1861 with a paternalistic governmental system. The territory welcomed this after the failure of pragmatic attempts to establish a workable frontier institution. Though the Nevada Territory barely even qualified as a territory, three years after it became a territory, Union sympathizers in the Civil War ensured Nevada's entered statehood.

Entering 1864, Abraham Lincoln and his party were intensely and reasonably in doubt about his re-election. The Republicans scrambling for every electorate vote decided to conjure the needed electorate vote from thin air, from Nevada—even though the 1860 census placed its population at 5,857, far short of the 60,000 ostensibly required for statehood. Nine days before the election, the Republican-controlled Congress ensured Nevada's participation in the 1864 presidential election in support of President Abraham Lincoln by granting Statehood to the Nevada Territory. Nevada provided the vote to elect Abraham Lincoln as the 10th US president and first-ever Republican President of the United States.

Nevada became the 36th State after telegraphing the Constitution of Nevada to the Congress days before the November 8 presidential election (the largest and costliest transmission ever telegraphed). With this telegraph transmission, Nevada's Statehood ensured three electoral votes for Abraham Lincoln's reelection and added to the Republican congressional majorities.

In total, Nevada sent 1,200 men to fight for the Union during the civil war. However, Nevada's main contribution to the cause was $400 million in silver from the Comstock Lode that the Union Army used to finance the war. The Comstock silver strike made Nevada also known as the Silver State.

In 1942, Nevada, the "battle born" state became the "battlefield" state when the United States established a defensive network in Nevada to repel a feared Japanese invasion of the West Coast. The Navy established its West Coast base outside Fallon, Nevada, and the Army Air Corps established several military bases throughout the state to contain the Japanese forces should it invade the West Coast of the United States. Nevada became the front line.

During World War II, the Wendover Army Airfield straddling the Nevada-Utah line became the training site for the 509th Composite Group, the B-29 unit that carried out the atomic bombings of Hiroshima and Nagasaki. At the Manhattan Project's Site K of the Wendover Army Airfield code name "Kingman," the 509th Composite

Group, code-named "Project W-47" assembled, modified, and flight-tested atomic bomb prototypes.

The 509th departed Wendover for Tinian, Marianna Island in late May, where it flew combat missions dropping "pumpkins" filled with high explosives over Japan, each aircraft dropping one bomb. The 216th base unit remained at Wendover, with two B-29 planes and crews and continued producing "pumpkins" and components for the atomic bombs until the end of the war.

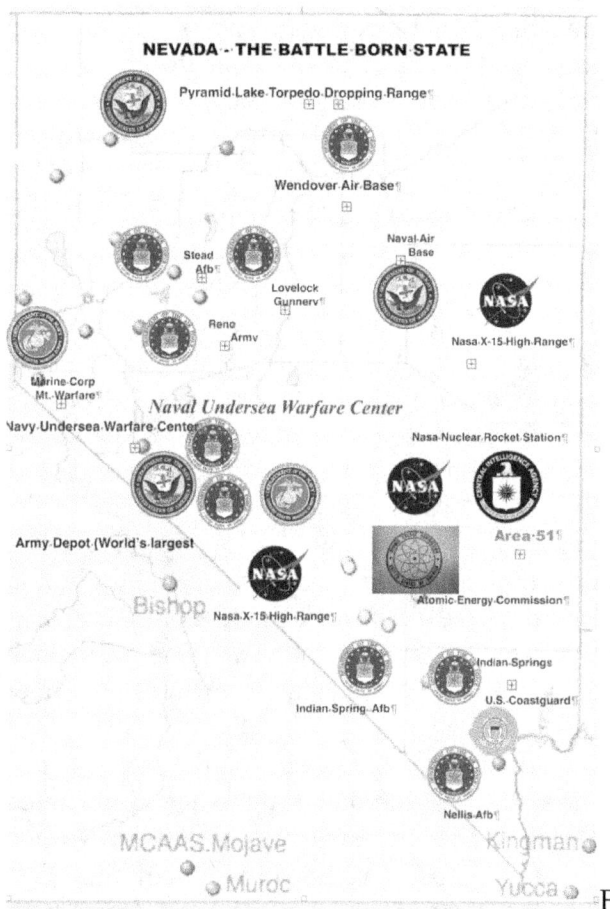

From the Wendover Army Air Field, Colonel Paul Tibbets took off in the "Enola Gay" from Tinian on the mission to deliver the Little Boy atomic bomb, dropped on the morning of 6 August 1945 at 0915 hours over Hiroshima, Japan. Three days later, the US dropped the "Fat Man" atomic bomb on Nagasaki, just hours after the Soviets invaded Manchuria.

Shortly after the bombing of Nagasaki, General LeMay had ordered Colonel Tibbets to dispatch Lieutenant Colonel Classen, the deputy group commander, and crews to Wendover AAFB with orders to stage the transporting of the Model 1561 Fat Man bomb assemblies to Tinian by special courier, Lieutenant William King. Washington halted the shipments due to the Japanese surrender.

Beginning with the Korean War, the Marine Corps Mountain Warfare Training Center near Lake Tahoe and the town of Gardnerville, Nevada provided cold-weather training for replacement personnel bound for Korea.

*The center continues operating today, occupying 54,000 acres of Humboldt-Toiyabe National Forest as one of the Corps' most remote and isolated posts. It maintains a staff of 250 Marines and 50 Civilian-Marines on permanent assignment to train units of up to 1,700 in mountain and wintry weather operations in the Sierra Mountains and provide ordnance for live-fire training at the Hawthorne Army Weapons Depot in Nevada.

Since 2001, thousands of trainees have completed the courses each year. They come from all branches of the US armed forces, as well as from nations such as Britain, Norway, Sweden, Chile, Peru, Israel, Argentina, Netherlands, Kyrgyzstan, Canada, and Germany. Instructors graduating from Mountain Leader courses have also deployed to train the army of Afghanistan.

When the CIA chose Watertown in Nevada for flight testing the U-2, it picked a state long-known as a military state, a military test venue where no one would notice yet another war activity. After all, the entire State had a population of only 237,000, with many of its residents depending upon the military and the Atomic Energy Commission for jobs.

Nevada the West Coast Line of Defense

Next door at Yucca Flats, Yucca Mesa, and Frenchman Flats, the Atomic Energy Commission exploded atomic bombs creating mushroom clouds that entertained the tourists watching from Las Vegas hotels and casinos. The fallout clouds traveled directly over Watertown before dissipating over Utah, or sometimes also traveling further and dropping radioactive rainfall over the eastern US.

At Jackass Flats, the National Aeronautics and

Space Administration (NASA) and the Atomic Energy Commission were developing a nuclear rocket engine for a crewed flight to Mars. NASA tested the NERVA (Nuclear Engine for Rocket Vehicle Application) there in the 1957 period. The nuclear engine moved on a railroad car. The "Jackass Railroad," said to be the world's slowest.

Watertown lay near the center of the 400-mile NASA High Range flight corridor to fly the X-15 rocket plane. The X-15 program produced eight astronauts.

The CIA's Groom Lake in Area 51 would soon become an issue between the CIA and NASA when NASA designated it an emergency landing site for the X-15 Project. The NASA High Range would soon support flights of the XB-70, the Lifting Bodies that became the space shuttle, and the CIA and Air Forces' YF-12. The NASA tracking station soon became one of the CIA's Seven Sisters radar sites for its flights of the YF-12, A-12, and M-21 / D-21 drone project at Area 51.

The CIA "weather research" project and the Atomic Energy Commission activities received scant attention because the US Army, Navy, Marine Corps, and even the Coast Guard were all present with most of the services conducting classified activities within the State. The US Coast Guard patrolled Lake Meade near Las Vegas, and the Army ran the largest weapons depot in the world at Hawthorne, Nevada.

The US Navy used the nearby Tonopah Test Range as impact targets for its Regulus tactical missile launched at sea. The Navy also operated a Naval Air Station at Fallon, NV, and the Naval Undersea Warfare Center in Walker Lake near Hawthorne, NV. More than 7,000 armed forces and civilians worked at the Hawthorne arsenal during the war, making it the busiest Nevada boomtown in a generation. In 1950, nearly 2,500 people lived in government housing at nearby Babbitt. In the 1960s, the Navy contributed more than 130 of the now-surplus Babbitt duplexes to the CIA for long term occupancy at Area 51.

The CIA's creating its Area 51 facility and combining its airspace with the adjoining Air Force Nellis gunnery range created the largest contiguous air and ground space available for peacetime military operations in the free world.

One might say the CIA chose Nevada because of its already hosting the unique and highly classified activities of four distinct worlds. These were the military world (Army, Navy, Marine Corps, Coast Guard, and Air Force), the white world (Atomic), the space world (NASA), and the black world (Area 51, the nucleus of black projects extending worldwide). A cloak of secrecy already shrouded Nevada's leadership in the national security of the United States, making it ideal for hiding the CIA's U-2 spy plane posing as a NASA aerial platform for weather research.

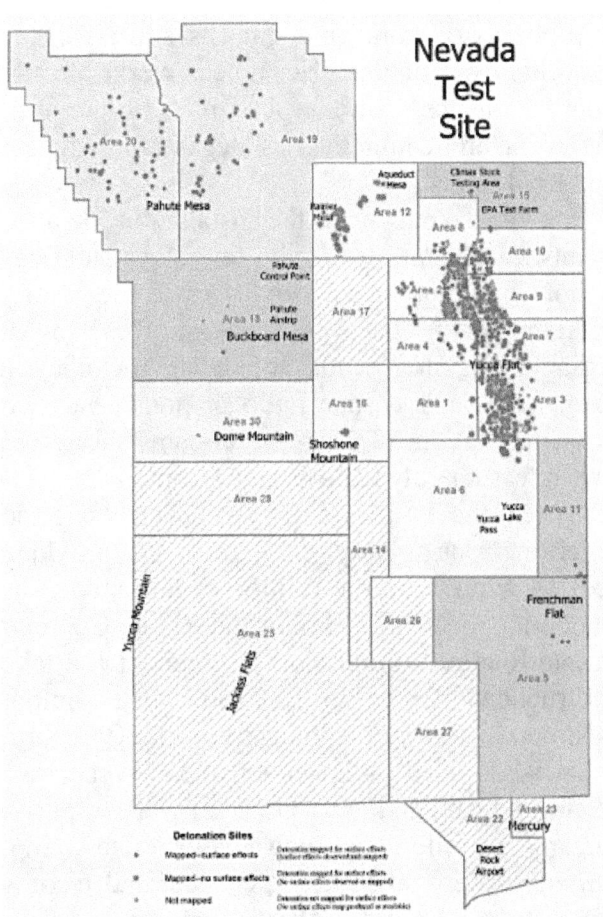

Above: Atomic test areas and detonation locations at the Nevada Test Site

The CIA did not build the Groom Lake facility in secrecy. The "facility that didn't exist" became public knowledge from the beginning in 1955 with the construction of the airstrip at Groom Lake, Nevada. On behalf of the CIA, the Atomic Energy Commission announced the construction in NASA's name and its use for weather research. *Nonetheless, despite numerous such announcements over the decades, Test Site insiders, government officials, military personnel, and the public have perpetuated the myth of the existence of the facility being a closely guarded secret.

Herbert Miller of the CIA Development Projects Staff used the cover of the Atomic Energy Commission to organize a team of construction crews after his issuing $800,000 in contracts for the building of the facility. Seth Woodruff, Jr., manager of the AEC's Las Vegas Field Office participated in the cover by announcing to the news media his instructing the Reynolds Electrical and Engineering Co., Inc. (REECo) to commence the preliminary work on a small, satellite Nevada Test Site installation. He noted work already underway at the location a few miles northeast of Yucca Flat and within the Las Vegas Bombing and Gunnery Range.

Woodruff announced the installation included "a runway, dormitories, and a few other buildings for housing equipment." He described the facility as "temporary." The press release went to 18 media outlets in Nevada and Utah, including a dozen newspapers, four radio stations, and two television stations. Under the circumstances, the news release hardly rated as significant.

Once the CIA and Lockheed selected Watertown during July 1955, the AEC provided the CIA with a secret facility by expanding its boundaries to include Groom Lake. The CIA and Atomic Energy Commission invented CLJ, a fake construction firm to oversee the initial construction done by sub-contractors. (CLJ were the initials of Kelly Johnson's name.) Workers for the firm began handling contracted work to put in hangars, a mile-long runway, concrete ramp, quarters, water well, sewers, etc. Official records referred to the facility as Watertown Strip. At first, the CIA paid for the work by placing a check in an envelope delivered to Kelly Johnson's home mailbox. Soon Miller's contract showed AEC paying $1,165,062 to REECo for construction and operation of Watertown through April 1956.

History has shown the CIA took the right approach at the beginning by announcing through the Atomic Energy Commission the construction of the facility. The CIA even identified it as a test site for the U-2. The agency used the cover story of the National Advisory Committee for Aeronautics (NACA) building the facility to conduct weather research. It provided enough information to satisfy the public's curiosity without revealing the CIA's involvement or classified operational details about the U-2's mission. The CIA hid in plain sight.

Naming the CIA's Secret Flight Test Facility

For decades, and even today, the debate has continued among Area 51 enthusiasts on how this small rectangular area adjoining the northeast corner of the much larger Nevada Test Site gained the name, "Area 51." CIA declassification of formerly top-secret documents has confirmed that it wasn't until June 1958 that Public Land Order 1662 withdrew 38,400 acres of land encompassing the Watertown base from public access. This rectangular addition to the Nevada Test Site was designated Area **51.** The CIA adopted the name depicted on the Atomic Energy Commission map dividing the Atomic Proving Grounds into areas. (Identifying the number 51 adjacent to the much lower area numbers and out of sequence, confirmed that the Atomic Energy Commission added Area 51 later in the sequencing.) This means that Area 51 did not exist during the CIA's U-2 era.

In any case, "Area 51" today conjures up images of conspiracy and mysteries dating back to the post-WW II Cold War era. Since the CIA's arrival in 1955, this non-existent Flight Test Center has acquired several identities: Groom Lake, Dreamland, Nevada Test Site, Nellis Test Range, Paradise Ranch, Area 51, Watertown Strip, and the Pig Farm. Its first mailing address was Pittman Station, a former one-room post office in a rundown area in Henderson, Nevada.

The pilots flying the North Range of Nellis today refer to Area 51 as "The Box" because the shape of the restricted area illustrated on their maps. In 1962, when the CIA formed the fifth discipline for intelligence-gathering, the Science and Technology Directorate, they would refer to it as Station D.

Bissell and Herbert I. Miller consulted with Dulles before asking the AEC to add the Groom Lake area to its real estate holdings in Nevada. AEC Chairman Adm Lewis Strauss readily agreed before Dulles approached President Eisenhower, who approved the addition of this box of land.

Thus, in 1955, the CIA identified the rectangular six by 10-mile (9.7 by 16.1 km) former WWII bombing and artillery practiced airfield and now a CIA test area as Watertown, unofficially naming the facility only for internal administrative purposes. Kelly Johnson did the same to give the new facility in the middle of nowhere a name for reference. He chose a name that sounded attractive to his workers.

When Johnson met with the CIA officials in Washington, DC, he discussed the progress made to the Watertown facility and the AQUATONE program. *(The CIA would later approve his proposal to unofficially identify the facility as "the Paradise Ranch" for the A-12 replacement for the U-2, an ironic choice that he later admitted a dirty trick to lure workers to the program. The workers soon shortened the name to "the Ranch," a fitting name for a range populated with stray cattle, burros, and wild horses.)

What's in a Name?

Since learning of Area 51 since around 1980, unknowing people have generated much speculation, debate, and skepticism about classified activities suspected of occurring there. In the late 1950s, the name Area 51 identified a block of the desert next to the AEC range for testing atomic bombs. Groom Lake was nothing more than the name of the geographic feature that defined the test facility.

The operating facilities developed in this desert valley used many unofficial names for shipping and administration purposes. The CIA bestowed the name Watertown Airstrip as the official name for the U-2 program. Per rumors, the name of CIA director Allen Dulles's birthplace of Watertown, New York inspired the Watertown name. However, records revealed the name referred to rainwater flooding the Groom dry lakebed by runoff from the nearby mountains. Workers related to the facility as the Watertown because of the way floods covered the airstrips. So, "Watertown" has several explanations.

The Nevada Test Site subdivision into numbered areas with the numbers seemingly selected at random made it convenient for the CIA to add another Area number to identify its creation at Groom Lake. Thus, the June 20, 1958, transfer of public land surrounding the facility used the "Area 51" designation specifically to identify the 38,400-acre block of land surrounding the airfield.

The area designations provided a means of calculating subsistence payments to the workforce. For example, in the early 1960s, REECo routinely paid its personnel different salary scales. The unionized company paid one scale for those working at Mercury, Nevada (the base camp for atomic testing). The company paid more if they worked in the forward Areas, whereas, Las Vegas-based personnel received no paid subsistence at all.

REECo time cards from this period contained a two-digit location code. For example, 01 identified the Nevada Test Site for atomic testing. Additional two-digit area codes identified the various areas that an employee worked in each week. Anywhere the access classification changed, the AEC installed a gate where security personnel checked all badges to ensure authorized access. Workers at Groom Lake sometimes drove through several numbered areas of the Nevada Test Site to travel to Area 51.

Consequently, Area 51 security badges contained several of these two-digit codes. The two-digit area code "08" identified Area 51 itself. (A different coding exists today)

Some explain that the Groom Lake facility acquired the Area 51 name from the "Project 51" code name for the building phase for "Project OXCART." Then, the new Science and Technology Directorate first identified Area 51 as CIA Station D. Like "Watertown," "Area 51" became a valid place name. Regardless of how the facility obtained the name as in the case of Watertown, the Area 51 name for the facility found its way onto official NTS maps, other public documentation and records, and into use by the media.

How the CIA's Lack of an Official Name Generated a Legacy

The cover story for the secret presence and activities in Nevada set a precedent that lasted for over half a century. Denial of the agency's presence and activities kept Area 51 unknown to the public until late 1979 when the Air Force assumed control of the operating facility.

To prevent associating the A-12 Project OXCART with the U-2 Project AQUATONE era at Area 51, the CIA would officially prohibit using the name "Watertown." Kelly Johnson would suggest naming the CIA facility the Paradise Ranch when the CIA reopened Area 51 for testing the A-12. The newly formed Science and Technology Directorate referred to and treated Area 51 internally as a station, "Station D." However, most old-timers today who worked on the A-12 OXCART project still refer to the site as "The Ranch." Generations of workers, some affectionately referred to themselves as "ranch hands" passed this nickname on.

Neither the CIA nor the Air Force ever classified the name Area 51 but treated the name as classified when used to identify specific classified activities. The Groom Lake operating facility managed deniability by avoiding it having a name that would identify and establish it as an official entity. With the A-12 program to replace the U-2, Area 51 would become a CIA station with an "operating facility." The Air Force support unit formed at Bolling AFB near Washington, DC, and according to records never left there. The unit owned an official unit patch. Area 51 never served as an Air Force Base during the CIA era.

The facilities at the generically named "Area 51" could not admit to an official name while operational because an officially named entity took away much of its ability to maintain secrecy. Had the CIA admitted the existence of Area 51, one can only imagine the problems this would have created. Long lines of attorneys, activists, environmentalists, labor unions, politicians, and even enemies would have used the US courts and frivolous litigation to impede progress and learn the secrets of not only this unnamed facility but also its various outposts around the world. Opening these doors would have changed the ability to develop a U-2 in 8 months, most likely changing the U-2 development timeline to that of the F-22 timeline spanning 33 years.

*In 2010, the CIA declassified its affiliation with Area 51, finally allowing the open use of the name to identify the area containing the facility. It remains doubtful that even now the facility managed by the United States Air Force operates under an official name.

Some would later identify the facility as Dreamland; a radio call sign introduced in the late 1960s for the facility, replacing the previous name, "Yuletide." The name referred specifically to a large block of airspace called a Special Operations Area surrounding Area 51, parts of the Nevada Test Site, and Nellis Air Force Range (now Nevada Test and Training Range).

The Dreamland airspace would eventually reduce in size to cover an approximately 24-square-mile box with Groom Lake at its center. The size and shape resulted in many pilots, mostly RED FLAG participants, nicknaming it "The Box" because the CIA and the Department of Defense defined it as a no-fly zone during the various exercises occurring out on the range. Some would call it "Red Square" because of the Soviet-built MiG planes flying out of Groom Lake. The Dreamland call sign still appears today in Air Force pilot's flight guides.

In some instances, people would refer to the Groom Lake test site by the designation of its operating organization. From about 1960 to 1971, USAF Detachment 1, 1129th Special Activities Squadron supported the facility for the CIA operations in Nevada during project OXCART. A closer look at this military unit reveals that the Air Force would activate the 1129th solely for the CIA's OXCART program at Groom Lake. Per Air Force records, the 1129th remained located throughout its existence at USAF Hq Command at Bolling Air Force Base, Washington, DC.

The pilots' Form 5 flight records illustrate their flights at Groom Lake as occurring at another Air Force or navy base—mainly at Bolling, Shaw, and March Air Force bases. The falsity applied to the CIA pilots returning to the Air Force at the end of Projects AQUATONE and OXCART as well.

When the Air Force would later form the 1129th SAS, Special Activities Squadron, it would do so strictly to support the CIA's Project OXCART. It did not have a unit name other than the 1129th SAS unit designation. Nor would the squadron have a unit patch. Consequently, the members being military would feel the squadron needed the traditional unit patch and would make their own. They would become the Roadrunners; a name that originated with two 1129th officers testing the installation of new radios in their vehicles. One of them remarked about the need for a call sign. MSgt O. B. Harnage in the command post heard the remark and said that they sounded like a couple of roadrunners, referring to the roadrunner birds abundant in the area and on the "Wile E. Coyote and the Roadrunner" cartoon. Today, the Roadrunner patch signifies anyone who worked on the CIA's U-2, A-12, YF-12, and M-21 / D-21 drone programs.

When the CIA personnel would later deploy on temporary duty to Kadena, Okinawa for OPERATION BLACK SHIELD, CIA pilot Jack Weeks would design a Cygnus patch for the CIA pilots. Consequently, the CIA pilots of BLACK SHIELD would often refer to the A-12 plane as the Cygnus.

The CIA would transfer responsibility for the test site to the Air Force in the late 1970s, under the management of the Detachment 3, Air Force Flight Test Center. Since the mid-1990s (and possibly as far back as the 1980s), the Air Force would refer to the Groom Lake facility as the National Classified Test Facility.

Thus, one can see how the CIA covered their classified activities in Nevada by not officially naming or acknowledging their existence. This practice became a lasting strategy that carried for half a century. No one casually spoke of the secret base even among themselves. Therefore, naming the facility never became an issue. Naming the facility meant that it existed, another excellent reason for it not having a name. Making the lack of a name an issue even in this book only adds to Area 51's mystique.

The Home Plate nickname dated to at least the 1960s and identified the facility in OXCART flight logs. Today, Home Plate still occasionally appears in documents and correspondence. Mostly, security personnel and aircrews used the nicknames Home Plate and C-Base in unencrypted radio communications rather than calling out the actual name of the facility.

Names were not a concern to anyone having lawful access to Area 51. Those working there for the CIA did not talk about their work to anyone. The workers never needed an official name. No one working at Area 51 spoke of the facility outside those having a valid and authorized need to know. The need for a name existed purely for the agency, Air Force, and corporate support documentation and routing purposes.

In any case, those working at Area 51 used pseudonyms to support cover stories for workers there. If asked where they worked, Air Force personnel said, "Pittman Station, Henderson, Nevada."

The name derives from the original mailing address at the now-defunct Pittman Postal Station (originally PO Box 121, later PO Box 52B). Civilian contractors at Groom Lake said they worked for "EG&G at the Test Site."

The CIA U-2 Flight Test Facility

The dry lakebed provided an excellent landing strip except for when it rained enough for the lake to collect rainwater runoff from the surrounding mountains. For this, the project managers provided a paved runway to allow testing during the times when the lakebed received rain. On 25 July 1955 when the first U-2 arrived at the flight test facility, the water did cover the lakebed. The C-124 landed on the new runway before it sealed and armored, "leaving deep wheel marks."

The CIA's facility requirements soon changed, however, calling for a permanent facility nearly 300% larger than Johnson's original design. Johnson estimated the construction of a larger Site I facility would cost $450,000. He expected a cost of $832,000 for building the same facility at Site II (Groom Lake).

The CIA's Flight Test Center became ready for occupancy during July 1955 with the Agency, Air Force, and Lockheed personnel moving in under the command of the CIA's station commander, Richard "Dick" Newton USMCR (Ret) who served from 1955-56, followed by Landon McConnell from 1956-57. The first CIA trainee pilots would arrive during January 1956.

By now, the CIA's fledgling Groom Lake facility possessed two dirt landing strips in an "X," scraped into the barren desert floor to the east and the 5000' asphalt strip near the SW corner.

At this same time, Earnest Williams, a REECo employee drilling a well for water hit a limited supply; however, trouble with the well still required trucking in water the same way ass the fuel.

Early flight from Groom Lake – no markings

The facility at first contained three "T" hangars and a few other buildings that offered rudimentary accommodations for the test personnel. By 3 October 1955, it established land-line communication with Burbank, also a security-cleared MATS shuttle flight. Testing of the U-2 shut down for 12 days to allow construction work by crews not cleared to view the plane or get any idea of the project.

AEC and REECo engineers together in Las Vegas spent 17-19 November 1955 laying out plans for aircraft parking aprons and tie-downs, dispensary addition, and other work. The 17 November crash of the C-54 shuttle on Mt. Charleston occurred at the stage surveyors were beginning to stake out the taxiway. As the search for the C-54 started, work at Groom Lake went on. Graders, cranes, and concrete mixer moved to the site. The control tower, fabricated at Camp Mercury, stood erected at this point, along with the security tower. The Quonset building and steel framing for the warehouse building began by 19 November 1955.

By late November approximately 60 construction workers billeted at the camp. Because temperatures were dropping to 6 F at night, it was necessary to use hot mix asphaltic concrete for paving the parking aprons and taxiway. The 12-day construction period in the last half of November produced the photo lab addition, classroom, preparation for 20 trailers, and installation of utilities, the compressor slab, plus rails and hoists in two hangars.

As of November 1955, the program had been seeing up to 30 flights a day by the Lockheed pilots and SAC Instructor Pilots. Lockheed provided firefighter protection for the base as it was proving difficult to obtain qualified civilian firefighters elsewhere.

The facility's few amenities eventually included a movie theater and a volleyball court in addition to a mess hall and fuel storage tanks. (Many today tend to refer to the Groom Lake operating facility in Area 51 as a base, as in an Air Force Base. During the CIA stewardship, Groom Lake remained a flight test center with no official name or designation as an airbase.)

*In 2007, the Federal Aviation Agency, FAA assigned the Groom Lake facility the airport identifier KXTA. The FAA describes two of its six runways as Runway 12-30, 5,420 feet by 120 feet, and Runway 14-32, 12,000 feet by 200 feet. That is not entirely correct. One end of 14-32 continues across the dry lakebed for another 11,000 feet with that portion closed and partially covered with blowing sand. Moreover, there are four additional runways marked in the sand of the dry lakebed.

The Ranch – In Their Own Words

Stockman: (In the first group of CIA pilots, January 1956) "We lived in trailers, three to a trailer as I recall. We couldn't write or call home from out there at Groom Lake. Probably fifty or so people on the site. There was a training building, which was also a trailer [right next door]. It was just all desert out there. To get to civilization, you were pretty dependent on aircraft. There was some road traffic, very carefully watched. Security people everywhere."

Jankowski: "It was very primitive, but the customer made up for it with a fantastic chow hall. We ate better than I ever ate in my life." - Roadrunner News, May 2010

Bevacqua: "We got our first glimpse of Groom Lake, commonly known to the trainees as 'the Ranch.' On the ground at the Ranch, my impression was that it was desolate country, truly cut off from civilization. There was only a large multi-purpose building where we took our meals, played pool and cards and watched movies. Groups of four pilots crowded into one Airstream trailer located adjacent to the flight line. Everything on that field was new since it was strictly a dry lake prior to it becoming a training base." - McIlmoyle p. 43

Ingram: (January 1957) "We boarded a C-124, destination unknown. I could tell we were flying east, and we landed on a dry lakebed an hour later; the lakebed designated as Groom Lake on a road map. Through the windows, I could see a paved runway, hangars, mess hall and barracks and on the ramp, six silver gliders with jet engines. That was my first glimpse of the U-2. I was learning fast that this area nicknamed The Ranch had tight security and even the name of the aircraft was classified Secret. Meals were excellent, and only cost $1 for a meal such as steak, salad, potato, and drink." - McIlmoyle p. 130-131

Powers: "As a place to live, it left much to be desired. As a secret training base for a revolutionary new plane, it was an excellent site, its remoteness effectively masking its activity" - Powers p. 15

Nelson: "A bus pulled up, and we climbed on board bound for a C-47 on the flight line. The aircraft held 27 people. I noticed most of the people on-board wore civilian clothes; some guys even wore Levi's. I was in fatigues. From my window, I looked down and saw huge black spots on the ground in Nevada, where the Atomic Energy Commission (AEC) had conducted atomic bomb testing. When we landed, I saw a mountain range in the distance. Our aircraft had landed on a runway at the edge of a dry lake."

"The back door opened, and I could tell we were in for an adventure. Two men in blue coveralls met us as we exited the aircraft with our bags. One of them said, 'Gentlemen, welcome to Groom Lake. You are on the part of the Atomic Energy test site. Please follow my colleague into the building.' No one was allowed on this facility without approval at the highest levels of security."

"This was a place of extreme security. We were told, 'What you see, you forget.' There was only one person authorized to talk to about our work, the man we worked with. Even mail was subject to inspection, and garbage cans were checked daily for any written material that was discarded."

"An armed guard was stationed at every building, and badges were carefully scrutinized. Finally granted access to the hangar, Pfeiffer stood aside and said, 'There it is.' It was like no aircraft I had ever seen, and I'd been around a few years. Pfeiffer said the aircraft, called the U-2, was classified Top-secret and a new language had sprung up to describe many aspects of the operation. For example, the U-2 was called an 'article' not an aircraft and pilots were called 'drivers.'" - McIlmoyle p. 173-176

Ritland: "We looked at that lake, and we all looked at each other. Kelly Johnson said, 'We'll build it right there, that's the hangar. We'll put the runway there. This will be the tower, right here.'" . Wild burros occasionally ventured across it, pebbles the size of peas blew around in the afternoon winds. " – Darlington, p. 113

"Its wilderness location brought some unusual neighbors, including a bobcat that set up house under one of the offices and a rattlesnake that turned up in the well of an airplane wheel." – Taubman p. 139

View of Watertown from Papoose Mountain

Air traffic control tower

C-124 delivering a U-2 plane to Watertown.

Base Operations
Headquarters flag at half-staff following Sieker crash

U-2 tail attached with three 5/8-inch bolts

Trailer area viewed from the tower

The first U-2 delivered to Watertown

View past shuttle to U-2 preparing for flight

Note guard shack on the tower, wind damage to the antenna

Looking W at the sunset
U-2 assembly work at night in T-hangar

Firetruck, control tower

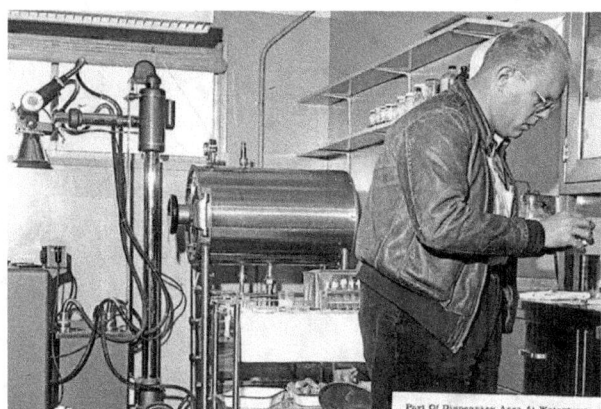

Flight surgeon in oxygen pre-breathing area

Aircraft assembly in T-hangar

U-2 final assembly in a hangar

Flight line looking east

Delivery of the First U-2

On 24 July, less than eight months after receiving clearance from Trevor Gardner, Kelly Johnson readied the first aircraft, known as Article 341, for delivery to the "Watertown" site. Lockheed completed the final inspection, flutter and vibration tests, control proof tests, and disassembled the aircraft for transporting to the Groom Lake facility for taxi and first flight.

Three days later, Lockheed transported the first top-secret AQUATONE prototype U-2, Angel 1, disassembled and wrapped in cloth, to the Watertown Groom Lake facility from a small

Lockheed factory at Oildale, California onboard a C-124 transport plane. The C-124 pilots delivering the aircraft did not know their destination. Their instructions were to fly (at night) to a point on the California-Nevada border and follow radio directions to their unknown landing site. The CIA security personnel required the crew to turn in their regular IDs on arrival and use aliases while at the Groom Lake facility.

The Lockheed U-2 #001, a lightweight, an unusual-looking plane with eighty feet wide wingspan resembling a glider, arrived at Watertown to commence flight-tests under the FAA designation N308X.

Among other equipment, it could now be armed with two telescopic-lens cameras with extraordinary capabilities—its high-resolution lens capable of focusing on a newspaper headline over the shoulder of a person on the ground from an altitude of fifteen miles. Johnson's aircraft provided the unprecedented potential for espionage, and vital for investigating the so-called bomber and missile "gaps" threatening the US national security.

At Groom Lake, Lockheed mechanics spent the next six days readying the craft for its maiden flight. Before "Kelly's Angel" could take to the air, however, it needed an Air Force designator.

Col Allman T. Culbertson from the Air Force's Office of the Director of Research and Development pointed this out to Lieutenant Colonel Geary. The two officers looked through the aircraft designator handbook to see their options for deciding if they should call the project aircraft a bomber, fighter, or transport plane. Not wanting anyone to know the CIA intended using the new plane for reconnaissance, Geary, and Culbertson decided to place it in the utility aircraft category between two utility aircraft on the books, a U-1 and a U-3.

Fuel Management Challenges

Johnson designed the U-2 using the Pratt & Whitney (P&W) J57/P-31 engine developing 13,000 pounds of thrust and weighing 3.820 pounds, giving it a power-to-weight ratio of 3.4:1. The U-2 first took to the air without these engines available because the Air Force needed the entire production to power specially configured Canberra RB-57Ds.

The first U-2s, therefore, used Pratt & Whitney J57/P-37 engines 276 pounds heavier and delivering only 10.200 pounds of thrust at sea-level with the resulting power-to-weight ratio of 2.7.1—20 percent less efficient than the preferred P-31 version.

The U-2 needed to carry a large amount of fuel to conduct long missions over hostile territory. To accomplish this, Kelly Johnson used a "wet wing" design for the U-2, which meant storing its fuel in the wings rather than in separate fuel tanks. Each wing divided into two leakproof compartments with the fuel pumping into all the cavities within these areas except for the outer 6 feet of the wings not used for fuel storage.

The U-2 carried a 100-gallon reserve tank in its nose. Later, in 1957, Johnson increased the fuel capacity of the U-2 by adding 100-gallon "slipper" tanks under each wing, projecting ahead of the leading edge. The need to maintain aircraft trim as it consumed fuel became one of the most important considerations in the U-2s fuel system.

The plane, therefore, contained a complex system of feed lines and valves draining to a central sump, making it impossible to provide the pilot with an empty/full type of fuel gauge. The first 50 U-2s would use mechanical fuel totalizer-counters instead of traditional fuel gauges. The pilots found this aircraft fuel control system very primitive. They referred to it as the "water spigot" because it was either on or off. It didn't seem to have an intermediate control, a crucial consideration with the early U-2 that required the pilot to maintain trim as it consumed fuel.

Before the start of each mission, the ground crew set the counter to indicate the total amount of fuel in the wings. A flow meter subtracted the gallons of fuel consumed during the flight. Once the U-2 began flying, the pilot kept a log of fuel consumption recorded by the counter and compared it with estimates made by flight planners for each leg of the trip.

As a double-check, U-2 pilots kept track of their fuel consumption by monitoring airspeed and time in the air, most pilots becoming quite expert at this. Even so, the planes often ran out of fuel or experienced flameouts because of the fuel control system. During the early part of the program, the pilots could average seven flameouts per flight, each time requiring the U-2 to descend into thicker air at 20-30,000 feet to get a restart.

Fuel Management in Their Own Words

Lopez: "Since the fuel was carried in the very long wings; great care had to be exercised to equalize the weight on both sides. One U-2 while refueling drifted out into the wingtip vortices of the tanker and shed a wing." - Polmar foreword p. vi

Cloke (RAF): "I dumped fuel and landed on the lakebed because the aircraft had developed an unusual heavy wing. At touchdown, the yoke was at 90 degrees to the normal position. Subsequent investigation revealed that one transfer pump had stuck in the 'on' position, continually transferring fuel to one wing." - McIlmoyle p. 386-7

U-2 flown by Maj Mike Hua forced down by fuel leak during a mission

Collecting and measuring unused fuel from the wings "After each flight, careful accounting is made of fuel consumption."

The Myths Spawned by the CIA Secrecy

News stories published as early as November 1955 described the Groom Lake base as "the super-secret 'proving ground within the proving ground.'" The name "Area 51" did not exist at this time. Later articles written in the 1960s described Area 51 as a "super-secret Nevada base" though without revealing the CIA connection.

Nonetheless, the world's best-known secret base did not become a household name while the CIA had it. The first public knowledge of the resumption and continued activities at Area 51 began circa 1979 after the CIA surrendered its management of the facility to the Air Force. As the CIA feared in 1955, the Air Force, with its chain of command and its practice of rotating personnel, was unable to maintain the secret existence of the facility. As word leaked about the installation, the conspiracy theorists wondered what the CIA did behind this secrecy for the past 24 years.

Few believed that an intelligence agency such as the CIA would have a secret place hidden in the Nevada desert as a venue to test the new aviation technology. After all, the CIA was supposed to be about spying and covert activities—cloak and dagger stuff. Almost overnight, Area 51 became known as a mythical Mecca for black conspiracies and cover-ups. Although the famous Roswell "flying saucer" event occurred in 1947, and the CIA formed Area 51 in 1959, ufologists around the world connected Area 51 to Roswell. That affiliation continues today.

Many believed the CIA constructed massive underground facilities for harvesting colonies of extraterrestrial alien species for their alien technology. For years, the media, movies, and authors have competed to generate conspiracy stories concerning the CIA conducting reverse engineering at Area 51. An endless number of diverse political and social groups throughout the world have used blogs to associate Area 51 with all sorts of trivial, superficial, and sensationalist phenomena. It is not because of what the media and some authors believe. Paranormal ufology and alien hypotheses sell books, movies, and memorabilia. After all, eighty percent of the world believes in flying saucers and little green men.

No such nonsense exists. Think about it. Area 51 became the most highly monitored spot on earth after the Russians launched their Sputnik satellites, knowing at the time about Area 51. Other than base support, security, and the special projects technical services team, Area 51 has always served transient occupants, here today and gone as soon as they complete their project. Countless low-skilled workers have come and gone at Area 51. No way could some of them have not said something over the past half-century had the stories been true. Therefore, the thought of the government keeping secret extraterrestrial aliens at Area 51 is ludicrous. Nonetheless, the state of Nevada named the Highway 375 access to the Area 51 region the Extraterrestrial Highway. Many businesses, including the Las Vegas 51s professional baseball team, have made a business of selling UFO memorabilia.

Contrary to what the world wanted to believe, Area 51 provided an operating technical laboratory for the advancement of military aerial systems, a business serving customers. Yes, secrecy cloaked Area 51, but not necessarily because of military activities.

In some cases, the secrecy existed to protect customers competing with one another via proof-of-concept systems or platforms undergoing testing for sale. Area 51 was where the military customer told the competitors to show what they had to sell. The competing vendors knew their trade secrets were safe at Area 51.

Chapter 8 - Organizing Project AQUATONE

The first Table of Organization, T/O, for Project AQUATONE, approved at the end of April 1955 by the Deputy Director, Support, provided staff for a headquarters office, a US field test site, and three foreign field bases. The staff totaled 357, consisting of 92 agency personnel, 109 Air Force officers and enlisted men, and 156 contract personnel that included tech reps, guards, and primary aircraft pilots.

The CIA, considering the changing requirements, revised the T/O within the month to delete the support aircraft crews, now an Air Force contribution. It increased the administrative support area, targeting clerical. It added a communications reserve cadre to permit retention of personnel while training on project equipment before their assignment to the field. For the four deployment bases, the T/O replaced civilian contract guards with staff security investigators and added a supply depot.

The Director of Personnel received a sterile version of the T/O, so he might produce agency candidates to fill the vacancies and provide support in keeping personnel records. The CIA assigned the highest priority to the project's requirements and made every effort to staff it with the best candidates. However, the CIA found it more difficult getting the actual bodies on board than getting approval to add them. For example, qualified physiological support personnel was in short supply.

The T/O called for scores of communications engineers and technicians and security investigators. Their offices handled the task by setting up their own recruiting and training programs to meet the requirements for personnel without depleting their staffs. The office reached a rapid decision disallowing accompanying dependents at either the ZI, (the part of the theater of war not included in the theater of operations), or foreign operating sites. Thus, the CIA chose single men wherever possible. To this end, the project made excellent use of Air Force enlisted men in clerical slots. The "no dependents" rule continued in effect until the end of 1957.

Watertown Support Aircraft

At Watertown, they also had a C-47 twin engine transport that was used for cargo and personnel transport. The Lockheed personnel traveled to and from Burbank in their twin Beech or in the scheduled Air Force C-54 (which later crashed into Mt. Charleston with fourteen AF crew, Hycon, CIA, and four Lockheed personnel aboard, all killed. Those of them in General Yancey's organization traveled to and from March AFB in one or more of the chase aircraft or the C-47. Normally, they would fly to March AFB on Friday afternoon, at the end of flying, and return on Monday morning.

Operations at Groom Lake went forward with one L-20A, one Beech Model 50 Twin Bonanza, four T-33 trainers, two C-47 transports, three C-54 shuttle transports (one crashed 17 November 1955 on Mt. Charleston – Chapter 12), and one C-124 transport. Several other C-124 and C-118 cargo planes brought in later for the deployment of the operational Detachments overseas.

Contract Personnel, Engineers, and Tech Reps

Maintaining security was a concern when the CIA contracted with each of the suppliers handling the furnishing of contract tech reps to maintain and service project equipment at Area 51 and the planned overseas operating sites. The CIA expedited security clearance and training by impressing upon the companies they need to draw current employees from their ranks for overseas assignments rather than sending recruits.

Each company maintained its policy regarding pay scales and other employee benefits. For those Lockheed employees planning to deploy, the company developed a plan whereby it withheld a part of the overseas pay that it would pay upon completion of an 18-month contract along with a bonus of $5,000 as an incentive to finish the contract. Any employee who failed to complete his term, or had his employment terminated for cause, received transportation home without their bonus and with the travel costs taken from the amount withheld.

Besides furnishing a five-person crew for each U-2 from Lockheed, the CIA executed service contracts or made other arrangements with Perkin-Elmer and Hycon for photo equipment. The CIA recruited Ramo-Wooldridge for electronics, Firewel for pilot equipment, Baird Atomics for the sextant, Westinghouse for side-looking radar, and Pratt & Whitney for the engines. It later signed other service contracts along with the same security terms for photo developing equipment. Many of these companies generated pseudo names for dealing with the CIA.

CIA Flight Testing Begins

Lockheed built the first U-2 planes used for training by hand. The same Lockheed people stayed with their U-2 aircraft at Watertown to provide maintenance and support. For each U-2, flight engineers served as what the Air Force would call crew chiefs.

Lockheed test pilot Tony LeVier, famous for flying the first flight in the U-2, only flew two low-altitude flights, whereas the Lockheed test pilots and the Air Force SAC instructor pilots flew the same and much more advanced missions. Colonel Phil Robertson, the operations officer, flew Article 345 to 74,500 feet. During October 1955. Tony LeVier would leave Groom Lake (never to return) to become Director of Flight Operations at Lockheed.

Not having a two-seat model of the U-2 made training difficult. The trainee pilots received their instructions on the ground before takeoff and over the radio once the craft became airborne. Almost 15 years would elapse before a two-seat U-2 became available for training new pilots.

The project's operating organization evolved slowly from January to April 1955, with most the individuals working on AQUATONE remaining on the rolls of their agency components.

On 2 March 1955, Dick Bissell discussed with the Deputy Director for Support, Colonel Lawrence K. White, his plan for the project's organizational structure. They discussed funding and staffing and agreed on charging personnel and operating costs to separate accounts and segregating both from those of other regular components—the "special project" concept.

Colonel White promised to name an administrative officer to help the project part-time on current regulatory matters and the development of an organizational plan. They planned to assign the officer full-time to the project later to manage administrative problems. During the last week of April 1955, the DD/S named Col Robert B. J. Hopkins as the administrative officer. Colonel Hopkins had just returned to duty from a recuperative leave following an illness. He stayed with the project only long enough to find it a "pressure" job. Two weeks later, he asked the Deputy Director to relieve him. The DD/S nominated Mr. James A. Cunningham, Jr., who proved a strong candidate and, in fact, held up under pressure for more than ten years.

Full Complement Achieved

Toward the end of April 1955, Project AQUATONE was now operational with 357 personnel divided among project headquarters, a US testing facility, and three foreign field bases. The CIA employees represented only one-fourth (92) of the total. The Air Force had committed 109 personnel to positions on the 1955 table of organization, not counting much of the other Air Force personnel such as the SAC meteorologists, who supported the U-2 project in addition to performing their other duties.

Contract employees made up the largest Project AQUATONE category with 156 positions in 1955. This class included maintenance and support personnel from Lockheed (five per aircraft), the pilots, and support personnel from other contractors for items such as photographic equipment.

The population of Watertown had increased in all categories to 444 at the end of 1955. With the staffing of Detachment "A" through the winter and spring of 1956 and the selection of cadres for two more detachments, the end of March 1956 found the population at 546.

By October 1956, both Detachments A and B had deployed as operational, and Detachment C awaited deployment. The AQUATONE population reached a high-water mark of 600 personnel. At this point, the operational pace slowed down because of the political stand-down of overflights of the Soviet Union. The CIA project and Watertown faced a reduction in force. Project Dragon Lady, the Air Force follow-on U-2 group, trained at Groom Lake from September

1956 into early 1957. Like AQUATONE, Dragon Lady would go on to produce its overseas U-2 detachments.

Watertown Goes Operational – In Their Own Words

Cordes: "My final U-2 checkout flight at The Ranch was a unit simulated overflight mission complete with cameras and designated targets. Takeoff was at night, and I was one of the first to use the sextant to measure my latitude with Polaris. This mission went west to California, north to Montana, east to the Dakotas, south through Colorado and west back to The Ranch. Total flight time was eight and a half hours. When I landed, I was certified, fully qualified to fly the U-2" - McIlmoyle p. 51

Bissell: "When the first field detachment of just over a hundred men and three aircraft approached a state of readiness, it was the Strategic Air Command training group that devised and supervised a simulated combat exercise involving a number of long-range reconnaissance missions over the United States. At the end of the exercise, they concluded the unit was combat-ready." - Bissell p. 97

Eisenhower: "We then tested the probability of the U-2s being discovered by the Soviets as it flew over the territory of that nation. So, a number of test flights were made over our own country. Even though our radar systems had been warned of strange airplanes flying over our national territory, the U-2 flights were either unseen or were tracked imperfectly." - Eisenhower p. 545

Security Concerns

On 29 April 1955, Richard Bissell had signed an agreement with the Air Force and the Navy, who exhibited interest at the time. The services agreed with the CIA's assuming primary responsibility for all security concerning the overhead reconnaissance of Project AQUATONE. From this period on, the CIA took responsibility for the security of overhead programs.

This responsibility placed a heavy burden on the Office of Security. The challenge called for establishing procedures to keep scores of contracts untraceable to the CIA. The responsibility included determining which contractor employees required security clearances and devising physical security measures for the various manufacturing facilities. The FBI watched the Lockheed factory to protect against foreign spying during the U-2 program.

Now with CIA detachments deploying overseas, the Office of Security found keeping the U-2 and subsequent overhead systems secret a time-consuming and costly undertaking. The most important aspect of the security program for the U-2 project called for the creation of an entirely new compartmented system for the product of U-2 missions: All imagery would be classified "Top-secret Chess" and the intelligence obtained handled in the strictly controlled 'Talent' system.

The terminology used to describe U-2 aircraft and pilots played a part in maintaining the security of the overhead reconnaissance program. The CIA reduced the chances of a security breach by always referring to its high-altitude aircraft as "articles," with each aircraft having its "article number" that the CIA used to identify the U-2 in classified internal documents. Similarly, the CIA identified the pilots as "drivers," a name that never set well with the former fighter pilots. (The prototype U-2, Article 341, never received a USAF serial number.)

Cable traffic referred to the aircraft as KWEXTRA-00 with the two-digit number identifying the precise aircraft; these figures were unrelated to the three-digit article numbers assigned by the factory. The CIA identified the pilots by a two-digit number identifying the specific pilot.

Thus, even if a message or document concerning overflight activities fell into unfriendly hands, the contents would refer to articles and code numbers without indicating the identity of the program.

Procurement of the aircraft components occurred secretly with even the plane's onboard equipment involving the CIA security planners. For example, Johnson ordered altimeters from Kollman Instruments with instructions for calibrating the devices to 80,000 feet. Johnson's choice raised eyebrows at Kollman because its instruments only went to 45,000 feet. CIA security personnel briefed Kollman officials and produced a cover story saying they planned to use the altimeter for rocket planes.

Security for the U-2 Project

Illuminated in an interview with Joe Murphy, former Security Officer with the program are the security concerns and practices of the CIA for the U-2 starting in 1955:

"Job requirements included remaining single for a number of years and being willing to spend a good deal of time overseas. Within a few months of my initial interview, I was in Washington, where, along with about 80 other men recruited for the same clandestine venture, I underwent polygraph testing and medical exams. All of the recruits were college graduates, and some had advanced degrees or government backgrounds."

"After our initial screening, a CIA security official briefed us on the U-2 program and told us that we were needed in the field as soon as possible. For the next five to six weeks, my new colleagues and I underwent basic training in physical security in the Washington area, including weapons training. Security and compartmentation were the order of the day."

"After completing the basic training, I joined a contingent of CIA, Air Force, and contractor personnel sent during February 1956 to a remote site near the Nevada Test Site at Watertown, Nevada, to prepare for our specific overseas missions. From Monday through Friday at 'the Ranch,' experienced Office of Security officials supervised security officer candidates in personnel security, weapons, access control, document control, and courier duties."

"In addition to protecting project aircraft and personnel, we were to escort operational film from the field to Headquarters. Security procedures established at the Ranch were the model for overseas detachments."

"Another important aspect of the security program for the U-2 project was the creation of an entire new compartment system for the product of U-2 missions. Access to the photographs taken by the U-2 was strictly controlled, often limiting the ability of CIA analysis to use the products of U-2 missions. To achieve maximum security, the U-2 program developed its own contract management, administration, financial, logistics, communications, and security personnel, and thus did not need to turn on a day-to-day basis to the Agency directorates for assistance."

Security Concerns, Then vs. Now

Today's security concerns at Area 51 have drastically changed from those of the CIA in 1955. Early on, the security concerns focused on cleared personnel discussing classified matters where those lacking a need-to-know might overhear them. About the only other concern in this area regarded mishandling of classified material. This concern lasted until October 1957, when the Russians launched their Sputnik satellite to introduce the United States to the space race. Suddenly, Russian satellites, referred to by Area 51 employees as "ashcans," appeared over Area 51, some seeking radio frequency (RF) transmissions, and others looking for infrared heat signatures.

Everything would change again when technology advanced into the computer age. In late 1961 and early 1962, satellite photoactivity by the Soviets was in its infancy, yet of grave concern to the CIA and its renewed operations at Area 51 to develop a stealth replacement for the U-2. Russian eavesdropping techniques would improve to where an ashcan (satellite) passing overhead could also listen in addition to detecting RF transmissions and infrared images. Security personnel at Area 51 would advise relevant parties of their overflight times over the area, at which time all air and ground operations would come to a halt, and all emissions, including audio and video transmissions, would cease at Area 51.

The introduction of computers would bring yet another level of security concern with the threat of hacking and cyber warfare. In addition to classified manuals and documents, security personnel would now have to contend with computer security, which included servers, modems and floppy disks, the latter advancing to CDs and USB flash drives, smartphones with cameras and SIRI, and even toys such as the Furby dolls that communicated with each other. Smart toys today connect to the Internet, making them suspect as electronic Trojans. All sorts of gadgets and personal devices would become potential security threats requiring security management and control. An iPhone, iPad, or computer with Google Earth would see and photograph the activities at Area 51 at will. Thus, Area 51 today disallows smartphones because of their cameras and Internet access.

Security badges have changed from the official Atomic Energy Commission badge using a number system to identify an individual's access authorization. Today, badge information includes colors and color barcoding to distinguish authorized personnel as sub-contract employees, customers, and government reps, etc. Color barcoding indicates the level of access. Yellow or orange designates top-secret access, red identifies a secret level, and green indicates an individual as having no security clearance.

Simply having a color-coded badge does not grant one access to Area 51. A top-secret clearance permits an individual to have access, on a need-to-know basis, to top-secret, secret, or confidential information as required in the performance of duties. The Defense Investigative Service (DIS) based top-secret clearances on the results of single scope background investigations (SSBI); however, typically, secret clearances based on national agency check (NAC).

A "Confidential" clearance permitted an individual to have access, on a need-to-know basis, to confidential information as required in the performance of duties. A "Q" clearance granted by the Department of Energy allowed an individual to have access, on a need-to-know basis, to top-secret, secret, and confidential, restricted data, formerly restricted data, National Security Information, or special nuclear material in Category I or II as required in the performance of duties.

Single Scope Background Investigation (SSBI)

Top-secret/Sensitive Compartmented Information (TS/SCI)

Access to Area 51 always required TS and SCI security clearances. Obtaining such clearances or for the AEC "Q" clearance entailed passing an SSBI or Single Scope Background Investigation. This involved the applicant submitting an executed Standard Form 86 (SF86) for the background check process where investigators or agents interviewed past employers, coworkers, and other individuals associated with the subject of the SSBI.

The National Agency Check checked on the subject and spouse/cohabitant using investigative and criminal history files of the FBI, including submission of fingerprint records on the subject, and such files of other national agencies (DCII, INS, OPM, the CIA, etc.) as appropriate to the individual.

Project Growth

As operations began at Watertown, growth came with a steady dribble of CIA, Air Force, and Lockheed personnel arriving. Earlier, during February 1955, Colonel Ritland urged the opening of a direct line to the DCS/P, Air Force Deputy Chief of Staff, Personnel to recruit the best candidates available and to expedite the paperwork required to transfer them to the project.

His urgings prompted the CIA's Military Personnel Division, headed by Col Jack Dahl, to set up procedures for handling the nominees separately from regular military assignees to other duty in the CIA. The Division placed requirements with the DCS/P liaison officer in the Pentagon who furnished the candidate files to Dahl for review by senior project officers.

Four months later, during June 1955, Colonel Dahl received word of the DCS/P, Gen John S. Mills, having concerns over the size and the phasing of project military personnel requirements. The Air Force acted reluctant to release so many extraordinary men from its critical categories.

The CIA overcame their reluctance by the signing of a joint agreement during August 1955. Nonetheless, these delays caused a sharp effect on training, equipping, and deployment. The reluctance of the Air Force to commit to the project justified the president's decision to have the CIA take the lead in training and deploying the U-2 detachments.

During May 1955, while the CIA dealt with security concerns and recruiting, Osmond J. Ritland, one of the first Air Force officers assigned to Project OILSTONE began coordinating with Richard Bissell the Air Force activities in the U-2 program. A month later, Ritland became Bissell's deputy, although Air Force chief of Staff Twining did not approve this assignment until the day after Air Force officer Lt Col Leo P. Geary joined the program during June 1955. Geary would remain until August 1966, longer than any of the other project managers. Geary used the Air Force Inspector General's office as a cover with the title of Project Officer, AFCIG-5 while serving as the focal point for all Defense Department support to the U-2, and later, OXCART programs. Geary provided a high degree of Air Force continuity during his 11 years with the overhead reconnaissance projects.

The Air Force attached its assigned project personnel to the 1007th Air Intelligence Service Group, Headquarters Command with a unit of MPD handling their records. It approached its selectees through a form letter describing the proposed assignment to the CIA as a sensitive activity overseas without dependents. It requested their Personal History Statement for use in a security office investigation and MPD's preliminary approval for administrative processing.

The candidate started the selection process upon receipt of orders to Washington. He completed the entry-on-duty processing, including physical and psychological examination, security briefing, and voluntary participation in a polygraph interview. Once the candidate received a final security clearance, the individual entered duty status and received a classified briefing on his assignment.

The first few months of this procedure saw a moderately high rate of washouts of military personnel for assorted reasons when subjected to agency tests. Despite efforts made to explain the necessity for it and to minimize the reaction to it, the CIA could do little to make this type of examination more palatable to senior Air Force officers. The career Air Force officers found it patently difficult to accustom themselves to civilian command with stringent security control over all their activities and movements. Only a modest number of problem cases came from the screening to give trouble later.

An excellent example of successful recruiting was the selection and development of SAC Instructor Pilot, Capt John Henry "Hank"

Meierdierck. Meierdierck, who would spend 22 years in the US Air Force and another eight years with the CIA, began his journey through the selection maze in mid-1955 when a colonel from the Strategic Air Command headquarters came to Turner Air Force Base recruiting pilots for a secret project. The colonel sought fighter pilots with experimental test time and lots of fighter jet time.

The colonel recruiting for the U-2 selected captains Louis Setter, an aeronautical engineer, and Meierdierck while providing no details concerning the assignment. Only after reporting in at March Air Force Base in Riverside, California did Meierdierck and Setter meet the other personnel making up this group.

Col William R. Yancey was the commander with Col Herbert Shingler as his deputy and chief of logistics. Col Phil O. Robertson was the operations chief, Maj John E. "Jack" Delap, the chief of navigation, and Maj Louis A. Garvin, a B-47 pilot and one of the experimental test pilots, along with captains Louis Setter and "Hank" Meierdierck. Several other officers received assignments to supply, base operations, air operation, and administration duties.

This unit was known as the 4070th Support Wing, also as the SAC Training Unit and the SAC Liaison Unit. It would have a life span of about 18 months until the three CIA detachments and the USAF U-2 follow-on group trained and deployed. At its peak, the 4070th used about twenty-three people.

Following a general security briefing, the 4070th staff flew to Watertown for an introduction to the secret reconnaissance aircraft. There, they met the Lockheed engineers led by Ernie Joiner and his staff. Settling in at Watertown, they also met Lockheed test pilots Ray Goudey, Bob Sieker, and Bob Schumacher, who did all the phase testing, testing the aircraft coming off the assembly line while also training the Strategic Air Command Instructor Pilots.

Arrival line for security check

MATS shuttle plane

Col. Herb Shingler

Pilot Jack Nole

Security check

T-hangars & flight line, looking WSW

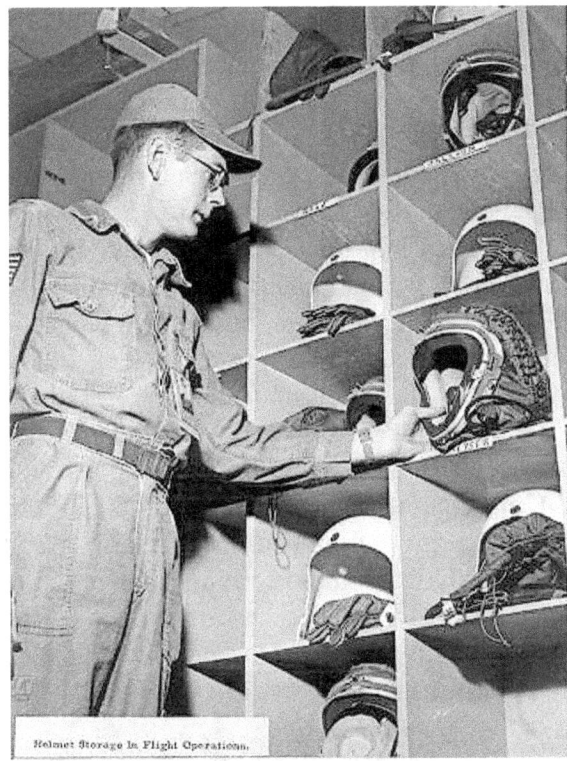
Helmet storage

Watertown weather office

Richard Heyser giving a briefing

Meteorologist's weather briefing

Colonel Perry, physically restricted from flying the U-2, had offered his resignation. He felt the commander, in the Strategic Air Command tradition, should fly the unit's aircraft, which he could not do. After much discussion, Perry stayed with the project only because of an obligation to those he disrupted to bring into the project. Nonetheless, he continued flying the T-33 used in training to simulate the floating tendencies of the

Pilot Ray Goudey breathing oxygen

Mission navigational planning

The U-2 accommodated only one person—the pilot—so pilots with their trainer flew the two-seater T-33 down the landing area, set 55% power, half flaps, and saw how long they could keep the plane a foot or less over the dry lake. This came close to the actual landing procedure of the U-2:

"Training of new pilots begins with the T-33. The pilot must be able to hold the T-bird inches in the air for the length of the lake so that he will be able to hold the Angel at the same altitude until its broad wings lose all their lift. This mastered, he graduates to the Angel for transition landings on the dry lake. The chase car and chase plane, both with radio, are used in this phase. The pilot makes at least three landings with the pogos installed—it's a little easier to handle that way. After the landings improve, the pogo safety pins removed, the new pilot is on his own."

Flying the chase aircraft for U-2 flights. (circa 1955)

At Groom Lake, during 1955, they trained three detachments of CIA pilots to fly the U-2, complying with a training program set up by Gen. William Yancey's 4070th Support Wing (SAC) organization for pilots already highly experienced and proficient.

One was instructor pilot Captain Hank Meierdierck had flown as a SAC F-84 fighter pilot, very proficient and trained in long-range navigation, including celestial navigation, which was a very rare thing for fighter pilots. The U-2 in those days was dependent on celestial navigation for long-range flights over very desolate and hostile territory. These were some of the most skilled pilots in the US Air Force.

So, although they checked out the CIA pilots in the U-2, came highly skilled and ready to solo the U-2 with a minimum of ground school and a few minutes' cockpit orientations to be sure they knew where all the switches were. Hitting the right switch with an inflated pressure suit could be a very trying task, especially at night. A single mistake could cost the pilot his life.

Kelly Johnson, head of the Lockheed Skunkworks, was at Watertown nearly every day, and had the final say about operations. He also had final say about anyone flying the U-2. If he didn't like the way the pilots flew the U-2, particularly during taxi, takeoff and landing, the pilot might never fly the U-2 again. On one occasion, the new (Lt. Col.) Operations Officer taxied his U-2 out on the lakebed, ground looped it, and came to a stop. Ground looping could do serious damage to the U-2, in most cases. Kelly saw this ground loop and ordered the flight canceled. This pilot did not fly the U-2 again, at least not at Groom Lake. Kelly was truly the boss.

This was for a good cause. The pilots were flying the very first prototype U-2 aircraft; planes that were considered "national treasures." The CIA and 4070th Support Wing (SAC) 4070th Support Wing (SAC) had to answer to President Eisenhauer, who wanted the U-2 to obtain photographs of Russian targets, rocket bases, and military bases-- very quickly before the Russians found ways to shoot them down. Every Wednesday, the President received a briefing on the test and training status at Groom Lake.

The instructor pilots had to fly a chase airplane for every U-2 training flight. The chase pilot would fly in loose formation with the U-2, checking for any smoke, fluid leakage, and proper landing gear, flap, and speed brake position. The problem was that the U-2 flew the traffic pattern at very low speeds, around 100 knots or less, and they found it difficult to find a chase airplane that could fly that slow, yet have enough power to stay anywhere near the U-2 in its initial climb. Just keeping the U-2 in sight, when it was above 50,000 feet, was very difficult. After painting the U-2s a dull black, it proved impossible to see at high altitude.

Over time, they used the T-33 jet trainer for medium altitude and high altitude chase up to 35,000 feet, and used various prop airplanes for chase in the traffic pattern, at speeds too low for the T-33. For traffic pattern chase aircraft, they tried numerous prop airplanes: a single-engine Navion, twin-engine Cessna L-27, Beech C-45, and finally, a modern Beechcraft Twin Beech. The Twin Beech proved to have plenty of power, the engines did not overheat, and it was quite easy to fly, so they preferred it.

For just plain fun flying, the favorite aircraft was the deHavilland L-20 Beaver. It was a well developed and very tough bush airplane.

Once the Air Force instructor pilots passed the T-33 and ground training phases, they set up in their pressure suit for their first flight. Four of the SAC instructor pilots received fitting with pressure suits: Robertson, Setter, Garvin, Meierdierck. Though they only ascended to 20,000 feet on their first flight, they went through the pre-breathing routine to dispel any nitrogen bubbles.

All the pilots, Lockheed, Air Force, and CIA alike found the U-2 a thrill to fly on a shorter takeoff roll of 500–1,000 feet compared to 8,000 to 10,000 in the B-47. They liked the straight-up climb to altitude.

On their second flight, the pilots flew to 70,000 feet to learn the idiosyncrasies of the coffin corner. The range of indicated airspeed in the U-2 there ranges around 100–105 knots or 115–121 mph indicated. This was where flying too slow the plane went into a stall buffet; too high it went into a speed buffet.

With time a deciding factor, Lockheed welcomed and accepted anything the Air Force pilots could add to the test phase. The Air Force pilots made a point of experiencing the correct flight profile to instruct the CIA pilots how to climb, how to handle the coffin corner, recover from flameout, land correctly, etc., and to dictate the best way to accomplish the mission.

At this point, the CIA appointed Col Richard Newton as station chief and base commander over the CIA, Air Force, and Lockheed personnel arriving at the Watertown test site.

The Inquisitive Angel   *T-33 with an instructor and new pilot U-2 with the chase vehicle*

****

Chapter 9 - The Pilots

Lockheed U-2 Test Pilots

The Lockheed test pilots for Project AQUATONE were Tony LeVier, Robert Matye, Ray Goudey, Robert "Bob" Sieker, and Robert Schumacher.

Anthony W. "Tony" LeVier began his aviation career with air racing in a Keith Rider racer dubbed "The Firecracker." Following World War II, he bought a war surplus P-38 Lightning that he modified, painted bright red, and used for racing. He began his career at Lockheed ferrying Hudson bombers to the RAF. In 1942, he became an engineering test pilot flying the PV-2 Ventura. At Lockheed, he made the first flight of the XP-80A. He flew the first flight of the XF-104 Starfighter and the first flight of the U-2. Kelly Johnson chose LeVier, then Lockheed's chief test pilot for the F-104, to fly the U-2 prototype. LeVier reportedly remarked, "I switched from flying the plane with the shortest wings in the world to the one with the longest." LeVier was a test pilot for Lockheed from the 1940s to the 1970s. During his flight career, LeVier survived eight crashes and one mid-air collision.

Robert L. Matye was a career experimental test pilot for the Lockheed Aircraft Company. By his being an accomplished fighter pilot in WWII, he received an assignment to the first jet-powered fighter group. After the war, Lockheed hired Matye as a test pilot, where Matye enjoyed an outstanding career lasting 26 years. He was the second pilot to fly the famous U-2 spy plane and the first person to take it to its maximum altitude capability. He flew every Lockheed aircraft produced during that time.

Ray Goudey first soloed on 25 September 1937. He enjoyed a career where he flew 258 different types of aircraft and accumulated 23,708 flight hours. Goudey flew single-engine jets, multi-engine jets, single and multi-engine turboprop and single and multi-engine reciprocating planes, helicopters, single and multi-engine seaplanes, and even 100 hours in gliders. Goudey trained civilian pilots for the Army Air Corps and was the US Navy acceptance pilot at Grumman, Chance Vought, Curtis, and the US Naval Factory. He flew in air shows for the flying circus and was the chief pilot for the Hank Coffin Flying Service.

From 1952 to 1990, Goudey flew as a test pilot for Lockheed, where he set several speed records, altitude records, and piloted several first flights. He flew 74 models designed and built by Lockheed. He was third to fly the U-2, and amassed 2200 hours in the U-2's early years through all phases of development. Lockheed would later schedule Goudey to fly the first flight in the Have Blue until realizing his age and replacing him with Bill Park.

Robert "Bob" Schumacher was a Navy dive bomber pilot assigned to the USS Bennington in WWII. He received the Navy's Distinguished Flying Cross for his part in sinking the Japanese Battleship Yamamoto. Schumacher joined Lockheed in 1953 as a test pilot. In 1956, he began testing the U2 spy plane. In 1965, he became the first pilot to land the U2 on an aircraft carrier.

Bob Sieker, one of the early Lockheed U-2 test pilots, flew the prototype stealthy U-2 in an attempt to defy the Russian high-frequency radar. The plane had a layer of what appeared to be a 4-inch foam layer on the bottom of the fuselage, topped with fiberglass. He did not know the technical approach, but aerodynamically, it added a layer of thick foam to the fuselage, causing the engine to overheat at high altitude.

Robert Sieker was the only Lockheed U-2 pilot fatality. The accident occurred when Sieker was conducting a test flight with the U-2 prototype, Article 341. Overheating caused the engine to fail and explode at high altitude near 70,000 feet. Sieker's pressure suit inflated, as planned, but his faceplate blew out, causing the sudden loss of all his oxygen. The face place blew out because of the failure of the latch, a steel latching pin that locked the faceplate closed. He lost all his oxygen almost immediately and probably lost consciousness and died at high altitude. The Air Force joined the Watertown pilots in a search lasting for days before finally finding the wreckage near Tonopah, Nevada. They found Sieker's body a few hundred yards from the U-2 wreckage. Kelly commented: "I lost a test pilot for the failure of a lousy 50 cent latch".

Lockheed test pilot Bob Mayte landed the twin Beech near the wreckage, on the desert.

There was no runway there. He made a safe landing with a twin-engine airplane with a full load of passengers, four people aboard. Someone radioed for Lou Setter to fly the L-20 out to the crash site to pick up the pilot's body and fly it back to the Ranch (Groom Lake). There was no room to carry the body in a body bag in the Twin Beech because it was already full of passengers. So he flew the L-20 to the crash site.

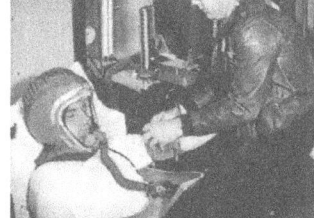

Lockheed's Bob Sieker Kelly Johnson, Bob Schumacher
Bob Sieker   Ray Goudey

Mayte had landed the Twin Beech on the desert near the U-2 wreckage—I could see his wheel tracks. So Setter landed in the same spot. When he touched down, it felt normal, until he felt and heard a loud "WHOOMP" Just after touchdown, Setter had rolled through a dry streambed. He got stopped OK, picked up the body, and flew back to the Ranch.

Air Force Instructor Pilots and Support

When Col William R. Yancey and the CIA project pilots left March Air Force Base to report to Watertown in Nevada, they learned their address for mail and emergency contact was a post office box in North Hollywood, California. None of them lived in Nevada.

SAC had selected Colonel Yancey to command the 4070th Support Wing pilots: Maj Louis Garvin and Captains John H. Meierdierck and Louis Setter, who received their training in the U-2 from the Lockheed test pilots. Majors Delap and Mullin and Colonel Robertson also served as SAC Instructor Pilots in roles of navigation, mission planning, classroom instruction, and operations. The Instructor Pilots (IPs) then trained the sheep-dipped CIA pilots to fly U-2 missions.

William R. Yancey was born at Parkin, Arkansas. He graduated from the University of Arkansas and commissioned a 2nd lieutenant in the AF Reserve during June 1938. He retired a brigadier general.

Col Louis A. "Lou" Garvin, born 1921 in Berryville, Virginia, attended Virginia Polytechnic Institute. His early assignments included Air Training Command and research and development for the Strategic Air Command. In 1955, the Air Force selected Garvin as development and flight test officer for the top-secret U-2. He became the first Air Force pilot to fly the U-2.

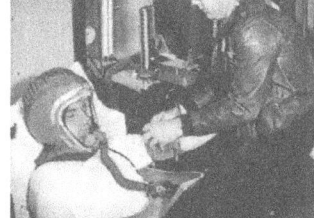

Meierdierck flight record, August 1956

Capt John Henry "Hank" Meierdierck, born in Newark, New Jersey in 1921 notably served as a SAC Instructor Pilot training the three detachments of the CIA pilots for the U-2. He would later retire from the Air Force as a lieutenant colonel to join the CIA as the Director of Operations of the A-12 program at Area 51. He would remain with the CIA and serve for a third time at Area 51 as Program Manager and Unit Commander of the 4027th Squadron for CIA Project AQUILINE, a stealthy propeller-driven, low altitude, anhedral-tailed plane by McDonnell Douglas. This was a UAV controlled by a data link from a high-flying U-2 for low-level electronic surveillance of the Chinese nuclear program.

Typical for all the pilots and flights at Watertown note how the Air Force showed Meierdierck's flights as occurring at March AFB,

California. Many of the U-2 flights show as T-33 flights this log shown here.

Colonel (then a captain) Setter before coming an instructor pilot at Watertown during the CIA Project AQUATONE had flown the F-84E and F-84G, air refueling and carrying atomic weapons while participating in the development of the "toss bombing" technique for atomic weapon delivery. Captain Setter, the fourth USAF pilot to fly the U-2 in 1955, assisted in developing and testing celestial navigation techniques for pressure-suit pilots. Setter would design the cruise control computer and plotter—techniques and equipment still used in U-2 overflights around the world today. Other aircraft using his computer included the F-84G, F-84F, T-33, and the T-39. After leaving Watertown, he continued with the CIA's U-2 program as the project's flight test engineer. He later assumed command of the CIA's North Base at Edwards Air Force Base, which included the duty of U-2 Ops Officer.

Lt Col (at that time a major) John E. Delap, born 1921 in Mukwonago, Wisconsin, and a World War II veteran, served at Watertown in 1955 as the mission planner and chief of navigation for the original SAC training cadre under Colonel Yancey.

Col Herbert I. Shingler, Jr., a native of Ashburn, Georgia, attended the University of Georgia before entering the US Air Force through the Aviation Cadet Training Program in 1940. From July 1955 until June 1957 he served with the 4070th Support Wing as Director of Materiel and later as the Wing Commander, stationed at March Air Force Base, California. On April 2, 1959, during a normal training mission, Col Shingler and his crew crashed while he attempted an emergency landing in Mountain Home AFB, ID. The B-47 he was flying suffered major problems, including engine fires, control problems and ultimately structural failure. During the flight engine #4 dislodged and fell off, and on a long final approach #5 rotated and lodged itself against the wing fuel tank, causing the plane to crash with no survivors. Colonel Shingler received burial at Arlington National Cemetery during April 1959 with his wife Frances interred with him on June 12, 1997.

The Air Force designated the unit as the 4070th Special Activities Squadron. The modus operandi called for working the civilians and the military personnel from Monday through Friday at Watertown and flying to Burbank or March Field for the weekend unless training required, they stay at Watertown through the weekend.

Pilot candidates who had never flown such an aircraft had to learn to land with no flaps, nor voice communication because of the secrecy of its missions.

The genesis of Watertown, later known as as Area 51 began in the summer of 1955, when Louis Setter, Ruth, and their three boys transferred from Great Falls, Montana, to an unknown classified program. He reported to Col Bill Yancey at March AFB, near Riverside, California. Their new office, quite small, was located at the rear of the base movie theater. There he met Col Yancey, their boss, Col Shingler, their logistics chief, Maj Art Lien (also logistics), and the other pilots were: Lt Col Robertson, their Ops Officer, Maj Lou Garvin, Maj Bob Mullen, and Capt Hank Meierdierck (an old friend from the 31st Fighter Wing). They were all from SAC (Strategic Air Command). Later he found out that this assignment that General LeMay (SAC four-star Commander) directed and insisted SAC pilots do the U-2 training. He was not allowed to interfere with the rest of the program; it was strictly under the CIA, and U-2 design, building, testing, and remote base operations were all under Lockheed, headed by Kelly Johnson. Col Yancey said that he and the two colonels plus the pilots were to report to the Lockheed Skunk Works building in Burbank the next day, to be briefed by Kelly Johnson, head of the Skunk Works and chief designer.

When they checked in to the Skunk Works, there was no Air Force security, only the CIA. Two CIA agents followed them and stayed with them. There were no badges and no security briefing. Col Yancey instructed them not to discuss any part of the program, nor where they were to work, or anything about the "article." It was not called the U-2 or even an airplane at this time. U-2 pilots were to be called "drivers." To this day, U-2 pilots are called "drivers." Their wives did not know what or where. They had no clue what they were doing.

In 2 or 3 hours, Kelly personally briefed them about the airplane, saying it was a single-engine, long-range single place airplane that looked like an F-104 fuselage with "Gooney Bird" (C-47 transport) wings. He would not let them in the manufacturing section; he didn't want any interruptions at all to work going on. They were on an extremely tight schedule, a schedule that they soon learned was nine months from the start of design until the first flight, a fantastic accomplishment. Kelly was the absolute boss of everything at the Skunk Works and at the remote location, which proved to be a remote dry lake bed located adjacent to the Atomic Test Range, west of Las Vegas, Nevada. They, the SAC training organization, Lockheed, Pratt and Whitney, and the CIA were the only occupants, all having Secret SAR (Special Access Required) clearance, which meant that absolutely no military, congressional, or media visitors were allowed.

The four pilots spent a week getting a physical exam at the Lovelace Clinic in Albuquerque, New Mexico, followed by the David Clark Company outfitted them with partial pressure suits, each tailor-made and very tight fitting. At the altitude chamber in Los Alamos, New Mexico, they had to pressure breathe in their suits for 45 minutes in a special chamber at an altitude of over 70,000 feet. They were essentially in outer space. That was not pleasant, but the four of them passed. Setter said that it felt like every breath he took might be his last. With their pressure suit fully inflated and squeezing them, they had to exhale high-pressure oxygen forced into their lungs. It left black and blue marks all over. Later, their CIA U-2 students went through the same tests. Not all of them made it.

At Groom Lake, a group of aeronautical engineers, maintenance personnel, a runway, and some trailers awaited them. As noted, other staff on the site included the LAC pilots, Bob Matye, Ray Goudy, Bob Schumacher, and Bob Sieker. The facilities included a kitchen and mess hall with little else.

Back at Watertown, they soon started getting checked out with an abridged ground school, mostly a brief rundown of the airplane systems, followed by a cockpit checkout by one of the Lockheed test pilots. The LAC pilots briefed the USAF pilots and Colonel Philip O. Robertson (ops officer) on the expected idiosyncrasies not yet learned.

There were no two-place trainers. Robertson, Garvin, Meierdierck, and Setter all flew their first low-level flight on the same day.

Louis Setter was the #10 U-2 pilot; the first six were the Lockheed test pilots, headed by Tony Levier, who flew the first flight, then only at low altitude. The test program proceeded very quickly. If Kelly didn't like what he saw a pilot doing, he immediately removed the pilot from the U-2 flying roster. Kelly could not afford to have his prototype aircraft damaged in any way, and they flew the prototypes before their first CIA students arrived, just a few weeks after the first flight, in the fall of 1955.

The U-2, with its single main landing gear and tailwheel, was truly a ground looping airplane, and still was to this day. That's the reason they used a Ford station wagon on the lakebed, so the instructor pilot could drive directly behind the airplane, giving instructions by radio to the U-2 pilot as to his height above the lakebed and if he was perfectly lined up to touch down safely. The airplane had to be landed in a full stall, with the tailwheel touching down at the same time as the main gear and perfectly aligned with the main gear. The pilot had to make control corrections very quickly to ensure a near perfect touchdown to avoid the possibility of a ground loop. Once a ground loop starts, it cannot be stopped, resulting in severe damage to the airplane. This made the U-2 the most difficult airplane to fly for any of them and was the cause of many accidents.

A stall landing was not smooth, but the airplane then stays firmly on the ground. Now the pilot must be extremely active with his rudder pedals to keep going straight until he was fully stopped. This was a difficult thing to do. The U-2 was probably the most difficult of all current aircraft to land.

Testing the first U-2 continued at Watertown while the Lockheed production crew kept moving additional U-2s off the line. Lockheed would produce the first six aircraft in Burbank, California before moving in 1956 to Bakersfield, California, 100 miles north of Los Angeles. The location became the venue where Lockheed assembled and functionally checked each plane before disassembling and trucking them to Bakersfield airport. They loaded each disassembled plane into a SAC C-124 for airlift to Watertown. The U-2s were assembled on-site by Glenn Fulkerson's team, and all Skunk Works people, a very skilled and capable group.

General Yancey

The CIA's Project AQUATONE produced specialists in the Air Force that would benefit it well into the future. USAF SMSgt Wayne S. Nelson (Ret) joined the 4070th during January 1955. Once he received his security clearance, he flew in the Gooney Bird to Watertown, where he maintained the flyaway kits for the U-2. ("Gooney Bird" was the nickname for the C-47 Skytrain, a military transport developed from the Douglas DC-3 airliner. The DC-3 was the first aircraft to land on Midway Island, previously home to the native long-winged albatross called the Gooney bird.

During April 1957, SSgt Nelson would transfer to Mira Loma Air Force Station, Riverside, California with four other military personnel, CWO Walt Moberly, TSgt Donald Northway, and A1C Winegard as part of an Air Force depot commanded by Col Robert Welch. The Skunk Works would use the depot for shipment of engines, avionics, aircraft skins and parts to Watertown. The depot would also store and ship parts for Lockheed, Perkin Elmer, Pratt & Whitney, and others. *(The Air Force later deactivated it in 1964 and moved everything to San Bernardino Air Materiel Area, San Bernardino, California to serve as both the U-2 and SR-71 depot support under the command of CWO James Molloy.)

SAC Instructor Pilots – In Their Own Words

Meierdierck, one of the SAC Instructor Pilots who trained the CIA pilots: "In mid-1955 a colonel from Hq. SAC came to Turner AFB. He was recruiting pilots for a very secret project. He also was looking for a fighter pilot with experimental test time and lots of fighter jet time. He took Lt. Setter, an aeronautical engineer who worked in my office of Wing Plans, and me. We were told nothing of what we were going to do or any other details. I was told to report to March AFB in Riverside, CA. Upon arrival, I met the other personnel who were to make up this group. Col. Wm. R. Yancey was the commander, Col. Herb Shingler was deputy and also chief of Logistics, Col. Phil O Robertson was the Operations chief, Maj. Jack E Delap was chief of navigation, and Major Garvin [a B47 pilot] was one of the experimental test pilots along with Capt. Louis Setter and myself. There were several other officers assigned for various duties and a few airmen.

"Robby [Robertson], Garvin, Setter and I were checked out for low-level flying and then went to David Clark Company back in New England to be fitted for partial pressure suits. We flew the same and sometimes more advanced missions as the LAC [Lockheed Aircraft Company] pilots. Time was a deciding factor, and anything that we could add to the test phase was accepted. We developed a complete program of ground school and flying. Of course, we had to experience the flight profile before we could tell a student how to fly, and more importantly, decide the better way to accomplish the mission. We trained three groups of CIA pilots, deployed them overseas, and then checked out the USAF pilots.

To illustrate how Kelly made necessary modifications to the aircraft, one day, Louis Setter had just landed on the lakebed after a short (2 hour) low altitude (below 50,000 feet) flight, which did not require a pressure suit. After he landed on the lakebed, while Setter was completing the airplane forms, he noticed one of the engineers using a template to mark a cutout area, about 6 inches by 12 inches long, on the tail skin of the airplane, with a penciled in rivet pattern. He asked what was going on. Kelly was standing there and said, "We are installing a black test box for flight test data. The airplane will be flying in half an hour". And it did fly in half an hour with the box installed." What Kelly said to do, they did, right now. At Watertown, they operated by Air Force or military or FAA rules. The FAA didn't care what happened above 60,000 feet, so initially, they navigated both within and outside of their test area with no radio calls except to their control tower. Later on, they found it was wise to make FAA position reports using "VFR on top." This way, they never reported their altitude, which was a legal position report. Once at high altitude above 60,000 feet, requiring pressure suits, they were strictly "on their own". Neither the FAA nor the Air Force knew what they were doing or even where they were.

"We gave them ground school on the aircraft systems, emergency procedures, flight planning, navigation, etc. etc. flying this airplane. Prior to our giving the ground school, we had to devise the procedures used in the air and preparation for a flight. This included preparing the route, fuel consumption, checkpoints, pre-breathing for 2 hours and all the myriad details necessary for an overflight of normally denied territory. Remember, this was a new type of aircraft flying in an environment where nobody had flown before." –

Nelson: "After receiving my security clearance, I was sent to go through U-2 pilot ground school. At the area, I met Hank Meierdierck for the first time and, as my instructor pilot, he started me out on a program designed to familiarize new pilots with some of the characteristics of the U-2. This involved the use of the T-33 as the lead-in trainer. I already had several hundred hours in the T-bird, so we got right to the different training."

"As I recall, the simulated U-2 configuration included no speed brakes, limited deployment of landing flaps, engine RPM no less than 80%. In my opinion, this was a very effective simulation of the U-2 performance in the landing phase. I must confess that even with all of Hank's excellent instruction, I still managed to overshoot my first U-2 landing, and I still recall seeing Hank and others standing near the lakebed landing strip waving me to go around. I worked a lot harder on airspeed control on the second approach and landing, and everything happened just as he told me it would."

http://roadrunnersinternationale.com/nelson_re

cruitment.html

Yancey, Meierdierck, Garvin, Shingler, Setter, Delap

McIlmoyle: "I learned that Captain Louis Setter of the 517th Strategic Fighter Bomber Squadron received an assignment and no one knew where. Setter was a regular officer and had not resigned his commission, another mystery. An engineer by education, Setter had designed a small handheld computer for use with F-84 flight planning. The Strategic Air Command (SAC) had jumped to put his invention into use." - McIlmoyle p. 58

****

During January 1957, Louis E. Dye would join the U-2 squadron as an electronics technician in the Commo shop. After thirty days of familiarization with the program, he would join the group in Giebelstadt, Germany. He would finish his commitment to the program at Detachment B in Turkey, responsible for maintaining communications and the ELINT systems on the U-2. This included testing the threat warning systems over the weapons simulators at Watertown to determine if the systems provided adequate notice and protection. The Squadron Director of Materiel, Colonel Thompson, analyzed the results before accepting a new U-2 and certifying that it was ready for deployment.

As early as August 1955 the CIA realized that even with Air Force support, establishing a secret flight test center in the Mojave Desert did not come easy. A shortage of supply personnel remained a recurring problem for setting up of the depot and the assembling of supplies for training the CIA's Detachment A scheduled to commence early in 1956. The shortage continued through the training and deployment of Detachment B.

In the face of this deficit, the SAC 4070th Support group, headed now by Col Herbert Shingler, carried the burden of getting Detachment A logistically ready at the time of deployment. From July 1955 until June 1957, while stationed at March Air Force Base, Shingler was the Director of Materiel for the 4070th Support Wing and later became the Wing Commander.

March AFB was not Colonel Shingler's first assignment regarding Russia. While a major, Shingler flew a secret mission to Teheran and met up with three British LB-30s that he escorted to Moscow. Major Shingler was the pilot of the only B-24 transporting General Maxwell, a Russian radio operator, and other passengers. One of the LB-30s carried British Prime Minister Winston Churchill on his way to meet with Joseph Stalin. Mr. Churchill was traveling with high-ranking American, British, and Russian officers. Flying time to Russia was 10 hours. At 0520 on 16 August, they returned to Teheran. It took 11 hours because of having to avoid German planes.

On 1 August 1943, Major Shingler at that time a squadron commander of the 415th, 98th Bomb Group participated in a raid on the Ploesti Oil Fields. Major Shingler flew a B-24 called "Fertile Myrtle" to lead the second wave following Col John Kane, better known as "Killer Kane." Maj Shingler completed his mission and brought his crew home safely. He received the Distinguished Service Cross for this mission.

By 1956, he was at March AFB and commuting to Watertown to train the CIA pilots. The shortage of aeromedical staff and personnel at March AFB and Watertown forced Shingler to borrow personnel to staff Detachment A at the time of deployment. Colonel Ritland reported the problem to the project director on 30 March 1956, saying,

"Because of the overall expansion and the lack of enough personnel, they drew on the Air Force commands to assume definite project responsibilities. Work proceeded rapidly, but by personnel outside of the project, making it an unsatisfactory situation not under the control of the project director and needing watching as the scope of the project expanded."

The CIA classified the tech reps assigned to overseas duty as Department of the Air Force civilians accredited to the Air Weather Service. Those allocated to the detachments deployed overseas all enjoyed the same benefits, privileges, and entitlements in a constant effort to equalize the treatment of all personnel and take care of the more significant complaints. The difficulty lay in the Watertown test site being part of the theater of war not included in the theater of operations, and each foreign field base's presenting different situations regarding billeting, messing, per diem, working conditions, recreation, etc. The Project Director described the cohesion achieved within these mixed task forces as follows,

"They put the detachment in the field made up of one-third CIA civilian personnel, one-third Air Force uniformed personnel, and one-third contractor personnel. These people must preserve the tightest security; they expected to achieve a standard of maintenance that three successive Strategic Air Command colonels fresh to the project admitted were above any achieved in a 100% military operation. To do this, they required a disciplined and hard-working organization. They must cope with the fact all three pay systems were different as were all sorts of conventional arrangements for fringe benefits (including most R&R leave) different. They average the regulations up to each of the three components receiving all the privileges under its union contract, plus all the privileges both other union contracts afforded. The cost was an expensive operation for the US government. I thought it worked as measured by maintenance standards achieved and maintained, and, by accomplishment. I believed that it worked well for relationships and morale."

For a highly skilled civilian-sponsored interagency organization with the highest security, the use of dedicated officers like Gibbs and Ritland was crucial for success.

L-20 Beaver Towing Skier on the Groom

Lakebed (circa 1956)

One day during their U-2 training of Detachment A, the first group of CIA U-2 pilots, they woke up to find a layer of snow, about four inches deep, covering everything, including the lakebed. The weather conditions canceled any U-2 flying operations. So, with nothing to do, Louis Setter climbed in the L-20 Beaver for some local proficiency flying. Those large tires had no problem handling a 4-inch snow. There was no wind at all, so drifting was not a problem. It was a beautiful clear day.

Setter was circling the dry lakebed at low altitude when he saw a single person on skis, slowly going across the lakebed. He waved him down, so he landed and waited for him to come up. It was the Detachment A commander, an ardent skier. He asked him if he could tie a rope to his wing strut so he could pull him across the lakebed. He said sure, so he went back to get the rope.

He tied the rope on his left wing strut, next to the fuselage, and slowly pulled away, not knowing how much speed the skier wanted. He could see him behind me, with lots of snow blowing on him from his prop. After a few minutes, they worked up to probably 50 or 60 knots (about 70 miles per hour), slowly traveling in a big circle around the lakebed, which was about 5 miles across. They were just below the L-20 takeoff speed. The skier waved Setter down and said he was having a great time, except the blowing snow from his prop made visibility difficult. So he said, "I can fix that", and they started off again, only this time he took off, at about 52 knots.

The skier was having a ball, now that the blowing snow was no longer a problem. Setter was cruising in a large circle around the lakebed, holding an altitude of about 20 feet and the speed at below 70, although at one point, it probably hit 75 knots (about 85 mph). He said later they probably broke the international speed for "schussing," which in Germany was done by a racing car towing a skier skiing in the ditch alongside the road. A wild ride.

Much later, it occurred to Setter, "what would have happened if the detachment commander had taken a fall at that speed, broken a leg or something else". With his key position as Detachment Commander, all hell would have broken loose, and for sure that would be the end of him on the U-2 program. But at the time they didn't think of anything bad happening. Again, maybe someone was watching over them.

Recruiting CIA Air Crews to Overfly Russia

Lt Gen Emmett (Rosy) O'Donnell, the Air Force's deputy chief of staff for Personnel, authorized the use of Air Force pilots and provided considerable assistance in the search for pilots meeting the standards established by the CIA and the Air Force.

The CIA won the battle between the CIA and the Strategic Air Command over control of AQUATONE, but under terms of the OILSTONE agreement, responsibility for pilot training lay with SAC. Accordingly, General LeMay won the right to name the staff that would accomplish the military support.

LeMay chose Col William F. Yancey to command the 4070th training wing out of March Air Force Base in California. Even before the CIA pilot recruiting effort was underway, the Air Force began developing the pilot training program. Col Yancey carried out this essential activity by commuting to Watertown.

The CIA screened and selected the mission pilots from the Strategic Air Command's F-84 pilots with long-range navigation training. The CIA organized and divided the pilots chosen into three squadrons known as Detachment A, Detachment B, and Detachment C. General LeMay named Colonel Fred McCoy to command Detachment A, Colonel Perry to command Detachment B, and Col Stan Beerli to command Detachment C. Each of the detachment commanders chose their staffs consisting of operations, flight planners, physiological trainers, and engineering officers.

F-84 Background of CIA Pilots

Lt Col Robertson recommended that AQUATONE use pilots from SAC's F-84 squadrons. "Some of these men had combat experience from the Korean War and had flown long overwater deployments in the single-engined fighter. Flying over hostile territory in the U-2 wouldn't be so different, he reckoned. The security staff at project HQ liked the idea." - Pocock 2000, p. 35

The US Air Force accepted the F-84 in 1951 and considered a first-line combat aircraft. Two of the U-2 instructor pilots, Setter and Meierdierck, had participated in SAC's development of overwater navigation for the F-84s. Because of the urgency of the Korean War, there was no time to ship F-84s by an aircraft carrier. Meierdierck and others led mass flights of F-84s across the Pacific Ocean to Korea, using the sextant. Kratt, later a CIA U-2 pilot in Detachment A, shot down 2 MiGs and a Yak-3 while flying the F-84 in Korea.

In addition to its combat role, was one F-84 secretly modified: "The aircraft, normally used for close air support, was fitted with a single K-39 camera and a 36-inch focal length lens. On February 11, 1951, the F-84 made an overflight of Vladivostok at 39,000 feet, without incident." - Peebles p. 14

The F-84G was adapted for nuclear-delivery capability. CIA pilot Overstreet of Detachment A spoke of his experience during 1952: "At Turner, we were learning to drop nukes. I got lots of flying time and crossed both oceans in single-seat aircraft." According to his memoir, CIA Pilot Powers received his nuclear-delivery training at Sandia, NM. Elsewhere in this book, CIA pilot Knutson is quoted about his nuclear training and his assigned target, Leningrad. SAC CIA Instructor Pilot Setter participated in development of the 'toss bombing' technique, enabling F-84s and other aircraft to attack Eastern-Bloc airfields from low altitudes with tactical nuclear weapons.

According to information from *History of Air Force Atomic Cloud Sampling*, four CIA U-2 pilots had earlier flown in the atomic testing program: Carey, Hall, Rose, and Stockman.

For the AQUATONE U-2 pilots, "experience with the F-84, with all its lousy handling and engine failure problems [there were thirteen F-84 pilot fatalities at Turner alone in 1955], would stand them in good stead for the Dragon Lady, where 'good stick-and-rudder men' would be needed." - Pocock 1989 p. 18

At the time of the early 1950s atomic tests,

modified B-57s were not quite ready, and only the F-84 was highly suitable for the assignment. The F-84 could operate for 2-3 hours at over 40,000' altitude for atomic sampling. The Air Force selected for the highly demanding job of atomic sampling pilots with stamina, trained reflexes, navigation ability, and skill at multi-tasking.

"Pilots with the ability to succeed in sampling missions had to possess the ability to receive radioed instructions, make taped recordings of instrument readings, be alert for excessive radiation and myriad other details simultaneously. Most pilots with less experience and proven ability were simply overwhelmed—so badly that they could not function satisfactorily—by the awesomeness of the cloud interior." – "Into the Mushroom Cloud" *Air & Space Magazine* Aug 2009

As mentioned earlier, before his role in the U-2 program, Col Osmond Ritland had organized and commanded the 4925th Test Group (Atomic) which was responsible for the development of Air Force nuclear capability. His test group, among other things, led the development of airborne sampling techniques for detection and analysis of atomic weapons tests. The 4925th operated within a double-barbed-wire fence at Kirtland Air Force Base.

During the Tumbler-Snapper series in Nevada (April-June 1952) the pilots rehearsed for the IVY series to take place in the Pacific late that year. Records state: Wilburn S. Rose and Hervey S. Stockman participated in IVY and then again at the Upshot-Knothole series in Nevada in 1953 where they could "lend advice." Air filters and gas canisters filled by F-84 sampler aircraft placed into lead containers and rushed to where waiting radiochemists could analyze the radioactive products.

Rehearsal of exiting an F-84 atomic cloud sampler aircraft after a mission

Outside surfaces were too "hot" with radiation to touch; the pilot stepped out onto a forklift

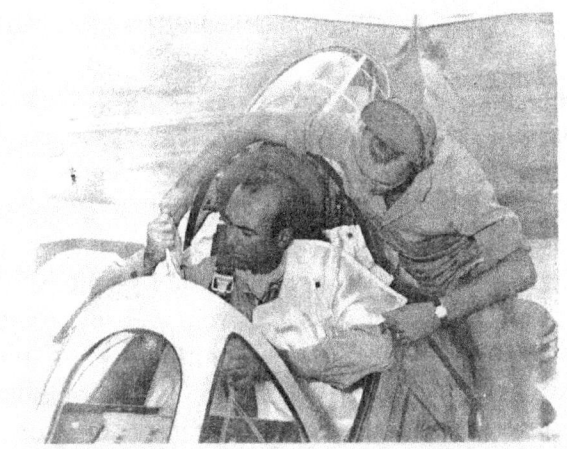

Figure 15.   Protective lead-glass cloth shroud being placed on
sampler pilot. IVY.

If an actual mission, the pilot wore lead-lined gloves and clothing

"On April 6, 1953, at Indian Springs Air Force Base, Nevada, a QF-80 (at left) and a two-seat DT-33 mothership head for the test site of Operation Upshot-Knothole on the first sampling flight by a jet drone."

Chapter 10 – Sheep Dipping the Pilots

The CIA faced complexities involved in having the pilots resign from the Air Force to hire on as civilians for the U-2 AQUATONE project. The problem arose where the Air Force pilots felt reluctant to leave the service and lose their seniority

The CIA overcame this by ordering handsome salaries to those who sheep-dipped from the Air Force. The Air Force promised that each pilot would be able to return to military service upon satisfactory completion of his CIA assignment. The Air Force would consider him for promotion along with his contemporaries who remained on Air Force duty.

Hervey Stockman flew the first flight of the U-2 over the Soviet Union on the 4th of July, which was also his birthday He was held for 2,093 days at a POW in North Vietnam before being released during Operation Homecoming in 1973.

The CIA U-2 pilots went through a very strenuous selection process. The first was to eliminate those too nervous or unstable to endure the rigorous strain of flying at extreme altitudes for extended periods. The CIA contracted the Lovelace Foundation for Medical Education and Research to conduct full physical and psychological screening of the pilots.

After passing the extensive physical and psychological tests, the first group of pilots for Detachment A sheep-dipped into a name change, separation from the service, and for practical purposes, anonymity.

How We Were Recruited – by Hervey Stockman

"At Bergstrom, I was preparing to ship out with my family on a PCS to Bangkok Siam (Thailand) to help the Thai air Force transition from F-8Fs to F-84s. The 42nd Air Division DO, Col Bill 'the Goat' Shick, a good friend, approached me outside the division headquarters and in his gravelly voice asked me if I would be interested in a new assignment. I said I had one. Shick kept talking as we walked towards the 27th Headquarters, 'Hervey, I really think you ought to look at what I'm talking about. I can't tell you anymore, but. ' You couldn't be at a fighter station without learning what's going on. Stockman, Kratt, Dunaway, and Grant were bailing out of the Air Force for another flying job. Hush-hush. Others approached turned it down. They had looked at the CIA contract, and the USAF headquarters had told them that they would be welcomed back on AD if they did the good work. We'd not lose our place in the 'golden ladder, trust us.' We received psychiatric examinations and lie detector test in out-of-the-way offices in the District. Our hotel rooms adjoined, and Company guys looked in on us periodically."

"From Washington, we headed for Lovelace Hospital via Denver. By now, we had a permanent watchdog. A tough little guy whose specialty was ballooning. Our connecting flight into ABQ was canceled, our hospital physical appointments were for 8 AM the next morning. Minor panic. The security man was useless and very goosey about having his wards doing anything out of the ordinary. But we did. We found a phone number for a pilot and six place bird for hire and cut a deal. He owned a very old twin Beech, prewar version of the C-45. It was a memorable ride. Jake Kratt took the right seat and held a flashlight on the console and flight instruments. Over the mountains around Raton, the pilot had to shut down the starboard, and we limped into Kirtland and the old terminal in the middle of the night pouring oil."

"Our old veteran pilot told us to press on. He could take care of the Beech. We booked into the motel across from Lovelace under our pseudonyms. We would use these almost exclusively for the next year except when encountering old pals who had known us on active duty. Concerning the Lovelace Clinic experience, I won't titillate you with what we suffered through. Tom Wolfe described the experience well in his book *The Right Stuff*."

"Glen Dunaway, Jake Kratt, Bruce Grant, and I hooked up at The Brown Derby in Hollywood.

We met up with CIA's John Raines and some other folks from security at a hotel on Hollywood and Vine. We proceeded to the Lockheed Air Terminal at Burbank where we boarded a C-54 and headed northeast towards Nellis AFB and Indian Springs in Nevada. Mt Charleston slid by, and we were in the restricted area. We dropped down and came onto a dry lake. Jake, Glen, Bruce, and I received a house trailer, which would be our home until we headed for England. Carmine, Marty Knutson, and Carl Overstreet were already in place."

"We met Col Fred McCoy, Hank Meierdierck, Lou Setter, and Maj Sam Cox (who was a weatherman). Detachment 'A,' as we were identified, completed training, and moved to the East Coast for a brief survival program and a short farewell to our families. We still did not have a comfortable cover story. We were NACA, the forerunner of NASA, pilots, Lockheed-trained, flying unspecified missions out of the country. My wife, worldly-wise and a tough trooper, said, 'horse shit' to me and kept quiet. It wouldn't be until after I had left the program that she would tell me how frustrated she found herself when asked by good friends, 'What's Hervey doing? What does he have to say?'"

Arrival at Watertown

The CIA contract pilots flew out of Burbank, California in an Air Force C-54, some of them thinking they were preparing all this time for a fast, exciting activity like space flight. At Watertown, the C-54 pulled up to an aircraft with long, glider-type wings, the U-2. When it was his turn, Marty Knutson and another pilot were reluctant to get off the C-54, saying "Thanks, but no thanks!" After some persuasion, the two joined the others.

The selected pilots had come from a fighter background, flying planes with the control sticks, or if a bomber pilot, with wheels or yoke-type controls. The fighter pilots in those days considered bomber pilots below them. They investigated the U-2 cockpit and saw the wheel/yoke control, and turned toward the C-54 saying, "No way." Again, after persuasion by the CIA and discussing it among themselves, they all decided to try it.

Checking Out in the U-2

The flying program was straight forward—T-33s and U-2s. To simulate the flat approach of the U-2, the T-33 flew with 10% flaps and a throttle setting of 65%. The pilots were not allowed to reduce power below that setting until the T-33 was on the lakebed.

Since they had no hard surface runway at that time, just the lakebed, the pilots never taxied the U-2. With the airplane parked in takeoff position on the lakebed, the pilots climbed in the cockpit, was hooked up to the oxygen and radio systems by the technicians. Somebody would close the canopy, and the pilot would lock it. Once the ground crew was clear of the plane, the pilot would push the throttle forward and take off straight ahead. Same on landing. Where the pilot stopped on the lakebed, he shut down, was unplugged by the suit technicians. They helped him out of the cockpit and drove him away in their "bread truck," a van outfitted with a big comfortable chair so the pilot could rest in a cooled area if there were a delay for any reason. The pilot's energy was saved very carefully because those long flights were so exhausting and dehydrating, with no water or food during the flight. He would typically lose 6 to 8 pounds on a flight. The autopilot seldom worked in those early days so they had to hand fly the airplane the whole flight. This required keeping the airspeed within plus or minus 5 knots of the scheduled indicated airspeed above 60,000 feet. Otherwise, the engine would flame out when the speed was too low, or the pilot got into compressibility if the speed got too high. A flameout required the pilot to glide down to below 35,000 feet altitude before air starting. It was a long way down, with suit fully inflated the whole time, and flight instruments mostly dead or not trustworthy. Without the pressurized suit, the pilot would die in a few seconds; their life depended 100% on their pressure suit working. If it failed, they died quickly. This happened. They were flying in very unfriendly territory in primitive conditions.

The pilot's airspeed gage—the only one he had—read indicated airspeed, which was a measure of the air pressure on the pitot tube in the nose of the airplane. At sea level, the indicated speed read very close to the actual true airspeed, but at 70,000 feet, the indicated airspeed read only one-fourth of the true airspeed, because the air pressure was so low there. So a rough correction was to multiply the indicated airspeed at 70,000 feet by a factor of 4, to make the true airspeed approximately 400 knots. The U-2 cruised at a true airspeed of 415 knots at high altitudes, which was nearly as fast as jet airliners, which cruised at half the U-2 altitude. The true airspeed, corrected for wind, gave the pilot his ground speed, although there was usually very little wind at very high altitudes (there might be much stronger winds at high latitudes, as in the polar regions). For the uninitiated, 415 knots true airspeed was approximately 478 statute miles per hour. So the U-2 at low altitudes was a very slow airplane, but it was not slow at very high altitudes; it cruised at nearly normal jet airplane speeds.

New models of the U-2 have modern GPS (ground position system) navigation systems and Mach meters, and many other navigation and control gages, so the pilot now knows precisely his speed and location at all times. Also, his autopilot is very reliable; the system at the time was not. They had to hand fly the old U-2s, and that took nearly all the pilot's time and energy.

The first U-2 flights were without the partial pressure suit and faceplate. The U-2 approached the landing threshold at 70-75 knots. To assist in depth perception and the desired landing attitude in the U-2, a station wagon with a UHF two-way radio would await U-2 approaches and speed alongside the landing U-2 while calling out height between the tandem wheels and the lake surface. (The pilots considered Capt Hank Meierdierck, a master caller.) Later the CIA pilots reportedly practiced landings without any help from the "Mobile" chase automobile. In fact, some later training and operational missions flown by pilots for the CIA (Kratt), Air Force (Maultsby), and ROCAF (Hua) had to undertake forced landings without any assist from the ground.

The already-experienced fighter pilots merely needed to become familiar with the plane and high-altitude flight. Aside from its extraordinary gliding ability, the U-2 proved to be a challenging aircraft to fly. The light weight that enabled it to achieve extreme altitudes also made it fragile. The plane was very sleek, and it sliced through the air with minimal drag. However, the U-2 design could not withstand high-speed G-forces, which made it very dangerous. Flying at high operational altitudes required the pilots to be extremely careful to keep the aircraft in a slightly nose-up attitude. If the pilot dropped the nose only a degree nose-down, it caused the plane to gain speed dramatically, enough for it to come apart in less than a minute.

Tremendous Lift, But Fragile – In Their Own Words

The two pogos, small wheels with steel spring struts holding up the wingtips during taxi and early takeoff, would drop out during the takeoff roll when the wings lifted and bent up a couple of feet. The wings were flexible and very lightweight. The airplane was so light and the engine so very powerful, the pilot had to retract the landing gear immediately after takeoff, or there was a danger of exceeding the landing gear redline airspeed. Initially, they took off using only 50% engine power (about 85% RPM), so the airplane did not accelerate so fast. This was nothing more than a large glider with a huge jet engine. Even at 85% RPM (50% thrust), you had to pull the nose up immediately, and steeply climb. At full power and lightweight, it was nearly vertical until reaching about 20,000 feet. Sometimes this climb was so steep that their attitude gyro would spill (start spinning around)—then the pilot had to fly visually and recover from the steep climb.

Powers: "The U-2 required very little runway for takeoff, a thousand feet would suffice. Within moments after the pogos dropped, you would begin climbing—at better than a forty-five-degree angle. (On the first couple of flights, you were sure you were going to continue right over on your back.)" - Powers p. 18

Michel: "We were airborne in no time, and Tucc said, 'Prepare for the pull' as the nose of the aircraft lifted nearly straight up. We were in a 15,000-feet-per-minute climb. Despite the high angle of attack, it was an extremely smooth ride." (Training U-2, note: rate = approx. 200+ ft. of climb/second) — https://www.usni.org/u-2s-still-flying-high

Fowler: "We had an early launch at a northern location, and the area was blanketed with dense fog. The aircraft was ready for takeoff. The pilot called the control tower for permission to take off, which was granted. Two or three minutes later, the tower called back to the pilot and stated he was given permission to take off. The pilot replied, 'I did take off, and right now I'm breaking 10,000 feet and climbing.'" - McIlmoyle p. 122

Warner: "This aircraft has an extremely sensitive center of gravity, and the weight and balance have to be figured on every mission. There is a lot of math involved; at home base, it was done by someone with a calculator." - McIlmoyle p. 262

May: "While in Hawaii, we were assigned to set up and receive our U-2s coming into Hickam. We were all told the bad news that one plane had a 'flameout' over the ocean about 500 miles out. The pilot was asked for his coordinates so any ships in the area could do a rescue. To their amazement, he said he was coming on in. Now, I call that gliding! He would only make one pass to land which he did with all of us watching." - McIlmoyle p. 168

****

To avoid security breaches, the CIA referred to the pilots as drivers and the planes as articles. Each plane had an article number. Cable traffic referred to the aircraft as KWEXTRA-00, the two-digit number identified the precise aircraft. The agency referred to the pilots as KWGLITTER-00, the last two digits identifying the pilot.

In 1955, the CIA selected 26-year old fighter pilot Marty Knutson and other elite military jet fighter pilots to fly an aircraft of undisclosed origin and told them that if they accepted the assignment, they might not live through it.

Knutson, from St. Los Park, Minnesota was typical of those chosen. As a youth, he enjoyed building model planes, electronics, and Ham radio. He attended the University of Minnesota, where he earned his Electrical Engineering degree under the Navy's Holloway Plan. After getting his degree, he separated from the Navy and enlisted in the Air Force, where he flew the F-86 in Korea against the Soviet MiG-15.

Following the Korean War, various levels of government decided that the CIA would get into the airplane business to see what was happening in Russia behind the Iron Curtain. The CIA developed an Air Division but had no pilots. The Agency set up criteria for the selection of pilots for the U-2. Prospective "Spook" pilots needed 1,000 hours of jet time, time in a single-engine aircraft, and to have flown them over oceans. Knutson met all the criteria, having accrued 1,400 hours of jet time. The CIA selected him along with six other military pilots for Detachment A.

Stan Beerli, a B-17 bombardier/navigator in World War II combat missions over Eastern Europe, flew after the war as a pilot and navigator on photo-reconnaissance versions of the B-17, B-29, and C-54 before joining AQUATONE in 1956. After training at Groom Lake, Stan commanded Detachment C in Japan until late 1957, when he took the same role at Detachment B in Turkey.

In mid-1959, he transferred to Agency headquarters as Director of Operations for a project now codenamed Project Chalice. Beerli was therefore in the major command positions during the last five U-2 overflights of the Soviet Union, including Gary Powers' unsuccessful mission on 1 May 1960. Stan devised the 'Quickmove' procedure for the rapid and covert deployment of the U-2 and support aircraft to staging bases.

Thus, the CIA and the United States Air Force initiated the development of high-flying aircraft capable of penetrating the airspace of the Cold War enemies of the United States. They screened personnel to engineer, test, and fly this new type of aircraft. Loyalty to the United States, physical and mental conditioning and exceptional skillset were the requirements in the selection of personnel to participate in these black, secret projects.

Initially, the recruiters used the criteria of the former OSS, seeking single men with little or no family ties. With the U-2 program, the Air Force and CIA found it more practical to recruit married candidates with family. The strong bonding and support within the family proved essential to the performance of their candidate. After that, with the A-12 Project OXCART, the screening process in most instances included the candidate's family.

In all cases, the individual only learned of

someone considering him after the FBI as well as various agencies wishing to recruit him completed meticulous scrutiny of his background. If the candidate failed to pass the examination, the individual never knew about the Agency even considering him.

The CIA U-2 pilots endured a rigorous selection process. Because of the strain involved in flying at extreme altitudes for extended periods, the selection process took painstaking efforts to exclude all pilots nervous or unstable in any way. The exclusions included those having excessive debt, alcoholism, or drug dependency, or being homosexual. The Lovelace Foundation for Medical Education and Research in Albuquerque, New Mexico, under a contract signed with the CIA on 28 November 1955, conducted the physical and psychological screening of potential U-2 pilots.

The CIA recruited a tough breed of pilots from the best. The CIA required pilots already having 1500 hours of flying time, with 900 being first pilot /instructor time. The CIA required they have experience in one-two and sometimes three aircraft, plus outstanding records, and a wing commander's recommendation. With all this, they might undergo an interview lasting two weeks.

The CIA's insistence on more rigorous physical and mental examinations than those used by the Air Force to select pilots for its U-2 fleet resulted in a higher rejection rate of candidates. The CIA's selection criteria remained high throughout its program and resulted in a much lower accident rate for Agency U-2 pilots than for their counterparts in the Air Force program.

Fifty percent or less of the prospective pilots made it past interviews conducted by the wing commander, his squadron commanders, and ops officers. Each of those chosen received orders to report to the 1007th Air Intelligence Service Group, Washington DC.

Even if those recruited possessed Q or top-secret security clearances, the CIA required they attended orientation classes, physiological tests, interviews, and the infamous lie detector examination. The CIA dismissed those not passing this phase of the recruitment after a strict debriefing. When the Agency requested the unsuspecting individual to appear for a personal interview, it invariably went something like this:

"Mr., Sergeant, Captain," (or whatever), the no-name person would say. "You've been recommended for a job of the utmost importance to your country. I have reviewed your records, and I have asked you and your spouse here today to request that you volunteer for this assignment. I will tell you both upfront; this is a dangerous job that you will not be able to discuss with your family or anyone else. At times, it may become a remote assignment unaccompanied by family. If you choose to decline this job, no one is to know what we discuss here today. I need you and your spouse to please sign this security agreement."

The recruiter advised the military recruits that they must resign from the military and become civilians. The recruit signed the security agreement, thinking this has something to do with an earlier application to become an astronaut, Navy Seal, a test pilot assignment, or whatever. At this point, the questions and answers went like, "Where is this assignment?" The recruiter responded,

"Sorry, that's classified. I can't tell you."
"What will I be doing?"
"Sorry, I can't tell you that either."
"Whom will I be working with?"
"Sorry, I can't tell you."

The man turned to his spouse and asked, "What do you think, Hon?"

"Just whatever you think. You know the kids, and I would stand by whatever you decide."

Recruiter: "Do you need some time to think it over?"

The recruit turned to the recruiter, "Negative, Sir. I'll take the assignment."

The recruiter smiled and told the recruit to return to his old job—he will hear from someone later.

The pilots recruited by the CIA went to March Air Force Base with no idea at this point, of why the CIA picked them or for exactly what purpose. All of them reaching this recruitment stage had volunteered. They resigned their Air Force commissions and signed a contract with the CIA. Only then did they report as civilians to Project AQUATONE Headquarters in a super-secure area of the Matomic Building at 1717 Street in downtown Washington. There, they met the CIA members of the team, including the project director, Richard Bissell, executive officer James Cunningham, plus the others recruited for their detachment.

This same procedure applied to the support team members except they retained their rank in the Air Force. Where the CIA project pilots received four times their Air Force pay with some withheld based on performance, the Air Force support members of the team received $7.00 per day per diem when deployed.

Most of the newly recruited rented apartments at 1600 15th Street, near the Russian Embassy, next door to the Cairo Hotel, and within walking distance of the Matomic Building.

At the time, no Dash-1 operating manuals existed for the U-2 plane. The recruited pilots reviewed notes and documents filed in loose-leaf binders. They studied the flying manual for the T-33 aircraft, which played a significant part in the U-2 training.

More Details of Recruitment

Cunningham remembers: "For every man we finally got, this represented about 20 people who had been considered in the first instance. The bulk of the pilots whom we had in the U-2 were those who had, in the first place, reserve officer status. They also had to be qualified jet pilots with as much experience as we could acquire. Physically, of course, they had to meet extremely rigid standards, and psychologically and from a security standpoint. Those who passed the first three screening interviews and who were subsequently approved by Security were then sent to the Lovelace Clinic for a specially designed series of physical exams before being vetted as qualified to proceed."

Those pilots selected for the U-2 knew of the plans to train them at Watertown in Nevada in preparation for deployment overseas. Even while waiting to report to Watertown for training, they used the unlimited access to "requisition" any office and flight planning materials they might need overseas. Also, as mentioned, they reported to the Lovelace Clinic.

Powers commented on some of the tests at

Lovelace: "I couldn't understand at the time why these were so important. It now became apparent. For safeguard, a special partial-pressure suit had been designed. Airtight, of rubberized fabric with almost no give or elasticity, it fit snugly around the body. a hermetic seal at the neck fastened the helmet into place. Once on, it felt exactly like a too-tight tie over a badly shrunk collar. On long flights, counting preparatory time, we would have to remain in the suit for up to twelve hours. Anyone with the slightest touch of claustrophobia would have gone mad. At Lovelace, we had our first washout. One pilot, though perfectly capable of flying for the Air Force, did not meet the rigid specifications required."

Records show that two contract pilots withdrew around early February 1956 and another resigned from the Detachment C group during training.

The high operating altitude and the partial cockpit pressure were equivalent to 28,000 feet pressure altitude. The pilot wore a partial pressure space suit to deliver the pilot's oxygen supply and provided emergency protection in case of cabin pressure loss. While in Washington, the pilots, in CIA fashion, met with Dave Clark of the Clark Clothing Company of Worcester, Massachusetts for measurement for their partial pressure suits.

David Clark Company Experience

Each of the pilots, Air Force, and CIA alike, went to David Clark Co. located in an old brownstone building in the old textile district of Worcester, Massachusetts, for the suit fitting. The pilots arrived at David Clark after the day shift had left. David Clark did this entire process 'in the black' with the CIA directing every aspect of the trip. The pilots went there with instructions not to bring a military ID card or dog tags. They brought only civilian clothes and no GI issue underwear because, for the fitting, they removed all clothes except undershorts and stood at a brace on the two-foot-high elevated round platform:

"A stoop-shouldered, wizened old man entered the room. With a tape measure dangling around his neck, he wore a white shirt with sleeves held away from his hands by armbands. This skillful man proceeded to measure every inch of my body, every inch. He even used a cup that he could adjust to determine the volume of my 'you know what apparatus'! He said, 'Okay, we are through. Come back in the morning. A car will come again for you. We will determine any adjustments made.'"

"The next day, I returned to the company. Again, I was instructed to strip down, don a set of inside out 'long johns' and then put on the pressure suit. The company people hooked it all up, showed me how it should be worn and proclaimed no adjustments were needed." - McIlmoyle p. 179.

Initial Testing of the U-2 by Lockheed

First Flight

"The Angel first moved from the surface of Groom Lake on August 2, 1955. However, it wasn't planned that way. These scenes show Lockheed chief test pilot Tony LeVier taking the Angel out for taxi tests."

"On the second taxi run, the U-2 popped into the air to 36 feet, and then dropped in so hard that it blew both tires on landing. During the rollout, the brakes caught fire, but were quickly extinguished. Two days later in a rainstorm, the Angel went to 8000 feet." - *The Inquisitive Angel*

(Note: official history differs, saying this first unintended flight took place on 1 August 1955)

Pilots wore street clothes for low-level flying

In what the CIA and Lockheed intended as a high-speed taxi test, Tony LeVier made the unofficial first flight in the U-2 on 1 August 1955. To everyone's surprise, the sailplane-like wings were so efficient the aircraft jumped into the air at 70 knots (81 mph; 130 km/h), amazing LeVier

who, as he later said, "Had no intentions whatsoever of flying."

As LeVier tried the ailerons, he became surprised to discover the plane airborne. The transition to flight had occurred so smoothly he did not notice. He cut the power and contacted the ground in a left bank of 10 degrees in a hard landing that blew both tires. The brakes were too weak and caught on fire, while at the same time the plane bounced back into the air.

LeVier brought it down for a second landing. He again applied the brakes with little effect. The aircraft rolled for a long distance before coming to a stop. Bissell, Cunningham, and Johnson saw the plane fall and bounce.

They leaped into a jeep and rushed to the aircraft while signaling to LeVier to climb out. They used fire extinguishers to put out a fire located in the brakes.

The prototype U-2 suffered only minor damage, blown front tires, a leaking oleo strut on the undercarriage, and damaged brakes. However, this unplanned flight foretold the airworthiness of the U-2. It also taught Lockheed test pilot Tony LeVier that the U-2 had a mind of its own when it came to it wanting to fly.

Taxi trials continued for one more day. The CIA knew now that the U-2 Angel nicknamed the Dragon Lady loved to fly and refused to land. At low speeds, it remained in ground effect and glided effortlessly above the runway for great distances.

Three days later, on 4 August 1955, LeVier piloted the U-2 at Groom Lake on its first intended flight, flying it to 8,000 feet. She flew beautifully, but the weather threatened. The plane, having only two landing wheels, would have to land like a bicycle. LeVier wanted to touch the rear wheel down first. But, Johnson insisted on touching the nose wheel down first.

LeVier disagreed, believing the U-2 would bounce if he attempted to land on the forward gear first. On this first flight, LeVier leveled off and cycled the landing gear up and down. He tested the flaps and the plane's stability and control systems before making his first landing approach.

As the U-2 settled down, the forward landing gear touched the runway, and the plane skipped and bounced into the air. LeVier made a second attempt to land front wheels first, and again the plane bounded into the air. Kelly Johnson watched from a chase plane, giving a constant stream of instructions as LeVier made three more unsuccessful landing attempts.

LeVier, with the light fading and a thunderstorm fast approaching from the mountains to the west, made one last approach using the method he first advocated, letting the aircraft first touch on its rear wheel. This time the U-2 made a near-perfect landing ten minutes before the thunderstorm dumped a rare 2 inches of rain, flooding the dry lakebed, and making the airstrip unusable.

Legend had it that LeVier climbed down while saluting Johnson with a "one-fingered" salute for almost getting him killed with his insistence on a nose-first landing. Johnson supposedly returned the "one-fingered" salute and yelled, "You, too!" The story spread that this was how the plane became known as the U-2.

The truth was that Geary and Culbertson decided to place it in the utility aircraft category along with two other utility aircraft on the books, a U-1 and a U-3. They dubbed it the "Utility, 2" to guise it as a civilian weather plane. Lockheed dubbed it the Angel, and the Air Force called it the "Dragon Lady." At Watertown, they often called whatever was flying "the bird."

Now with the first problems in flying and landing the U-2 worked out, Kelly Johnson scheduled the "official" first flight for 8 August 1955. This time, outsiders present included Richard Bissell, Col Osmond Ritland, Richard Homer, and Garrison Norton, all there to witness LeVier, using the call sign ANGEL 1, make the "VIP" official flight in Article 341. Bob Matye flew chase in a C-47 with Johnson on board as an observer.

The U-2 ascended to 32,000 feet and performed well, meeting Kelly Johnson's eight-month deadline. LeVier made an additional 19 flights in Article 341 before moving on to other Lockheed flight test programs in early September. This first phase of U-2 testing explored the craft's stall envelope, took the aircraft's maximum stress limit (2.5 g's), and explored its speed potential. LeVier was soon flying at its maximum speed of Mach 0.85.

Little did the CIA know that the plane expected to fly for two years would still be flying today, more than half a century later. In eight months and under budget, the CIA produced at Watertown the most capable and reliable high-altitude intelligence, surveillance, and reconnaissance (ISR) platform ever, even compared to any system flying today manned or unmanned.

Tests continued, with the U-2 ascending to altitudes never attainable in sustained flight. On 16 August, LeVier took the aircraft up to 52,000 feet. During preparation for this flight, the 42-year-old pilot completed the Air Force partial-pressure suit-training program, becoming the oldest pilot to do so.

Ray Goudey demonstrating the "training stand" (cockpit mockup) for the sextant/drift sight integrated unit—lined up here below the sextant unit on the U-2 that appears as a 5-inch bulb on the nose in front of the cockpit.

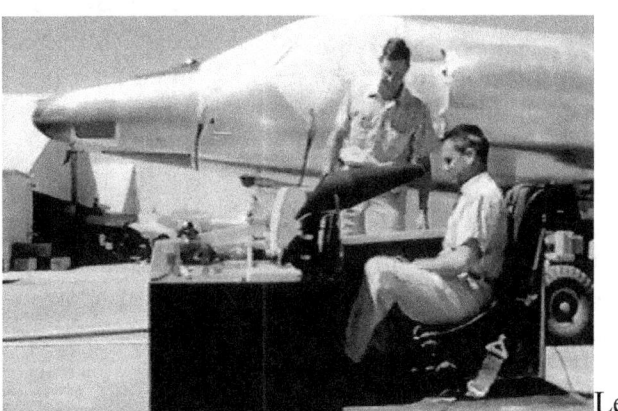

Le Vier completed contractor testing. This included taking the U-2 to 50,000 feet, achieving a maximum design speed of Mach 0.85, and making a successful dead-stick landing.

Eventually, every day the U-2 broke the world's altitude record of 64,000 feet, but due to the secrecy, they could not tell anyone. The Lockheed test pilots Bob Matye, and Ray Goudey expanded the altitude envelope to 70,000 feet. The two pilots replaced LeVier shortly before the second U-2 (Article 342) arrived at Watertown.

Landing – In Their Own Words

Schiff: "The U-2 must be close to the normal touchdown attitude at the height of one foot before the control wheel is brought aft firmly to stall the wings and plant the tailwheels on the concrete. The feet remain active on the pedals, during which time it is necessary to work diligently to keep the wings level." —

http://www.barryschiff.com/high_flight.htm

U2 Being Prepared For Flight As Seen From Cockpit Of Shuttle.

View of U-2 from the cockpit of the shuttle

Stratton: "I passed Ed [Instructor Pilot in chase vehicle], and he started calling out my altitude above the runway: '. 3, 2, 1 foot, hold it off, hold it off,' I kept sucking back on the yoke. The bird started to shudder, so I kept coming back and it 'flopped' to the runway. I figured this was a disaster and then Ed called out, 'Nice landing!'" – McIlmoyle p. 61-62

Espinoza: "As I was told before one of my tryout flights, 'Landing the U-2 is a lot like playing pool. It's not so much how you shoot as how you set up your shot.'" - Cholene Espinoza, *NY Times* op-ed, May 7, 2010

Wood: "Approach and landing were normal with tail wheel down as the Mobile pilot had instructed. After touchdown, I brought the yoke back. The next thing I knew, I was airborne again at a very steep angle! The power was in idle, and my left wing started to drop. I hit the power to maximum to recover from a most difficult position and to save the aircraft, to say nothing for saving my own butt."

"My thoughts at the time were that I had just 'bought the farm' and this was it. I was holding back the yoke, trying to pick up the left wing to keep it from hitting the ground and the aircraft turned approximately 45 degrees to the runway. So, in this nose high, left wing down position, I

said to myself, 'Where in the F--- is all this power that I have been hearing about?' Just in the nick of time, before impact, the power did kick back in, and the aircraft started to fly. Final landing of the day was normal. Debriefing comments by General Russell were, 'About average first landings, Captain Wood. You have a nice weekend.'" - McIlmoyle p. 101-102

Rehearsal of exiting an F-84 atomic cloud sampler aircraft after a mission

Landing short

SAC Instructor Pilot Meierdierck's memoir describes one U-2 crash taking place in 1956, that of CIA pilot Vito, who damaged his plane in landing on the desert, although not injured (Article 342, March 21, 1956):

"The engine was still running, and fuel was pouring out of the broken wing, and the pilot was unconscious. I jumped up on the wing to pull him out. I was afraid to shut the engine off, for fear of causing a spark and an explosion, so I dragged him out, but the seat pack attached to him and the wire wrapping was holding me back. I had to go back in the plane and release the seat pack and then drag him across the desert until the fire trucks and ambulance arrived."

CIA pilot Strickland experienced a forced landing in Article 344 on June 1, 1956, due to running low on fuel. The movie *The Inquisitive Angel* shows a U-2 salvage scene and gives the following description and comment:

"Rescue crews rush to the end of the runway where an Angel has landed short. The pilot here was uninjured. But emergency crews take no chances with leaking fuel. Salvage operations mean that this fallen Angel will soon fly again to rejoin its sister ships already in the air. This project has had fewer mishaps than is normal with new aircraft. Yet the unique ground equipment, designed solely for the Angel, operates as well at this crash scene as it does on the flight line."

(Note other U-2 in the air)

Training the CIA U-2 Pilots

The Air Force cadre instructor pilots had received only a few U-2 flights of their own before the agency (Central Intelligence) pilots, all F-84 qualified, arrived from Turner, Bergstrom, and Malmstrom Air Force Bases, where they had been recruited. Typically, they had plenty of combat experience in the F-84 and had participated in long-range navigation using the sextant, dead reckoning, and extensive pre-planning. The F-84 fighter had seen plenty of service in the Korean War. They flew them en masse across the Pacific Ocean to Korea using sextant navigation.

Major Delap, a USAF navigator, devised a system of flight planning and navigation while the cadre officers learned the systems with help from the Lockheed and other contractor people. Using this learn-as-they-go process, the cadre officers used their flight experience to brief the student CIA pilots in ground school and subsequent flights.

Though the CIA trainee pilots arrived at Watertown as experienced military pilots, the Air Force instructor pilots, nonetheless, began their training with the ground school on the aircraft systems, emergency procedures, flight planning, navigation, etc., before allowing them to fly the U-2.

Before starting ground school, the instructor pilots devised procedures to use in the air and to prepare for a flight. The preparation included the route, fuel consumption, checkpoints, pre-breathing for 2 hours, and the myriad details necessary for an overflight of denied territory in a new type of aircraft flying in an environment never flown before.

To ready these pilots to fly the U-2, the instructor pilots first verified the CIA contract pilots' flying experience and qualifications before allowing them to fly the U-2, the first million-dollar airplane. To do this, the instructor pilots first placed the trainee pilots in the back seat of a two-place T-33 jets.

The instructor pilot took control of the airplane at the end of each T-33 flight. He climbed to 10,000 feet, put the gear, flaps, and speed brakes out, and shut the engine off. The lakebed contained a painted cross that marked the pilot's target to land as close to as possible. After a few practice flights, most trainee pilots touched down within 100 feet of the mark.

Even with the T-33 trainer, the Air Force used a souped-up Mercury station wagon for mobile control and to chase the plane on takeoff and landing. On landing, a driver, and an instructor pilot raced alongside and to the rear of the aircraft while calling out his altitude above the ground, "Two feet, two feet, one foot, one foot, OK, ease her on down."

Even for experienced pilots, landing the U-2 proved challenging and required flying in a two-mph range to make a safe landing. Knowing the exact stalling speed of the T-33, the instructor pilots instructed the pilots to fly at 2 knots above the stall speed, 2 feet above the surface of the dry lake. The instructor pilots induced contact with the lake to let them develop the correct recovery procedures. When satisfied with the pilots' capability of transitioning to this new airplane, the instructor pilots checked them out in the U-2 with Carmine Vito and Marty Knutson the first CIA pilots to complete this stage.

Flight training began with the T-33, then several low-level U-2 flights with the pogos still attached. Next, the CIA pilots practiced takeoffs and landings with pogos released. From there, training proceeded to high-altitude flights, night flights, navigation, camera operation, and other mission-readiness preparation.

A single J57 (non-afterburning) engine powered the early U-2s. The J57 was the first 10,000-pound thrust engine for the US, and requested for the B-52, F-100, RB-57, and other designs, obtaining them became a challenge. All training for the U-2 used the J57/P-37 engine. The initial Lockheed design however was to use lighter and more powerful J57/P-31 engines.

The P-31 engines used primarily on the KC-135 tanker powered the U-2 on actual operational missions from 1956-58. The heavier, less powerful P-37s used in U-2 training were also unfortunately prone to flameout at high altitudes, causing over 900 flameouts during the flight testing at Watertown.

In late 1958 the CIA began replacing the J57/P-31 engines with the newer J75/P-13 non-afterburning engine, offering more power and 2500 feet higher mission cruise altitude for the U-2. This model was known as the U-2C. As described by Lockheed pilot Ken Weir: "That was the C, the hot rod. The Cs had a higher rate of climb; their thrust-to-weight ratio was a lot better." With the J75 engine, "It would just leap up. It would rocket to altitude."

For simplicity, reliability, and lightweight, the U-2 had no hydraulically boosted flight controls. The U-2 used the hydraulic system only to operate the landing gear, flaps, speed brake, and the fuel boost pump. As a result, the plane had slow roll rates with normal pitch and yaw rates. Normally, the plane seldom indicated over 220 knots, and then only in "gust" position, which moved flaps few degrees, greatly reducing tail loads. The limit load was 2 Gs. Typically, cruise speeds at altitude were in the 120-95 knot IAS range, stall speeds between 52 and 65 knots, depending on gross weight and flap settings. Because of the high thrust to weight ratio in training flights, the instructors started training using only 85% RPM (50% thrust) to avoid excessive climb angles.

Groom Lake had a 5,400-foot macadam strip at the SW corner, with continuation markers (Los Angeles and Pasadena traffic markers) going northward on the clay lake surface for about five miles. The paved runway was for night takeoffs and for when the lake surface covered with water. There were no two-place U-2s at that time. Initial training involved flying shallow power-on approaches and landings (long float distance) in the T-33. Each student, already highly proficient, made 20 to 40 T-33 landings.

The U-2 had an idle engine thrust at sea-level of approximately 600 pounds. At fairly-light weights, 12,000 to 15,000 pounds, the airplane

drag with gear down, speed brake out, and 15 degrees of flaps was very near 600 pounds, so when the pilot pulled the throttle back to idle on final approach, the plane decelerated very slowly—around one or two knots per thousand feet of runway. Speed control on final was critical. It was not uncommon on a hot day to float 8,000 feet down the runway.

During this time, envelope expansion was going on, restricting the pilots from using full flaps, which would probably have reduced some of the float. Final approach speed was 10 knots above stall speed. The pilots could make a go-around in idle simply by retracting the speed brake.

The student pilots spent many hours in T-33s practicing precise airspeed control because it was so critical in the U-2, both at high altitudes and in landing. At cruise altitude, the pilot had a certain "window" to maintain plus or minus a few knots from cruise speed. Going over that the pilot got into compressibility quickly. Going under that speed, the engine flamed out with no warning. (Full throttle thrust at altitude was about the same as at idle on the ground, 600 pounds, burning 90 to 100 gallons per hour. For each pound of fuel burned, the U-2 climbed one foot.) Maintaining weight as low as possible was the only way the U-2 could operate at high altitudes. The Lockheed engineers tracked equipment weight down to the ounce, literally.

After a flameout, the pilot glided down to 35,000 feet with the pressure suit deployed and the canopy frosted over. Very few instruments worked to attempt an air start. Most air starts were successful, but not all. Louis Setter said, "Getting squeezed by that primitive pressure suit was something you never forgot." Regarding flameouts, Hervey Stockman said, "It became quiet during a flameout. The cockpit canopy and windscreen frosted over; cockpit instruments failed. The cockpit altitude of 30,000 feet begins to climb as the pressurization from engine bleeds air."

The pilots avoided crosswind landings on the lakebed simply by always landing into the wind. Later, the students conducted crosswind landings to prepare for operating on long, hard runways at their deployment bases. Since the U-2 had to land in a stall, the pilots found the wing down technique with opposite rudder was the best technique. As the speed slowed the pilot was not always sure when the stall would occur, so the technique used by some pilots of crabbing into the wind, wings level, and kicking it out just before touchdown did not work. The U-2 had a sharp stall with an unforgiving single main gear with a tailwheel for landing. A ground loop would normally collapse the main gear, doing considerable structural damage.

To make the problem worse, the U-2 pilot had this unfriendly plane to land after a mission which was long and very fatiguing, with both frost and oil vapor on the inside of the canopy. Louis Setter flew one mission for 10.5 hours with no autopilot and lost the feelings in his arms and hands. He was totally dependent on the chase station wagon driver to talk him down on that landing. (Setter achieved an altitude of 72,500 feet on that flight.)

The U-2 should be stalled with tail wheel touching down first or at the same time as the main wheel, with the yoke full-back and the pilot looking out both left and right sides of the canopy like a T-6 or P-51 to keep the plane straight down the runway. The pilot could not see over the nose in a stalled attitude.

Sometimes, with turbulence on touchdown, the pilot would hit both stops on both ailerons and rudder. Under these conditions, pilot inputs had to be large and quick. Other times, the airplane would stall straight ahead and show no tendency to yaw after touchdown.

The engine oil vapor would get into the cockpit through the pressurization system because the engine's seals were unable to do the job at high altitude. A fine mist of engine oil spread a layer of oil on the canopy. During the flight, since there was nothing to look at outside the cockpit anyway, no one cared. However, for landing, part of the landing checklist was to wipe the inside of the canopy. The pilot carried a long stick with a gauze pad (a women's sanitary pad) on the end of it to clear an area on the inside of the canopy of oil vapor or frost so he could see to land.

Because it was so easy to get into a ground loop and so difficult to judge one's altitude over the lakebed, the pilots required a chase aircraft and a chase vehicle (a station wagon with UHF radio) on every landing. The ground chase pilot literally talked the U-2 pilot down, giving him running report as to height, pitch attitude, and rudder inputs to keep it straight. The U-2 was a ground looping aircraft worse than the T-6. After landing, if a wing touched the ground, the pilot could not bring it back up with the aileron because of the remaining fuel rushing toward the wingtip.

At Watertown, the CIA and Air Force tried several aircraft as chase aircraft; initially a Navion, then a C-45, a Twin Bonanza, and finally a Twin Cessna L-27 for traffic pattern work and T-33s for the high altitude chase, particularly during stall testing.

Testing at even higher altitudes continued, and on 8 September 1955, the U-2 reached its initial design altitude of 65,600 feet. The pilots' final U-2 check out flight occurred at night with the pilot using a sextant to measure his rough latitude from the reading of the North Star Polaris. The mission headed west to California, north to Montana, east to the Dakotas, south through Colorado, and west back to Watertown. Total flight time was usually eight and a half hours.

On 22 September 1955, the U-2 experienced its first flameout at 64,000 feet more than 12 miles up, with Lockheed pilot Robert Schumacher at the controls. After a brief restart, the J57/P-37 engine again flamed out at 60,000 feet, and the aircraft descended to 35,000 feet before the engine relit. Engineers from Pratt & Whitney set to work on the problem.

Test pilots Robert Sieker and Robert Schumacher had joined the U-2 test team during October 1955, the same month the Atomic Energy Commission, under a request by the *Las Vegas Review-Journal*, released a statement regarding progress on the "Watertown Project." It stated, "Construction at Watertown installation a few miles north of Yucca Flat announced last spring continued. Data security to date indicated a need for limited additional facilities and modifications of the existing installation. The Reynolds Electrical and Engineering Company, Incorporated performed this additional work under the direction of the Atomic Energy Commission's Las Vegas branch office."

During the month, Lockheed test pilots Bob Matye, and Ray Goudey replaced LeVier. Afterward, Ray Goudey expanded the altitude envelope to 74,500 feet, and the second U-2 (Article 342) arrived at Watertown.

The P-37 model engine demonstrated poorer combustion characteristics than the preferred, but unavailable P-31 version. The P-37 tended to flame out at high-altitudes. The combustion problems became apparent as the U-2 began the final part of its climb from 57,000 to 65,000 feet. The pilots referred to this area as the "badlands" or the "chimney." According to Hervey Stockman, high altitude flameouts and associated decompression led to Bruce Grant's having to leave the program.

The flameouts bedeviled the U-2 project until enough numbers of the more powerful P-31 engines became available in the spring of 1956. Louis Setter claimed that they'd had over 900 flameouts when he left the program, nearly all with the P37 engine. Meanwhile, with the airworthiness of the U-2 airframe proven, Lockheed set up a production line in Oildale, California. Nonetheless, the delivery of even the second choice J57/P-37 became a major problem. The CIA learned of Pratt & Whitney contracting with the Air Force for its full production capacity of engines for the next year for its F-100 fighters and KC-135 tankers.

Colonel Geary, with the help of a colleague in the Air Force Materiel Command, managed to arrange diversion of several these engines from a shipment destined for Boeing's KC-135 production line, making it possible to continue building the U-2s.

The new J57 / P-31 engine had a 16-stage compressor with nine stages in the low range and 7 in the high-pressure chamber. The Pratt and Whitney engine operates at full power for the duration of the flight. At sea-level, this unit gulps nearly 9000 pounds of fuel per hour. At 70,000 feet this drops to 700 pounds per hour. At 74,600 feet, the engine will quit from oxygen starvation. In the initial stages of the program as many as six flameouts occurred on a single flight. With the new fuel system and turbine design of the -31 engine, flameouts have ceased to be a critical problem." – *The Inquisitive Angel*

Though the USAF would eventually fly the U-2, the CIA had majority control over the project, code-named Project Dragon Lady. Despite the Strategic Air Command chief LeMay's early dismissal of the CL-282, the USAF in 1955 sought to take over the project and put it under the Strategic Air Command until Eisenhower repeated his opposition to military personnel flying the aircraft. Nonetheless, the USAF substantially participated in the project; Bissell described it as a "49 percent" partner. The USAF agreed to select and train pilots and plot missions, while the CIA would handle cameras and project security, process film, and arrange foreign bases.

It took a face-to-face meeting of Dulles and the top Air Force officials to bring the approval of the joint agreement entitled "Organization and Delineation of Responsibilities - Project OILSTONE" and signed by General Twining for the Air Force and by Mr. Dulles for the CIA during August 1955.

The agreement gave responsibility for the general direction and joint control of the project to the DCI, and the Chief of Staff, USAF. Subject to guidance from higher authority, the CIA appointed a project director, and the Air Force appointed a deputy project director responsible for the project through all its phases.

President Eisenhower's decision was crucial: "I wanted this whole thing a civilian operation," the president wrote. "If uniformed personnel of the armed services of the United States flew over Russia, it became an act of war—and I do not want any part of it." With the issue of control over the program settled, the two agencies soon worked out the remaining details.

The OILSTONE pact gave the Air Force responsibility for pilot selection and training, weather information, mission plotting, and operational support, the lion's share of the work. The CIA retained responsibility for cameras, security, contracting film processing, and arrangements for foreign bases. Also, the CIA kept a voice in the selection of pilots. All aeronautical aspects of the project, including the construction and testing of the aircraft, remained the exclusive province of Lockheed.

Because of this agreement, the CIA remained in control of the program; Richard Bissell later remarked how the Air Force became a supporting element, and to a major degree wanted a role more than supplying half the government personnel. He said the Air Force held, "if you wanted precise numbers, 49 percent of the common stock."

Chapter 11 - High Flight

Adjusting to High-altitude Flight

Hervey Stockman: "We needed to wear our partial pressure suits in the cramped U-2 cockpit in the event the engine quit—or in jet vernacular—flamed out. Where commercial aircraft passenger compartments pressurize to about 5,000 feet, the U-2 cockpit pressurized to 30,000 feet when flying at 70,000 feet. The pilot breathed 100% oxygen delivered into our helmets under just enough pressure to inflate our lungs without our breathing in. We had to exhale against the incoming flow. It did not take long to adjust to "reverse breathing." Above 55,000 feet, nitrogen suspended in the blood came out of suspension in bubble form, causing the blood to boil and instant death. The partial pressure suit prevented this phenomenon by effectively squeezing the body and its blood vessels sufficiently to keep the nitrogen in suspension, squeezing the body back to sea-level where atmospheric weight and density were livable."

Putting a man into high flight required a change that spawned several experiments with both the pilot and the plane throughout the aviation industry. The Air Force undertook high-altitude bailout experiments from balloons in the autumn of 1955 to determine if the suit designed for the U-2 pilot protected him during his parachute descent once he separated from the life-support mechanisms inside the aircraft.

Like earlier X-1 rocket plane pilots and Air Force RB-57A pilots who wore T-1 partial pressure suits, the pilots flying the U-2 helped pioneer the development of protections against the hazards of traveling to high altitude. Not only is there not enough oxygen to breathe, the lack of air pressure on the body causes dissolved gases to expand in the body's tissues, called decompression sickness. Initial symptoms such as disorientation can quickly go on to include brain injury. Without protection above 62,000 feet, Armstrong's Limit, fluids in one's throat and lungs will boil away.

The U-2 pilots used a procedure known as pre-breathing to avoid getting the "bends" during descents from the long flights. They breathed pure oxygen, starting at least 90 minutes before takeoff to dissipate nitrogen gas from their bloodstream. They used a portable oxygen supply while entering the aircraft.

The pilots found eating, drinking, and urination a significant problem while wearing their suits. The first model of the pressure suit used by Lockheed test pilots made no provision for urination. At first, they catheterized the pilot for permitting urination during flight. By the autumn of 1955, a pilot brought a condom and a football bladder to the Ranch, which led to a new external system that worked and made catheterization unnecessary.

The pilots reduced their stimulation by eating a low bulk, high-protein diet on the day before and the morning of each mission. They drank sweetened water to prevent desiccation during the lengthy missions, a condition exasperated by their breathing pure oxygen. They accomplished this by providing a small self-sealing hole in the face mask to allow the pilot to push in a straw-like tube attached to the water supply.

Due to the long mission duration, the CIA project pioneered in the development of the ready-to-eat foods in squeezable containers used today. They chose bacon or cheese flavor mixtures that the pilot squeezed into his mouth using the self-sealing hole in the face mask. Despite all these precautions, the U-2 pilots lost 3 to 6 pounds of body weight during an eight-hour mission.

The pilot carried his food and liquid tubes in a pocket on the leg of the flight coverall. (The loose-fitting coverall served mainly to protect the expensive pressure suit from rips and snags. No leaks in the pressure suit permitted.)

Flight Testing the U-2

When the first Air Force U-2 squadron trained at Watertown in 1957, the Air Force intended to activate an Air Force U-2 Squadron at Turner Air Force Base, Georgia. Colonel Yancey sent Captain Meierdierck to Turner Air Force Base to advise them on the placement of the ground approach control (GAC) vans on the base.

At Turner, the division commander refused to take suggestions from a lowly captain, until Col Gerald Johnson, who knew Captain Meierdierck, intervened.

Training and flight-testing progressed on

schedule at Watertown while the first detachment trained and prepared to deploy to the UK. At the time, the Watertown support aircraft consisted of one C 47, 4 T-33s, one B-25, a Twin Bonanza, and two Navions. Meierdierck went to the Sacramento Air Depot to pick up one of the Navions, only to learn, "It belongs to General LeMay."

In those days, the wing and base commander signed all flight clearances, except the pilots in the unit. Meierdierck lacked authority for his taking General LeMay's plane. Nonetheless, he signed for it and headed to Watertown with no known repercussions.

The U-2 encountered some developmental and technical problems as did all-new aircraft design. Among early issues reported by U-2 test pilots: sun in one's face, cold cockpit, the layout of the controls, need for autopilot, rudder and aileron response, stall characteristics, tendency to ground loop.

One such issue involved an oil film often appearing on the windscreen, which clouded the forward visibility and increased the landing problem. During the interval, until Kelly Johnson solved the problem, the pilots used a sanitary napkin on the end of a stick to clean enough of the windscreen for them to see to land.

U-2 Instructor Pilot Work (circa 1955)

There were four of them military instructor pilots, and the Lockheed test pilots sometimes would also act as instructor pilots. They all worked together.

The U-2 IP (Instructor Pilot) was much more critical to the safety of his student than an IP was for probably any other airplane, because the U-2 IP had a vital role in the safe landing of the aircraft. The IP rode in a car equipped with a radio so he could talk to the pilot. He drove down the runway directly behind his student, who was landing the U-2. They had no two-place U-2 trainers, so the student was on his own in his airplane, even on his first solo flight. The U-2 was easy to fly but difficult to land. It was quite different from any Air Force or civilian airplanes that the pilots had flown, because basically, it was a very lightweight glider, having no hydraulically boosted controls, with a massive jet engine. This mismatch makes little sense until you flew the airplane at very high altitudes, then you see that that big engine was struggling to get enough air down the intake, and was putting out about the same thrust at full power as it was at idle power at sea level.

Before the first solo flight, the IP or Ops Officer signed a training form attesting the student trained and ready to solo. The student had cockpit familiarization and knew where all the switches were. The student pilot knew the emergency procedures, and he had completed a course of T-33 training with his instructor, simulating the very low and slow "drag in" approach necessary for the U-2. A "drag-in" approach meant the airplane was flying only about 5 knots above stall speed and quite low to the ground. This "drag-in" approach suggests the reason for calling the U-2 the "Dragon Lady." The "Dragon Lady" was a cartoon character in the comics of the 1950s.

The crucial things were memorized because at high altitude, with a flameout, there was no time or opportunity to get out a checklist. Things happened too fast. Emergency procedures for the U-2 were much simpler than for a typical Air Force airplane because the systems were few and simple. Flight controls were mechanical—no hydraulic boost with only the landing gear, flaps, and speed brake hydraulically actuated.

In the early days, they used to chase their students during their mission in a T-33, cruising at 35,000 feet and trying to keep the student's U-2 in sight at almost 70,000 feet. When the U-2s were unpainted, they could usually keep them in sight, but later, when painted a dull black, they couldn't see them. They stopped the T-33 chase requirement.

But they did chase each U-2 in their traffic pattern in a slower prop airplane, utilizing the twin-engine C-45, the single-engine Navion, and finally a leased Twin Bonanza. Only the Twin Bonanza worked out well because the older planes did not have enough power at the high altitude of their remote site, and many times had engine overheating problems. The prop chase airplane would follow the student U-2 down final approach, at about 80 knots, and the mobile control would pick up the U-2 as chase when the U-2 came over the edge of the lakebed. The Ford Station Wagon worked quite well as a chase. The instructor pilots spent a lot of time on the lakebed in their Ford mobile control. One time,

For the long navigation flights, the students were really on their own. They began pre breathing 100% oxygen in their pressure suits in the medical facility, starting at 0300 or 0400 hours. Then a pressure suit technician completed dressing them. He hooked them up to a "walk-around" oxygen bottle and assisted them in getting into the "bread truck" and drove them out to their U-2. Another U-2 pilot had already done the preflight walk around to save the energy of the mission pilot. He had no food or water on his long flight, which often lasted 9 hours, and if his autopilot did not work, he had to hand fly his airplane within plus or minus 5 knots indicated airspeed the full time at high altitude. If he got over 5 knots slow, the engine would flameout suddenly, and his suit would "blow up," he would have to pressure breath and glide down to 35,000 feet or lower to attempt an air start. Or glide to an emergency airport if he could not glide home or get an air start. Glide distance was about 200 miles. It was a very long, cold, miserable glide.

Lockheed test pilots and the SAC Instructor Pilots took the lead in working out solutions and producing a pilot manual. Weir commented: "They built so few of these airplanes that they were really considered to be hand-made, and they all had idiosyncrasies. If one had its wing attached at a slightly different angle than the other, that caused the airplane to have lateral trim difficulties, it causing the airplane to have peculiar stall characteristic.; It might roll off to the right or roll off to the left, or it wouldn't stall straight through, and there was always something that you had to be conscious of. And it was mainly by word of mouth that you learned about the difference. One pilot would tell you, 'well, be careful of this. This plane's going to do such and such.' Lockheed did everything it possibly could to eliminate those differences."

U-2 flameouts and recovery (circa 1955-56)

The pilots experienced hundreds of flameouts during the test and training program. Some claimed that they got used to them. Setter had three flameouts gliding back at the same time,

one on final, one circling, and one while still at altitude. He had to land on two occasions. One, he managed an air start and resumed his mission. His pressure suit saved his life three times.

Sometimes the engine would quit without warning even if the pilot was holding the correct airspeed. Any slight wiggle of the airflow going into the air ducts was sometimes enough for a flameout. At 70,000 feet, the correct indicated airspeed was 104 knots (415 knots true speed). The pilots had to hold airspeed within plus or minus 5 knots. Flying five knots indicated speed too high would put the pilot into compressibility, and 5 knots too low caused a flameout. These flameout problems originated primarily with the Pratt and Whitney J57-P37 engine; the later J57-P31 engine was a significant improvement and had few flameouts. The U-2 used the P31 engine for all of the Russian overflights.

Louis Setter oft-stated that his first flameout was embarrassing. It occurred during his first pressure suit ride, for orientation. He had leveled off at 45,000 feet to do the standard stall series: stall with no flaps, stall with flaps (15 degrees was all they were authorized to use at this time), gear up, gear down, left turn, right turn. He did this at idle rpm, about 80% rpm. Suddenly the engine flamed out! Setter was quite surprised because 45,000 feet was not considered to be in the flameout zone, but when he checked his altitude, he found that he had climbed, at idle rpm, to 53,000 feet! That was a good lesson—this airplane would climb in idle! So his suit inflated. he glided down to 35,000 feet and made a good airstart. Setter was only a few miles away from the lakebed, so He continued his training mission. At that time, they could have many flameouts in a flight and still continue the mission after a good airstart. Later, one flameout was cause for mission abort; the medical people found that one flameout at high altitude could cause damage to the pilot.

The pilots compared a flameout landing in the U-2 to landing a T-6. There was no idle thrust—you just flared and landed with no float. Idle thrust at low altitude was around 600#, which was almost exactly equal to the airplane drag. In other words, the airplane could cruise at low altitudes at idle thrust. One of the student pilots made a go-around in idle power by merely retracting the speed brake. Landing the airplane at idle thrust was difficult—it didn't want to land. Their approaches were very low, airspeed control was precise, and still, the aircraft would float for thousands of feet across the lakebed before it would stall, or had to be force stalled to land.

CIA pilot Gary Powers tried to make a wheel landing, and the airplane kept bouncing back in the air. He didn't like to stall the airplane but learned he had to.

The early U-2s did not have wing spoilers, as later models had, so slowing to stall speed at around 55 or 60 knots was a precise and challenging thing to do. Doing so required an instructor pilot with radio in the station wagon, following right behind you. The IP would tell the pilot if he was in a yaw, or not, and how high his main wheels were above the ground. The airplane typically floated 1000 feet across the lakebed for each knot above stall speed. The stall was abrupt and was a definite hard bump; the pilot knew when he was down. The tail wheel had to hit at the same time, or slightly before, the main gear for the airplane to stick without bouncing, and the pilot needed not to touch down in a yaw position. Doing so could lead to a ground loop, and once begun, could not be stopped.

The pilot had to be very active with his rudder. Ground looping was the primary reason for damaging or destroying many U-2s. Ground looping was a significant problem requiring a mobile control on the runway to assist the pilot to land. After a very long flight, the pilot was exhausted and dehydrated. His canopy was at times frosted over, so he had trouble seeing the runway, and he may have a crosswind.

In the early days, the U-2 pilot had no water nor food during the mission. He would typically lose 6 to 8 pounds of weight on an extended mission.

After landing, they were grounded for about 48 hours to get rehydrated. It was a difficult mission to pull off but unlike anything seen before. The pilot saw the curvature of the earth, and felt like he was "cruising in space, above the earth." There was little or no wind there and no turbulence. Boring but beautiful. His autopilot never worked at high altitude, so it was all hand flying, requiring high concentration and physical effort. One time Louis Setter was on final approach and could no longer feel in his arms or hands. Setter had to talk to himself while on final approach, "wheel 2 inches left, now 4 inches back__". He could not get out of the cockpit. Two people had to unhook him and lift him out and carry him to the "bread truck", working on getting the circulation going in his arms. He recovered OK but learned that day that he was operating at his limit.

Setter once experienced a flameout on a high altitude flight that he never forgot. He was at around 70,000 feet when the engine suddenly flamed out, for no good reason. His pressure suit inflated, so Setter was pressure breathing, the canopy immediately frosted over, so Setter could not see out, most of his flight instruments had suddenly quit, and there he was, gliding and in trouble. He could not attempt an airstart until he got down to 35,000 feet, which was a long way down—about 30 minutes of gliding. His suit squeezed him quite hard, but Setter was mostly concerned about getting the next breath of oxygen. He called the base and told the tower that he was flamed out and gliding back (from about 100 miles away). It took a long time to get down to 35,000 feet, the first time he could attempt a restart of the engine. It started OK, so he climbed back to high altitude and finished his mission. This was on a Friday afternoon, and he knew his friends were waiting for him to land so they could fly back to Riverside for the weekend. When he finally landed on the lakebed, they hurried him into the Twin Beech, and they flew to Riverside. Setter had gotten out of his pressure suit but was still in his long underwear and flying suit. And he probably smelled pretty sweaty at this point. When he got home, he tried to sneak into the shower so Ruth wouldn't see him, but she saw him and yelled, "what was that---?" She saw all the black and blue pinch marks on his shoulders, arms, and back from getting squeezed by his pressure suit after a high altitude flameout, (called "patequii), and was alarmed. (She was a nurse.) He couldn't tell her what caused the black and blue marks. She never learned until years later what caused them. She was not happy.

Flameouts at altitude proved to be a significant problem because it depressurized the cockpit and inflated the suit, thus preventing the pilot from talking on the radio. It forced oxygen into the faceplate and mouth at a fast rate while the aircraft slowly descended. Also, the neckpiece of the helmet often popped out, requiring the pilot to hold it in as far as possible with one hand. Besides making it necessary to fly with one hand, suit inflation stiffened the arms. The procedure created a dangerous predicament. It did not take Mr. Johnson and the Pratt & Whitney engineers long to help solve the flameout problem using a bleed valve—making the U-2 much less hazardous to fly at altitude.

Flameouts – In Their Own Words

Setter (SAC Instructor Pilot who trained the CIA and AF U-2 pilots): "When the engines flame out the crewmember's suit freezes, and the aircraft instruments quit working. There was almost no air up there, and the engines were starved for air and would quite often flame out. Without a pressure suit, I would have died within six seconds after flaming out. those tests were just scary as hell." — from past article at edwards.af.mil

Woodhull: "On takeoff, the acceleration was so powerful it was like being launched by a gigantic rubber band. The steep climb out and departure went fine. After reaching 60,000 feet, I got busy with the mission's activities. BANG! The airplane produced a violent, high-frequency vibration, with an immediate sensation of deceleration. On the cabin altimeter, the needle that indicated the atmospheric pressure in the cockpit spun rapidly toward the same altitude as the airplane. Simultaneously with the engine flameout, the capstans of my partial pressure suit inflated, squeezing my torso in their grip and forcing me into a stiff, hunchback posture." - Woodhull, Jr., Richard G., "Above & Beyond: I Have a Flameout," *Air & Space Magazine*, September 2008

With approval from the National Advisory Committee on Aeronautics (NACA)'s director Hugh Dryden, Bissell's team at the CIA developed the cover story for the U-2 that described it as used by NACA for high-altitude weather research. To support the story, U-2s several times took weather photographs that appeared in the press. The civilian advisers Land and Killian disagreed with the cover story, advising that in the case of an aircraft loss, the United States forthrightly acknowledge its using of U-2 overflights "to guard against surprise attack." The CIA did not follow their advice, and the weather cover story led to the propaganda disaster following the May 1960 U-2 loss.

The weather research cover story turned out to be accurate when the CIA sent a U-2 flown by Captain Meierdierck to Alaska to check out the high-altitude, high-latitude winds. Meierdierck kept the aircraft on course despite high crosswinds by using the drift sight, a device allowing the pilot to look under the plane to check his course and determine any drift. A C-54 flew along under the U-2 in case he encountered any difficulty.

On another flight, Captain Meierdierck departed from Watertown and flew over the Pacific Ocean to check the wind patterns when he developed engine problems, and the engine quit. The pressure suit inflated, forcing him to descend to 30,000 feet to attempt an air start. It worked, and he climbed back to altitude only to have the same thing happen again. The engine malfunction and suit inflation occurred 15 times during his return to base.

Meierdierck was sent to the David Clark Company to solve the communication problem that occurred when the engine flamed out, and the suit inflated. Clark built a new suit for him containing a bladder that placed pressure on the chest to enable the pilot to talk with his suit pressurized. To verify this, the CIA sent Meierdierck to the altitude chamber at Wright-Patterson, where they simulated an altitude of 120,000 feet. The suit inflated, and Meierdierck shouted to see if they could hear him. They heard him, solving the inability to communicate during a flameout.

A U-2 pilot always took care not to hit the front gear first when landing as it caused the plane to bounce and "porpoise" at near the stall speed. This problem occurred once on an occasion when a pilot hyperventilated as he was coming in for a landing. Lou Garvin, the mobile control officer, drove to the end of the runway to assist the landing by calling out the plane's altitude above the runway as he raced alongside in the souped-up chase vehicle.

In his hyperventilated condition, the pilot heard Garvin calling, "Two feet, two feet, one foot." The pilot said, "Screw you," and jammed the front gear onto the lakebed with the nose a foot off the ground and the tail high in the air. He conducted this impossible maneuver by leaving the power on and continuing two or three miles before he pulled the power off and let it settle to the runway. When the flight surgeons met him, they grounded him until they could evaluate his condition. The grounding set him off again. He gave them all hell for a little while until the effects of too much oxygen wore off and he returned to normal.

Despite the difficulties involved in training U-2 pilots, Colonel Yancey's cadre of six qualified Air Force U-2 pilots was ready to begin to train the CIA pilots by September 1955.

Until then, the US did not have pilots trained to fly a plane at a high altitude for the long duration needed successfully to overfly the Soviet Union. Nor were many qualified to fly a plane whose landing gear resembled a bicycle, its wheels aligned under the center of the aircraft. Not many had experience flying a plane as lightweight as a glider. None had experience with a plane that refused to land. (The RB-57D would also prove unwilling to land.)

The U-2's being a mixture of glider and jet made training pilots more difficult even for the qualified fighter pilots chosen for the overflight program. The purpose of the training program was to teach the fighter pilots to fly the delicate U-2, a revolutionary new plane design concept with large wings, and tremendous lift, but too fragile to survive the stresses of loops and barrel rolls to which the pilots were accustomed. Moreover, flying the original U-2s at sea-level limited their speed to a mere 190 knots in the smooth air or 150 knots in rough air, according to the placard placed in the plane.

At operational altitude, the much less dense air limited their speed to Mach 0.8 (394 knots). Speeds exceeding this limit would cause a shuddering called "Mach buffet." Then the tail section (attached with only three bolts) became extremely vulnerable

Airspeed remained a critical factor for the U-2

throughout the CIA program. The U-2's limiting Mach of 0.8 at optimum altitude was very close to the stall speed. The pilots called this narrow range of acceptable airspeeds at maximum altitude the "coffin corner" because, at this speed, the U-2 was always on the brink of falling from the sky. Slowing the aircraft to below the stall limit caused it to lose lift and fall with enough stress that it tore off the wings and tail. A little too much speed led to buffeting, which resulted in the loss of the tail, or wings as well.

U-2 pilot Weir later commented: "We were looking at plus or minus two-and-a-half knots at one particular point in the climb schedule between critical Mach and stall speed, so that's very, very small."

The Pratt & Whitney representative often joked that the roaring inferno of the engine at sea-level was no more than a Zippo flame at altitude. These flying conditions required a U-2 pilot's full attention when he used the autopilot, so much so that Kelly Johnson added a vernier adjustment to the throttle to allow the pilot to make minute alterations to the fuel supply. Normal hand action would precipitate either engine overheat or flameout.

Weir observed: "You got into the climb-Mach schedule, and you engaged 'Mach hold'; you had a small Mach trim wheel so you could tweak it just a little—more Mach or less Mach. And it maintained that climb schedule. It was almost impossible to 'hand-fly' it—disengage the autopilot and try to fly it up through there with the yoke."

U-2 Navigation circa 1955

In the first CIA training program for their U-2 pilots, they used dead reckoning (magnetic heading and ground speed) and pilotage (looking out the window) plus celestial navigation at high altitude. Neither GPS nor Inertial Navigation was available in those days. The airplane had a viewport directly in front of the pilot so he could view what was below him in reasonable detail, through a small bubble on the bottom of the fuselage. He could slew the optics around to look forward, back, right, or left. If he pulled a handle a mirror flips, so he was now looking up through his sextant bubble, which was located on top of the fuselage directly in front of the canopy.

In the old days, a navigator on a ship or large airplane took a four-minute or longer shot of a star or the sun or the moon. The pilot then used two fat navigation books to compute his position was at the time of the shot. This took a long time and required a good steady autopilot. Without an autopilot, celestial was not possible because the airplane moved around too much, and the movement was not repeatable, as with an autopilot.

The U-2 pilot, now cruising at high altitude on autopilot, referred to his precomputed celestial shot readings. He set his sextant to the given azimuth (direction) and elevation, and adjusted the sextant manually so the star (or sun or moon) was precisely in the center of the sextant bubble (looks just like the bubble in a carpenter's level in your workshop), reading the exact time, and kept adjusting the sextant to keep the star in the bubble over a one minute period. With this sextant reading, he then looked at his map, on which the navigator had drawn lines of position for that particular star at that exact time. The pilot marked a line on his map the average elevation his sextant had shown for the one minute shot. One shot did not show a position—it showed a line of position, or LOP, and he was somewhere on that line at the time of the shot. If his shot was over his left or right wing, the LOP line would run parallel to his flight path and would show if he was left or right, of course. Say the LOP indicated he was 30 miles left, of course. Then he would compute a small heading change to the right so that one hour later, when he took another shot left or right of his airplane, he should have corrected back to course. If not, he would make another course correction. This procedure continued for the remainder of his flight—a celestial shot every half hour.

If the shot azimuth was straight ahead or over the tail, the LOP showed whether he was ahead or behind his planned ground speed. He then recomputed his arrival time at his position one hour from now. So every half hour, the pilot took another sextant shot, alternating with a speed shot (over the nose) and a course shot (over the left or right wing). Before takeoff, a navigator had precomputed all these sextant shots for the complete mission, assuming the pilot taking each

shot at the time shown in his log. An on-time takeoff was mandatory. Otherwise, the celestial data was unusable.

The wind at roughly 70,000 feet was often calm or less than 10 knots at low latitudes (i.e., California). Reports were that near the Arctic Circle, winds could be much higher, up to 50 knots or more. With very low winds and the strict maintenance of cruise speed by the pilot and autopilot, flying time and heading were usually very accurate. Another problem at high latitudes (near the North Pole) was that magnetic compass variation changes rapidly, causing significant errors in true heading if the pilot was not aware of this fact. Most U-2 flights were accomplished with high navigation accuracy.

According to records, the SAC Instructor Pilots participated in the development of the integrated drift sight/sextant, the Mach control, TAS (True Air Speed) computer, time and distance computer, and a "how-goes-it" curve for fuel consumption.

The old "bird dog" direction finder was the main navigation aid other than the integrated drift sight/sextant. Baird Associates developed the sextant for making celestial fixes during the long U-2 flights. With clouds often covering navigational points on the earth, the sextant became the pilot's principal navigational instrument during the first years of deployment. Otherwise, the periscope 'hole in the floor' proved accurate for navigation. During U-2 development tests, pilots found they could navigate by dead reckoning with an error of less than one nautical mile over a 1,000-NM course.

'Porpoising' – In Their Own Words
Meierdierck: "The one thing that you had to be very careful of was not to hit the front gear first as that would make you porpoise; there was a great possibility of breaking the plane."

Bogash: "The airplane could fly at Idle Thrust (actually it could CLIMB at Idle Thrust) and didn't want to land. Landing on the front wheel caused a porpoise."
- http://rbogash.com/U-2/U-2.html

Stratton: "My Instructor Pilot (IP) Ed Dixon

repeatedly emphasized the fact that the bird would not land unless it was held off the runway until it stalled and 'dropped' to the runway. Dixon drove home to me the fact that if I touched down above stall, I would be airborne again and 'bullfrogging' down the runway." - McIlmoyle p. 60

Martinez, on the training of the Republic of China pilots: "This particular time, his Chinese student was listening intently to Eddie's explanation. Eddie said, 'I don't want you to start "crow hopping" from the main gear to tailwheel, or you're going to be in real trouble. Do you understand?' The ROC pilot answered with a puzzled look, 'What do you mean, "crow hop"?' Then Eddie proceeded to demonstrate with his hand bouncing up and down on the table. We all looked at each other and had a good laugh." - McIlmoyle p. 62

(porpoising = bullfrogging = crow hopping)

Weir: "When you just barely touch the main gear on the runway, the weight of the engine will make the tail go down and immediately you generate an enormous amount of lift. The nose comes up, the lift increases, and you are back in the air. The tendency is to push the yoke forward to decrease the lift and put it back on the runway. Well, you're always a day late and a dollar short when you do that, and you're behind it, and the airplane is going down when you think it should be going up, and it's going up when you think it should be going down, and you get a big porpoise going. Then it bounces down the runway, and it stalls out, and it crashes. And that's where so many guys got in trouble and beat up the airplane." - Kenneth Weir interview, *Air & Space*, March 2012

Congress Briefed on AOUATONE

Although a guarded secret within both the CIA and the Eisenhower administration, DCI Dulles decided to tell a few key members of Congress about the U-2 project. On 24 February 1956, Dulles met with Senators Leverett Saltonstall and Richard B. Russell, the ranking members of the Senate Armed Services Committee and its subcommittee on the CIA. He shared with them the details of Project AQUATONE and asked their opinion on informing some members of the House

of Representatives.

Because of the senators' recommendation to brief the senior members of the House Appropriations Committee, Dulles met later with Representatives John Taber and Clarence Cannon.

Official Congressional knowledge of the U-2 project remained confined to this small group for the next four years. The House Armed Services Committee and its CIA subcommittees did not receive a CIA briefing on the U-2 Project until after the loss of Francis Gary Powers' U-2 over the Soviet Union during May 1960.

Celestial Navigation – In Their Own Words
Meierdierck, U-2 instructor pilot: "Prior to the central Atlantic crossing, [Cocoa Alpha flights] I was the Program Manager in charge of the USAF program to test celestial navigation for jet fighters. We would pre-compute what our sextant shots should be on our route, make good a takeoff time and then when I took a shot of the sun or moon, during daylight hours, I could determine by my numbers, whether I was right or left of course or faster or slower than planned. I had proven the theory by taking off from Turner AFB in an F-84, going under the hood immediately and flying instruments all the way. At a preset time, I took my sextant shots, made a determination as to my position, made a 90-degree right turn and let down. At 5000 feet, I took the hood off, and Bergstrom AFB in Texas was right in front of me."

Setter, U-2 instructor pilot, from his résumé: "1952 flew F-84G across the Pacific Ocean using celestial navigation" (via Roadrunners Internationale)

McIlmoyle: "Contemplating flying the U-2 on 3,600 nautical-mile flights without navigation aids really got my attention. I understood that celestial navigation would be critical to successful mission completion. We knew that the U-2 on autopilot was seeking the set airspeed and heading. This 'seeking' had a sine wave curve of approximately two minutes' duration. We had to keep the star centered in the bubble for two minutes and then average out the elevation readings recorded by centering and aligning the sextant's indices. . the best navigation approach was a three-star fix with one of the three stars as close to my line of flight as possible. this would give me a good speed line. The other two stars should be preferred so that their Line of Position (LOP) formed as close to an equilateral triangle as possible. The small triangle derived from dropping a perpendicular from the apex of the triangle to the opposite side would be my most probable position and close to my real location." - McIlmoyle p. 88

Brown: (missions out of Alaska) "The worst thing about flying north and using grid navigation was the need to outguess the gyro precession rate and in which direction. Would they proceed in the direction forecasted for the next rating period? Also, remember, once over the extreme northern latitudes, every direction is south. The gyro should indicate the correct south longitude heading for the next flight route segment, and not one headed south. The navigator/mission planner pre-computed all my fixes to be three-star fixes. It made for extra work. For the next three-plus hours, I was constantly writing down the C-121 transmitted numbers, evaluating their position information, taking three-star fixe,s and evaluating one or the other for my position. I was using the Rudd Star Finder to plot my fixes, so it cut down on time to plot and evaluate my position." - McIlmoyle p. 302-304

Maultsby: auroras foiled sextant readings on missions out of Alaska - McIlmoyle p. 337

Herman: "This route went directly over the magnetic North Pole and required Grid Navigation. We neared the magnetic pole, and the aircraft went into a slight bank to hold heading due to rapid changes in variation. Rather than upset the gyro, we went to free run on the slaved gyrocompass and used that heading as our reference. Our takeoffs were planned around 7:00 a.m. so we could land in daylight and tail in on the ice-covered runways and taxiways. We relied on the sextant for navigation, so at the star,t we hoped for a good shot of a well-known body." - McIlmoyle p. 308

 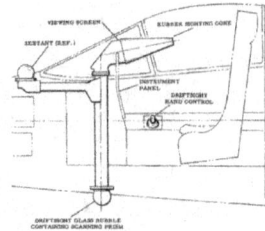

Sextant mockup Sextant / drift sight system

McIlmoyle: "Thirty minutes went by, the limit upon which preplanned celestial readings were considered useable. later I followed Steve, also minus the celestial charts. Steve, Chuck and I dead reckoned our way across the beautiful blue Pacific Ocean. " - McIlmoyle p. 347

The Impact of the Air Force Project GENETRIX Balloon

While the CIA made its final preparations for U-2 overflights, the Air Force began a reconnaissance project causing considerable protest around the world and threatening the existence of the U-2 overflight program before it even began.

Project GENETRIX, a project originating from the 1951 RAND Corporation study, involved the use of camera-carrying balloons to obtain high-altitude photography of Eastern Europe, the Soviet Union, and the People's Republic of China.

By the end of 1955, the Air Force overcame numerous technical problems in camera design and recovery techniques and manufactured hundreds of balloons and cameras for use in the project. President Eisenhower gave his approval on 27 December 1955. Two weeks later, launches began from bases in Western Europe. The Air Force launched 516 balloons by the end of February 1956.

Project GENETRIX proved far less successful than its sponsors hoped. Once launched, the balloons flew at the mercy of the prevailing winds, missing the prime target areas, which lay in the higher latitudes. Instead, many drifted toward southern Europe, across the Black Sea, and the desert areas of China.

Scores of balloons failed to cross the Soviet Union and China because of hostile aircraft shooting them down, or for prematurely expending their ballast and descending too soon.

Of the 516 balloons launched, the Air Force

recovered only 46 payloads (one more than a year later and the last not until 1958). The camera malfunctioned in four of the payloads and contained no intelligence value in another eight. Thus, only 34 balloons succeeded in obtaining useful photographs.

The GENETRIX balloons encountered more severe problems than the low success rate. The balloon flights provoked a storm of protest and unfavorable publicity, even with the Air Force issuing a cover story concerning using the balloons for weather research connected with the International Geophysical Year,

The East European nations protested to the United States and international aviation authorities, claiming the balloons endangered civilian aircraft. The Soviet Union sent a strongly worded protest note to the United States and to the nations from which the balloons launched. The Soviets also collected numerous polyethylene bags, cameras, and transmitters from GENETRIX balloons and put them on display in Moscow for the world press to see.

According to a summary memo to Dulles, General Cabell asked for a stop to GENETRIX flights because of the "additional political pressures being generated against all balloon operations and overflights." The amount of publicity and protest led President Eisenhower to conclude the balloons were giving more grounds for irritation than providing benefits. He ordered the project halted.

On 7 February 1956, Secretary of State Dulles informed the Soviet Union of the United States' deciding not to release any more "weather research" balloons. He did not offer an apology for the overflights. Note that this was in the same timeframe that Detachment A was training full time and preparing to become operational in Europe. Also during February, personnel from the SAC 26th Strategic wing staged to Thule, Greenland ahead of 16 RB-47E photoreconnaissance aircraft about to penetrate the Northern USSR in the bold covert operation called Project Home Run.

Despite this and the furor caused by GENETRIX, Air Force Chief of Staff Twining proposed yet another balloon project only five weeks later, in mid-March 1956. This project employed even higher-flying balloons than GENETRIX. President Eisenhower put a stop to the idea, however, informing the Air Force that he was "not interested in any more balloons."

Although Project GENETRIX gained limited quality photo intelligence, it still rates as some of the best and most complete photography obtained over the Soviet Union since World War II. It was used as "pioneer" photography because it provided a baseline for all future overhead photography, including by the U-2. Even innocuous photos of forests proved valuable in later years when the U-2 and satellite photography revealed construction activity.

The data obtained by NATO and US radars as they tracked the paths of the balloons at an average altitude of 45,800 feet over the Soviet Bloc proved of still greater importance to the U-2 program. This data provided the most accurate record to date of high-altitude wind currents that knowledgeable meteorologists later used for determining optimum flight paths for U-2 flights.

One fortuitous development coming from Project GENETRIX had nothing to do with the cameras. It involved a steel bar serving a dual purpose. The bar secured the top rigging of the huge polyethylene gasbag with the camera payload. By sheer chance, the length of the bar—91 centimeters—corresponded to the wavelength of the frequency used by Soviet radar. Known by its NATO designator as TOKEN, the Soviet forces used this S-band radar for early warning and ground-control intercept.

The bar on the GENETRIX balloons resonated when struck by TOKEN radar pulses, making it possible for radar operators at the US and NATO installations on the periphery of the Soviet Union to locate several unknown TOKEN radars. These radar findings, coupled with other intercepts made during the balloon flights, provided extensive data on the Warsaw Pact radar network, radar sets, and ground-control interception techniques.

Analysis of these intercepted findings revealed the altitude capabilities and the accuracy of Warsaw Pact radars. It also revealed the methods used by the nations to notify each other of the balloons' passage (handing off) and the altitudes at which Soviet aircraft could intercept the balloons. All this information applied to future U-2 missions.

These positive results from Project GENETRIX did not outweigh the political liabilities of the international protests. The amount of ill will generated by balloon overflights concerned the CIA officials. It soured the Eisenhower administration on all overflights, including those by the U-2, now ready for deployment.

Therefore, DDCI Cabell wrote to Air Force chief of Staff Twining during February 1956 to warn against further balloon flights because of the "additional political pressures being generated against all balloon operations and overflights, thus increasing the difficulties of policy decisions which would permit such operations in the future."

In addition to its concern for the future of the U-2 program, the CIA feared President Eisenhower's anger at balloon overflights might result in the curtailment of the balloon program of the Free Europe Committee, a covert Agency operation based in West Germany to release propaganda pamphlets over Eastern Europe.

During December 1955, Kelly Johnson wrote, "Our anniversary. We have built four flying airplanes, have the ninth airplane in the jig, and have flown over the design altitude any number of times. We have trained crews and are developing the Bakersfield factory. It has been quite a year."

The program progressed with the Lockheed Aircraft Co, Air Force, and CIA student pilots learning more concerning the airplane each day. The Air Force instructor pilots were soon able to declare each student pilot mission and combat-capable and the detachments ready to deploy.

Finalizing the U-2 Cover Story

The rage and controversy over balloon flights continued into February 1956. Feeling the heat from Project GENETRIX and with the U-2 completing its final airworthiness tests, Richard Bissell and his staff realized the need for a cover story for overseas operations. They knew the glider wings, and the odd landing gear was bound to gain curiosity. They decided on NACA weather research as a plausible reason for deploying such a plane.

A weather research cover story would need the approval of all concerned: Air Force Intelligence,

the Air Weather Service, the Third Air Force, the Seventh Air Division, the Strategic Air Command's U-2 project officer, the Air Force Headquarters' project officer, and NACA's top official, Dr. Hugh Dryden. Bissell also consulted with the CIA Scientific Advisory Committee concerning the coverage plan. The CIA's Photo Intelligence Division grew to prepare for the expected flood of photos.

With approval from the NACA's director Hugh Dryden, U-2s on several occasions took weather photographs that later were released to the press. By the end of March 1956, senior CIA officials and the other agencies involved had approved the definitive version of the overall cover story. This left the project staff to work on contingency plans for the loss of a U-2 over hostile territory.

Bissell advised the project cover officer to produce a document setting forth all proposed actions. He wanted all press releases to use the cover story. He wanted the suspension of operations and at least an indication of there being some proper diplomatic action. The cover officer prepared emergency procedures based on the overall weather research cover story, and Bissell approved these plans and ensured one final high-level look at the cover story on 21 June 1956.

This high-level look was on the day after the first U-2 mission over Eastern Europe. Bissell met with General Goodpaster, James Killian, and Edwin Land to discuss the pending overflights of the Soviet Union. Killian and Land disagreed with Bissell's concept, including proposed emergency procedures, and made a much bolder and a more forthright proposal in the event of the loss of a U-2 over hostile territory:

Rather than to deny responsibility, the United States should state that it was conducting the overflights "to guard against surprise attack." They all put this proposal aside for further thought, which it, however, did not receive. Bissell's weather research cover remained the basis for statements made after a loss.

The project staff continued to prepare several different statements for use in various scenarios. Included was what to say or utilize in the event of a captured pilot. Even in such a case, the proposed policy remained where the United States would stick to the weather research story, a course of action proving disastrous in the May 1960 shootdown of CIA pilot Francis Gary Powers.

A committee of Army, Navy, Air Force, CIA, NSA, and State Department representatives created lists of priority targets. The U-2 project received the list and drew up flight plans that enabled the committee to provide a detailed rationale for each plan for the president to consider as he decided whether to approve it.

By January 1956, everyone working on Project AQUATONE saw the U-2 nearing operational deployment. The aircraft met all criteria established in late 1954. Its range of 2,950 miles was enough to overfly continents, its altitude of 72,000 feet was beyond the reach of all known antiaircraft weapons and interceptors, and it carried the finest camera lenses available.

The primary targets for the U-2 lay behind the Iron Curtain. Bissell and his staff considered this and looked for operational bases in Europe. The CIA felt America's closest ally, the United Kingdom, as a logical choice for a first U-2 base location.

Bissell saw the British as having something more to offer than merely their bases. In a remarkable act of lèse-majesté, Bissell found a way to carry out the U-2 missions without consulting the president. He approached the British intelligence service and the Royal Air Force about undertaking U-2 flights on their own. When the British asked if they would need Eisenhower's permission, Bissell replied, "Not at all. That's not the play."

On 10 January 1956, Bissell flew to London to discuss the matter with the Royal Air Force (RAF) and MI-6 officials. Initially, they responded favorably. Nonetheless, they referred Bissell to a higher level for approval of the proposal. Bissell reported his findings to DCI Dulles, who arranged to meet with Foreign Secretary Lloyd in London to explore the possibility of winning the British government's approval for the project.

Dulles presented his case to Lloyd on 2 February, and, by early March, Lloyd approved the basing of U-2s in the United Kingdom, suggesting the U-2s use Lakenheath Air Force Base that the USAF Strategic Air Command already used.

During March 1956, Colonel Ritland returned to the Air Force as the deputy project director followed by Col Jack A. Gibbs. During March 1956, Colonel Landon McConnell took command of Watertown, and the CIA director Allen Dulles visited the flight test facility to meet the first training class.

Colonel Gibbs, who retired a brigadier general, was born in Canton, Ohio in 1912. In 1936, he earned his BS in mechanical engineering with a specialty in aeronautics from the Oregon State College. Gibbs received the prestigious W.E. Boeing Scholarship to attend Boeing School of Aeronautics, obtaining a multi-engine transport license in 1937. Rather than joining United Air Lines, he accepted a commission in the US Army Corps of Engineers with an assignment to Fort Lawton, Washington.

In 1938 Gibbs applied for US Army Air Corps flight school and graduated in the Class of 39-B. He received an assignment to the CIA as the deputy project director of the U-2 under Richard Bissell. He would also become the CIA's engineering manager for R&D efforts under the Project RAINBOW to reduce the radar detectability of the U-2 and assess the feasibility of developing a new aircraft with lower RCS characteristics under Project GUSTO. He would receive the Legion of Merit in 1958 for his performance in this assignment. In 2010, the NRO would posthumously award him The Pioneer of National Reconnaissance Medal for a career in reconnaissance spanning from 1942 until 1966.

As of April 1956, the fleet consisted of nine aircraft and the CIA contract pilots undergoing flight training at the Ranch.

For a time before this point, Lockheed's own DC-3 made almost daily flights to Watertown with a hand-picked crew of flight line mechanics. As construction proceeded, other contractor personnel and advance personnel became ready to report in preparation for training of the first CIA detachment; activity increased, a major logistic problem arose: how to transfer more project employees from Burbank to Watertown without arousing much curiosity.

The project staff decided the simplest approach was to fly the essential personnel to the site on Monday morning and return them on Friday evening. This was a decision that would almost wipe out much of the key management of the U-2 program.

U-2 at the Ranch - NACA "weather plane" markings on the wing

Chapter 12 – Mt. Charleston Crash

By 3 October 1955, landline communication between the Groom Lake facility and Burbank was established, also a MATS shuttle flight using a C-54 with a security-cleared USAF crew. Flight testing of the U-2 was shut down for twelve days to allow construction work in the second half of November 1955. (Construction personnel lacked a security clearance to view the U-2 and its operation.)

During October 1955, Lockheed added test pilots Sieker and Schumacher to the flight-testing program at Watertown. Their presence demonstrated the considerable progress made in preparing for the first Agency U-2 pilots.

The civilian version of C-54 was called the DC-4n used by President Roosevelt like today's Air Force One

Lockheed was now scheduling frequent flights to bring in supplies and visitors from contractors and the CIA headquarters. Following the decision to commute the workers to Watertown, the Air Force initiated a scheduled Military Air Transport Service (MATS) flight using a USAF C-54 aircraft. James Cunningham dubbed this activity "Bissell's Narrow-Gauge Airline." During November, less than seven weeks after it began, Douglas MC-54M Skymaster, 44-9068, c/n 27294/DO240 was one of the transport planes attached to the 1700th Air Transport Group, of MATS, at Kelly AFB, Texas.

On November 17, 1955, C-54 Skymaster 44-9068 with five military personnel departed Norton Air Force Base, California carrying also nine civilians who it picked up from the Lockheed "Skunk Works" at Burbank, California for a routine flight to Watertown. Their mission was the testing of the U-2 spy plane under construction at Lockheed. It was a top-secret project that would allow the US to fly undetected over enemy sites to photograph the installation of missiles capable of carrying nuclear weapons.

Because of the secrecy involved with the Lockheed U-2 project, it was routine for the C-54 crew to maintain radio silence and not to contact Air Traffic Control. At this hour, the passengers were most likely asleep when a few miles northwest of Las Vegas, Nevada and as the plane approached the Spring Mountain Range, a severe blizzard blew the plane off course.

From Burbank to the California/Nevada border at Goodsprings, Nevada, the plane went with winds blowing towards the east due to a strong Low-pressure center in the atmosphere over southern Nevada. At the Goodsprings beacon their flight path turned north (left). Heading north they flew parallel to the Spring Mountain Range on the west side of the range, heading toward Watertown. At Goodsprings, to avoid tracking while flying in and out of Watertown, military orders required they shut off all navigational instruments, drop below a 10,000-foot elevation ceiling, maintain radio silence and fly only by visual means. This created a dangerous situation since the Spring Range varied from 10,000 feet to nearly 12,000 feet at its peak – Mt. Charleston.

When they turned north at Goodsprings, the prevailing wind became a crosswind that was 80% stronger than forecast for the Las Vegas area. Without realizing it, the crosswind pushed them further east toward the Spring Range. Without navigational instruments and due to the early winter storm, USAF 9068 drifted off course and found itself on the east side of the Mt. Charleston peak, rather than on the flight plan which should have taken them safely west of the Spring Mountains.

In a blinding blizzard and lost in the clouds, an error in plotting the position of the Skymaster over the Spring Mountain range caused the plane to crash at 0819 hours only 30 feet below the crest of an 11,300-foot ridge leading to the peak of Mount Charleston. All on board died instantly.

It may be of interest to note that the same Sheriff's Mounted Posse that retrieved the bodies of USAF 9068 on November 20, 1955, had on January 16, 1942, recovered the body of Carole Lombard at Mt. Potosi, Nevada. While the posse was on Mt. Potosi Lombard's husband Clark

Gable waited in the bar at nearby Goodsprings, Nevada. Mt. Potosi is also in the Spring Mountain range just 30 miles south of Mt. Charleston.

When the medical staff at Nellis AFB examined and identified the 14 persons involved in the C-54 accident, the examining doctor noted that one the watches from the victims had stopped at 0820. Lt Col George H. Pittman, Jr. observed the same and determined the crystal crushed on the hands of the watch, stopping their movement. Since the accident occurred between 0800 and 0830, he suggested that 0819 hours and 10 seconds might be the exact time of the crash. The accident took the lives of key CIA and Hycon personnel and dealt a blow to the fledgling U-2 project. Killed in the crash were:

Pilot	USAF	George M. Pappas, Jr., 27
Co-pilot	USAF	Paul E. Winham, 24
Flight Mech	--	USAF Clayton D. Farris, 26
Flight Attendant	--	USAF Guy R. Fasolas
Unknown	--	USAF John H. Gaines, 23

Civilians lost included:

Harold Silent, 59	Hycon
Fred Hanks, 35	Hycon
Rodney Kremendahl, 38	Lockheed
Richard Hruda, 37	Lockheed
James F. Bray, 48	CIA
Terence O'Donnell, 22	CIA
James W. "Billy" Brown	CIA
Edwin J. Urolatis, 27	CIA
William Marr, 37	CIA

When the plane failed to land at Watertown, the CIA and Air Force frantically initiated a search. Searching for a secret aircraft, carrying secret documents and personnel to a secret facility proved delicate. No one could identify the type of plane, who was flying it, who was on board, where it originated, or its destination.

During the afternoon of Thursday 17 November 1955, as USAF Capt Hank Meierdierck, SAC Instructor Pilot at Watertown relates: "We heard of the crash and sent all our airplanes on the search mission. I happened to be the one that discovered the site, very close to the crest of Mt. Charleston."

The Spring Range was experiencing the worst weather conditions seen in years. Those conditions were bad enough for the US Forest Service to predict the crash site would remain inaccessible until the middle of June. The CIA found this, a seven-month delay, unacceptable for the secret U-2 project.

Two Air Force parachute rescue teams arrived from March Air Force Base, California with the intention of offering first aid to anyone who survived the impact. The Air Force aborted the paratrooper drop because of the intense wind conditions still in force at the top of the mountain.

Instead, the Air Force deployed the 42nd Air Rescue Squadron, a trained mountaineer team from March Air Force Base under the command of Col Frank Schwikert.

After several days of struggling through the deep snow and sub-zero degree weather while using skis and snowshoes, this team aborted their attempt to climb the north side of the mountain in time to save any survivors. The team became bogged down in the deep snow as it returned to base camp.

Las Vegas Sheriff Butch Leypoldt also deployed his mounted posse to assist in military rescue efforts that included two paratroopers from the original March AFB team. Now, no one realistically expected to find any survivors. Nonetheless, providing first aid to any possible survivors remained a necessary priority. Recovery of classified documents and body recovery was a secondary priority for the Air Force and CIA.

The Sheriff's Mounted Posse had no idea of the top-secret documents scattered at the crash site, or the identity of the crash victims for that matter. It would stay that way for nearly half a century.

The first rescue party, the paramedic group from March Field, California, headed by SSgt Donald S. Pipes and including SSgt Kenneth Woods, SSgt Derald Parks, A/1C Robert Taylor, and A/B Gilbert Seeburger, left Friday morning on foot and wearing snowshoes.

One of the posse members became severely ill during the climb and returned down the mountain, aided by the other three members of the posse. The two parachute rescuers continued their climb. They reached the crash site and radioed base camp, confirming no survivors and their intent to remain near the site to wait the arrival of a horse party headed up the mountain.

Ranger Hank Hoffman, Deputy George Dykes, SSgt Walter F. Adkins, and A/1C Gordon Bailey of the 2nd Air Rescue Unit, plus Deputy Sheriffs Melvin Scholl and Vernon Bosserman headed out on foot early Saturday morning as a second rescue party.

At six am, Sheriff Leypoldt headed out with another group of Sheriff's Mounted Posse members. This group included two Air Force colonels along with deputy sheriffs Roy Neagle and Floyd Hayword. Others in the sheriff's posse were Eldon and Harold Ballinger, Earl Murdel, Ray Gubser, Sr., Frank Scott, Merle and Vivian Frehner, Charles Steel, Bill Stratman, Ed B. Taylor, Dr. Robert Clark, J. E. Williams, Russ Walters, Newell Knight, Hagen Thompson, Duain Titus, and Pat McDowell.

The top secrecy of this accident posed many problems for the CIA and Air Force. The CIA was extremely concerned about the top-secret documents and equipment needing recovery before the civilian rescue party approached the wreckage. Securing the area now became the mission.

To this end, Colonel Schwikert from March AFB and Colonel Pittman from Norton AFB met up with Sheriff Leypoldt, 15 members of the Sheriff's Mounted Posse and 17 horses at the base camp at the bottom of the mountain. They briefed the sheriff to ensure that he and the posse stood down while they attended to classified matters. Schwikert recalls, "AF Headquarters was in touch with me concerning the classified material aboard. They impressed on me the necessity of speed in securing the area."

At 0600 hours, on a cold November Sunday, the group departed base camp, taking a longer route to the south of the mountain. They traveled up a switchback trail with snowdrifts so deep the horses' feet never touched the ground. The riders at times dragged their feet in the snow behind their saddle or dismounted to slide on their stomach while hanging onto the horses' tails. Conditions became worse with the rescue team realizing they had no water to drink and had only a few cans of Spam for food.

Sheriff Leypoldt led the rescue party up this old hiking trail in snow varying up to six feet deep. In temperatures below zero and drifts twenty feet deep, the rescue party worked its way up the mountain in increasing wind velocities that covered them with powdered snow. The horses often slipped off the narrow trail, at times, pinning their riders beneath them in the snow. The rescue party, traveling on skis and snowshoes, overtook and joined the first Army rescue team six miles from the plane¾finding them so cold and miserable, they ordered them to return to the base camp.

The rescuers traveled along the ridge near the summit to the crash site in the freezing wind. The shallow snow turned to ice that the horses' hooves broke through, cutting their legs. The rescue party arrived at the wreckage at 1300 hours, where they stood down, shivering in the blizzard wind and snow to allow the sheriff and Colonel Schwikert time to inspect the crash site and the colonel to collect classified material. When finished with this, Colonel Schwikert turned the scene over to the sheriff.

The pilot appeared to have seen the mountain at the last moment and tried to climb over it. It was too late. The plane disintegrated against the mountain, causing cargo and ten passengers to erupt through the top of the cabin and scatter forty or fifty feet in all directions. The plane's motors lay 20 or 30 feet from the aircraft and its nose and wings lying downslope in front of the fuselage tail.

The sheriff's posse had to break the frozen bodies to bend them for tying them onto the saddles of spooked horses for the trek back down the mountain. The first group of five rescuers departed at 1300 hours, on foot leading the horses through a cold blizzard, down slippery slopes. Several times, the horses carrying the recovered bodies and lunging through the snow slipped and rolled down the slopes.

It became necessary for the rescuers to clear the trail with a shovel, again breaking the frozen bodies to keep them from extending to the sides of the narrow trail and catching in the brush. In one incident, one of the horses slipped. It fell and rolled down the mountainside. The rescuer managed to return his horse on its feet but was too exhausted to put the body back on the horse.

The sun set with the blizzard howling around

the rescue party with no flashlights. They groped in the darkness, hanging onto the saddle to prevent their staggering and falling off the steep mountainside. They finally met up with the half-frozen snowshoe troops from March Field. The March Field troops were in their third day of encampment in small pup tents on the side of the frozen mountain.

The 20-hour rescue and recovery effort ended with the Sheriff's Mounted Posse volunteers taking an oath of secrecy concerning everything they had seen. Following this, the greatest single loss of life in the entire U-2 program, Lockheed assumed the responsibility of transporting personnel to Watertown using a Lockheed-owned C-47.

The Air Force Rescue Party Members:
MSgt Kenneth Woods (First-Sergeant)
Sgt Donald Pipes   Pararescue
SSgt Derald Parks   Pararescue
SSgt Walter Adkins   Pararescue
A/B Gilbert Seeburger   Aeromed
A/1C Robert Taylor   Aeromed
A/1C Gordon Bailey   Pararescue
  The first party:
  SSgt Donald S. Pipes (Officer in charge) 42 Air Pararescue
  SSgt Donald D. Parks 42 Air Pararescue
  MSgt Kenneth Woods 42 Air Pararescue
  A/1C Robert L. Taylor 42 Air Pararescue
  A/1C Gilbert Seeburger 42 Air Pararescue

The second party:
Ranger Hank Hoffman
Sheriff's Deputy George Dykes
Deputy Sheriff Vernon Bosserman
Deputy Sheriff Melvin Scholl
SSgt Walter F. Adkins, A/1C Gordon C. Bailey --42nd Air Rescue Sqdn
The third party:
 Lt Col George Pittman
Lt Col Frank Schwikert
Clark County Sheriff Butch Leypoldt
Deputy Sheriff Roy Neagle
Deputy Sheriff Floyd Hayword
 Volunteer Vivian Frehner
Members of the Mounted Posse:
Eldon and Harold Ballinger

Russ Walters
Earl Murdell
Pat McDowell
Ray Gubser, Sr.
Frank Scott
Merle Frehner
Charles Steel
Ed B. Taylor
William Stratman
Dr. Robert Clark
J. E. Williams
Newell Knight
Hagen Thompson
Duain Titus

Technical Challenge to High-Altitude Flight

AEC and Reynolds Electrical and Engineering Company (REECo) engineers spent 17-19 November 1955 in the Las Vegas AEC office laying out plans for aircraft parking aprons and tie-downs, dispensary addition, and other work at Watertown. The accidental crash of the C-54 shuttle on Mt. Charleston on 17 November understandably affected work schedule at the base. By 19 November 1955 surveyors were staking out parking aprons and a taxiway. The control tower, ladders and security tower, fabricated at Camp Mercury on the adjacent AEC nuclear test site, were up at this point, the Quonset building erected and the warehouse started by about 19 November. Larger parking pads with tie-downs also laid to accommodate three C-47s, four T-33s, and one each C-54 and C-124 transport.

With numerous other installations made, by the end of November, approximately 60 contractor's personnel billeted at the camp. Lack of housing accommodation prevented use of a larger construction crew while U-2 testing shut down for the infrastructure addition.

The Lockheed engineers at Watertown encountered and solved many problems while getting the U-2 aircraft ready to deploy. One such issue involved obtaining a fuel that would not boil off and evaporate at the plane's high design altitudes.

Retired USAF General James H. Doolittle, long involved in overhead reconnaissance and more recently as a member of the TCP, solved this problem. Now a vice president of the Shell Oil Company, he arranged for Shell to develop a special low-volatility, low-vapor-pressure kerosene fuel for the craft.

Shell produced a dense mixture, known as LF-1A, JP-TS (thermally stable), or JP-7, aka "Cough Syrup," with a boiling point of 300°F at sea-level. Manufacturing this fuel required petroleum by-products that Shell used to make its "Flit" fly and bug spray. Shell produced several hundred thousand gallons of LF-1A for the U-2 project in the spring and summer of 1955 by limiting the production of Flit, causing a nationwide shortage.

Another problem involved special tanks and modifications to the aircraft's fuel control and ignition systems because of the new fuel's density.

Even more important than the issue of boiling fuel became the problem of boiling the pilot's blood when reaching altitudes that exceed the Armstrong limit of around 62,000 feet—where fluids in the human body vaporize unless the body kept under pressure. Furthermore, the reduced atmospheric pressure placed considerable stress on the U-2 pilot's cardiovascular system and deprived the pilot adequate blood oxygenation.

Keeping the pilot alive at the extreme altitudes required for overflights, therefore, called for a different approach to personal environmental equipment. Pilot survival during high-altitude flights required a system that maintained pressure over much of the pilot's body. The system prevented hypoxia, a pathological condition, from depriving the body or a region of the body of adequate oxygen supply. The solution required

technology enabling U-2 pilots to operate during extended periods of reduced atmospheric pressure. *(The solution later played a significant role in the manned space program.)

The CIA sent out to experienced Air Force doctors, Col Donald D. Flickinger and Col W. Randolph Lovelace for help on high-altitude survival.

Dr. Lovelace, co-inventor of the standard Air Force oxygen mask, had begun his research on a high-altitude flight before World War II. In the early 1950s, he and Flickinger had made daring parachute jumps from B-47 bombers to test pilot survival gear under extreme conditions. Flickinger, at this point, had been the medical advisor on the subject for a decade.

Flickinger and Lovelace suggested the CIA ask the David Clark Company of Worcester, Massachusetts, a manufacturer of environmental suits for Air Force pilots, to submit designs for more advanced gear for the U-2 pilots.

NACA pilot in Clark T-1 suit

David Clark Company

In 1941, David Clark Company pioneered the development of air and space crew protective equipment that supported every manned high-altitude/space program since that date. In peacetime, the company manufactured posture-correcting corsets for women.

Establishing an aero-protective division in World War II for the manufacture of the first standard anti-G valves and suit hose connectors for allied fighter pilots, this wholly-owned subsidiary of David Clark Co., Inc. now would provide the U-2 AQUATONE and the following A-12 OXCART programs with specialized suits, connectors, and related equipment for its pilots.

The polycarbonate "bubble" pressure helmet epitomized the extraordinary quality and reliability of the company's products, continuing with visor assemblies developed for the Apollo spacesuit program later and the Space Shuttle astronauts and

International Space Station spacewalking astronauts today.

David Clark expert and U-2 program manager Joseph Ruseckas had joined David Clark Company in 1948 after an early career as a professional pilot and member of the Army Air Corps. At David Clark, he developed a complex life-support system, the first partial pressure "spacesuit" for keeping humans alive for lengthy periods at ultra-high altitudes. He helped in pioneering the development of pressure suits for the early rocket-powered X-plane test pilots.

The effort to provide a safe environment for pilots at high altitudes also involved the Firewel Company of Buffalo, New York, which pressured the U-2 cockpit in creating an interior environment equivalent to the atmosphere at a 28,000 feet altitude. Firewel designed a system where the pilot's suit inflated if the interior cockpit pressure fell below the 28,000-foot level. In either case, the pilot obtained oxygen only through his helmet.

*The CIA declassification of the AQUATONE and OXCART projects at Area 51 has brought forth stories of startup companies and obscure companies fronting for other enterprises in the interest of secrecy and national security. Firewel was one.

Pressure Suits – In Their Own Words

Birkett: "Physiological Support Division (PSD) technicians assisted the mission pilot dressing in the partial pressure suit. The U-2 pilot was fully pressurized from the neck up and physically constrained from the neck down by a skin-tight suit like a corset. The transition between the two suits was accomplished by a rubber bladder the pilot pulled down around his neck and tucked back under, it then resembling a turtleneck sweater."

"Positive oxygen pressure against the tight bladder combined with body moisture formed the seal for the tiny pressure chamber for the pilot's head. The full-body partial pressure suit was required because blood boils at 63,000 feet altitude, and we were subject to pressurization failures, such as flameouts, canopy seal problems, etc. above 70,000 feet." - McIlmoyle p. 292

Powers: "Airtight, of rubberized fabric with almost no give or elasticity, it fit snugly around the body, so snugly that the slightest movement—bending a knee or arm, turning the head—would rub the skin, leaving bruises. Wearing long johns helped, but not much." – Powers p. 16 - 17

Firewel

In 1946, two Fairbanks Aviation alumni glanced around the basement of their home at 135 Aurora Street in Lancaster, New York and decided it would be a good place to launch a company that they named Firewel. Philip Edward (Ed) Meidenbauer, Jr., was the president, Donald Nesbitt, a ceramic engineer, was the vice president.

Meidenbauer, a self-taught mechanical engineer, had been the Director of Oxygen Research at Fairbanks Aviation, where he developed the original Air-Pak. Meidenbauer's brother, Clifford Meidenbauer, a Signals Corps officer during World War II, joined Nesbitt and him as Firewel's financial officer.

Operating from the basement amidst Lois Meidenbauer's home-canned peaches and jellies, near the old furnace and around the corner from her laundry room, the company began building furnace burners to convert old coal furnaces to oil or gas. Workers trooped down the cellar stairs to hammer out the burners. Buyers and suppliers conducted business with the company from the living room sofa, charmed by Meidenbauer's youngest daughter who loved to toddle to them and crawl into their laps. By 1947 the company advanced from conversions to a line of new furnaces.

Outside events now were occurring that would eventually drastically change the company. During March 1946, the first US-built rocket left earth's atmosphere, reaching an altitude of 50 miles at about the same time Meidenbauer was leaving Fairbanks Aviation. A month later, during April, the US Navy revealed it had created an 8,000-horsepower aircraft rocket engine. On August 27, a pilot ejector seat tested successfully at Wright Field.

During October 1947, the Air Force became independent of the Army, and on 14 October 1947, Air Force test pilot Chuck Yeager broke the sound barrier in a Bell X-1 rocket plane. Yeager's X-1 approached Mach 1.06, or at his altitude about 700 mph. On 28 February 1948, Yeager exceeded the speed of sound again in a Bell XS-1. On August 8, 1949, pilot Frank Everest climbed to 63,000 feet in the Bell X-1.

Because of their technical background, Firewel's principals decided to add a corporate focus on the future need of high-altitude military flight, where innovative kinds of breathing apparatus would be needed. The Firewel executives saw this as a means of expanding their business. They contacted various individuals they knew during the war. Their original development contract was for $80,000, followed by an additional $70,000 extension.

In 1951, David Clark Company contacted Ed and Firewel to solve problems with the regulators for a prototype full-pressure Navy space suit that David Clark Company and the BF Goodrich Company were developing.

Thus Firewel advanced its designing and manufacturing of small valves, regulators, and systems to the field of survival technology for high altitude flight. Firewel designed an aircraft-mounted regulator to pressurize early partial pressure suits, and soon was supplying components to David Clark and Goodrich.

The partial pressure suit provided counter pressure to the torso and facepiece of a pilot if exposed to the barometric pressure at altitudes of 40,000 feet or higher. Firewel equipped the spacesuit with the instrumentation and controls and oxygen breathing and ventilating systems that automatically protected the wearer from the hazards of reaching extreme height. The suit design covered such hazards as blood-draining acceleration, blood-boiling low pressure, cosmic rays, and extreme temperatures.

The Model 1-A, a prototype stratospheric suit design, included a Firewel regulator that supplied oxygen and air mixture on-demand or pure oxygen under pressure. Oxygen delivery occurred through a mask attached to a standard crash helmet.

In 1952, seven years after laying out much of the patented technology groundwork, President Ed Meidenbauer died. Donald Nesbitt succeeded him with Clifford Meidenbauer continuing to handle financial matters. By 1956 Firewel employed 140 people and was a multi-million-dollar corporation.

By the mid-1950s, Firewel's specialty of high-altitude regulators exceeded the production of all their larger competitors. The company's aeronautical division, with only one-third of the employees, accounted for almost 70 percent of the firm's business. In part, this could be attributed to Firewel's new concepts in oxygen equipment and survival system integral to the pilot's flight gear and part of the state-of-the-art military aviation at the time

Firewel designed new valves and incorporated them into the pilot's parachute pack rather than the instrument console and miniaturized the oxygen hoses and radio communications wiring. Firewel developed delicate silicon diaphragms strengthened with nylon filament.

The switch from regular to bailout oxygen became automatic rather than manual upon pilot ejection, making the new concept in oxygen equipment and survival systems an integral part of the pilot's flight gear.

Firewel first developed the miniaturized mask-mounted oxygen system, thus becoming a major player with the Air Force and Navy in the development of breathing apparatus for high altitude flight.

The Firewel-built survival kit came as a soft pack interfaced with the aircraft oxygen supply and connected to the pilot by two hoses. One hose provided breathing oxygen from the miniature regulator to the facepiece and chest bladder — the other connected to the capstans of the partial pressure suit. Integrated into the survival kit hoses were a microphone and earphone wiring harness as well as the power to heat the facepiece of the partial pressure suit.

The emergency oxygen supply in the survival kit manually actuated by pulling a cable that terminated in a green ball, aka the "Green Apple." The U-2 used this soft pack survival kit because the earlier versions of the aircraft did not have an ejection seat. The soft pack survival kit after that evolved into a rigid hard-shell kit used with all century series aircraft.

In the mid-to-late 1950s, the Firewel production line went from an oxygen regulator, oxygen bottle, and mask to now producing anti-g gloves, a "Global Survival Kit" for pilot ejection seats, pressure regulators for flight suits and helmets, and a backpack oxygen manifold. The company developed a "Get Me Down" oxygen supply¾back-pack oxygen assembly, couplings, gauges, and pressure reducer. It was during this period that the company became involved in the design of breathing apparatus and life support for the U-2 at Watertown. Firewel's field representatives trained for deployment from Watertown with the original three overseas detachments of the program in 1956.

Earliest partial-pressure suits came in 12 different sizes, but for the more advanced U-2 suit, each one would have to be individually tailored. The suit for the U-2 "was a cross between the original S-2 and the modified MC-1." Marty Knutson was the first CIA U-2 pilot fitted with this new suit. - Jenkins p. 154

The U-2 pilots found the early models of the suits (called MC-2 and MC-3) uncomfortable because the heavy rubberized fabric was very snug, so much so that pilots could not gain or lose more than two lbs., to be able to fit. The pressure in the early models of their suits excluded the feet and the pilot's gloved hands. The attachment areas for the gloves and the heavy helmet chafed the wrists, neck, and shoulders. The helmet was prone to fogging.

This life-support system worked countless times and saved the lives of pilots during the numerous high-altitude flameouts experienced during the program. Problems with the pilot life-support system, on the other hand, are believed to have contributed to Ericson's and Sieker's U-2 crashes.

Ground support shoehorned the pilot in his bulky suit into his seat in the small cockpit, something they could not do to remove him in an emergency. The early model U-2 lacked having an ejection seat, and even after Lockheed included an ejection seat, the pilots felt reluctant to use it in fear of losing their legs below the knees when blown from the cockpit.

The U-2 design saved weight by keeping the pilot's seat simple with no height adjustment mechanism. The shorter pilots adjusted the seat by inserting wooden blocks beneath the seat to raise it. Kelly Johnson added an adjustable seat in later versions.

Chapter 13 – CIA's Watertown Goes Operational

Continuations flying of the U-2 continued at Watertown throughout the fall and winter of 1955–56 to test all the various systems. The CIA and its Air Force training unit, the Strategic Air Command 4070th Support Wing, were also coming together for their work at the Ranch.

This effort would include everything needed for the restricted overseas deployment of tactical units that included personnel, finances, base facilities, communications, parts, transportation, fuel, photographic supplies, oxygen trailers and equipment, aircraft maintenance, flight planning, weather forecasting, security, and many other provisions.

During October 1955, the third Agency aircraft arrived at Watertown, along with the SAC officers and enlisted men to assist Lockheed in determining the operational capabilities of the U-2 and training three CIA pilot detachments.

By January 1956, the U-2 so impressed the USAF that it decided to obtain the aircraft. The USAF purchased 31 U-2s through the CIA; the transaction code name, Project DRAGON LADY, was the origin of one of the plane's nicknames.

On 30 January, DCI Dulles agreed to have the CIA act as the executive agent for Project DRAGON LADY. Ironically, the concept of the Air Force supporting the CIA at Watertown switched to the CIA assisting the Air Force in obtaining the U-2 plane, and later, the Mach 3 YF-12 interceptor, the D-21 stealth drone, and ultimately, the famous SR-71 Blackbird.

The Air Force maintained secrecy by transferring funds to the CIA, which placed an order with Lockheed for 29 U-2s in configurations determined by the Air Force. The Air Force later bought two more U-2s; bringing the total to 31 aircraft purchased by the Air Force as the follow-on group, FOG.

With the Instructor Pilots trained and ready, AQUATONE managers now concentrated on checking out the complete U-2 system: planes, pilots, navigation systems, life-support systems, and cameras. On 15 December 1955, the SAC 4070th Support Wing issued its Operational Plan for training and deploying all detachments. On 11 January 1956, the first three contract pilots for Detachment A arrived at the test site to begin training.

The 4070th Support Wing produced a technical pamphlet used to score the Unit Simulated Combat Mission (USCM) performed after each detachment completed formal training. Some of the areas of the evaluation were briefings, quality of maintenance and supplies, on-time takeoff, accuracy of celestial readings, accuracy in following the flight lines, photo results.

For example, "Each sortie must be airborne within one minute of the scheduled time." One to two minutes early or late meant only an 80% score. Two to five minutes early or late meant 70% and so on. More points awarded were for staying close to the flight line, with points decreasing to a 14-mile error, then no points for any further away from the flight line. Sextant readings were similarly graded. The tracking camera confirmed actual positions. It was not all up to the pilot; if a camera or sextant failed, this reflected on the performance score for the entire detachment.

Detachment A finished formal training 3 April 1956. The follow-up report states: "During the week of 9 – 14 April a USCM was scheduled. Project Headquarters issued daily operational orders, Detachment 'A' flew the missions as directed; the exposed film was shipped to the contractor for processing who, in turn, delivered the finished product to Project Headquarters for review."

Eight overflights of the US flew using the A-2 model cameras. The USCM tested all systems according to the grading pamphlet, in conformity with the standard Air Force Operational Readiness Inspection as well. Colonel Yancey and his staff, plus some visitors from Washington observed as Detachment A flew the full missions. During the weeklong exercise, US radars were unable to track these flights.

At the end of the exercise, Yancey's 4070th group reported the detachment ready for deployment. He briefed a high-level Pentagon panel that included the secretary of the Air Force and the chief of air staff. Both concurred with Yancey's determination concerning the readiness of the U-2 for deployment.

In late April 1956, Detachment, A begun to deploy, with Detachment B scheduled to arrive at the Ranch for training around 16-25 May. The CIA wanted to resurface the runway during ten days between these classes, but compromised to a proposal that covered parking U-2s and towed U-2s far onto the lake for flying, as the construction crew lacked security clearance.

As with any new aircraft design, the U-2 encountered problems for Lockheed and other relevant support vendors to resolve. One such problem was fuel control. A second was the propensity of the engine to dump oil into the cockpit via the ventilation system. Some seals had a hard time holding oil, so oil consumption on an early mission could vary from 3 to 17 quarts (out of 56 qt. capacity). Yet another challenge was restarting the aircraft due to flameout, which required descending to 35,000 feet or lower, making it vulnerable to Soviet interception.

Finding this unsatisfactory, Pratt & Whitney concentrated on the J57/P-31 engine to provide the added payload and altitude capability. One hurdle to overcome was a short supply of the J57 engines in general, due to high demand from other programs.

J57/ P-37 and P-31 Engine Availability

Kelly Johnson only accepted the J57/P-37 as a temporary engine used for testing and training. It was too heavy for his design and lacked thrust for long high-altitudes missions; he said he might as well 'tow the aircraft partway to altitude.' Even so, it was hard obtaining P-37s for the U-2 due to short supply.

Air Force officials at the Air Development Command during October 1954 appealed to General Putt that diverting J57/P-37 engines to Lockheed would jeopardize the X-16 and RB-57. The P-37 was in demand for other Air Force programs as well.

By December 1954, Pratt & Whitney indicated to Kelly Johnson that they could make a lighter, more powerful J57 engine for the U-2. The oil tanks would be smaller, and the blades for the compressors forged rather than cast. It would be called P-31.

Once the J57/P-31 became revealed, other aircraft designs scrambled for it. As of early 1956,

the USAF had a procurement program for 70 P-31 engines, and according to Col. Gibbs as of 20 April 1956: "They do not contemplate procuring additional -31 engines for the U-2 program, but with the SAC's concurrence -31 engines might be diverted from the C-135s." As the U-2 was secret, it could not have its own publicly known engine contract.

Within the U-2 program, Col. Gibbs also reported: "Col. Berg interposed, and stated that he did not go along with the idea that our project got number one priority over the Air Force because we had articles equipped with the -31 engines to the detriment of the Air Force's U-2s. he still indicated that he would not say yes to our project having any priority over the Air Force U-2."

Politics aside, the pilots found it essential for safety and workload to have a reliable engine and an autopilot. "Mach control" was critical as the aircraft drove into the Coffin Corner where maximum speed and stall speed converged, threatening a loss of control. The Lear Company produced an autopilot that, after numerous test hours, Lockheed installed in its production aircraft.

As the flight envelope began to expand, flameouts became a frequent occurrence with the P-37 engines. Lockheed test pilot Bob Matye experienced the first flameout on his third high-altitude flight. The experience proved that the pressure suit, regulator, and the emergency oxygen system worked.

During the USCM for Detachment A, the U-2 once again demonstrated its unique airworthiness. On 14 April 1956, James Cunningham received a call from Watertown while in his Washington, D.C. headquarters office. The caller informed him of a westward-bound U-2 experiencing a flameout over the Mississippi River along the western border of Tennessee.

After restarting his engine, the pilot reported a second flameout and engine vibration so violent it prevented him from getting the power plant to start again. The pilot was still gliding at high altitude with no power and would need to make a dead-stick landing.

Early in the program, Bissell and Ritland foresaw such an emergency and, with the cooperation of the Air Force, arranged for delivery of a sealed order to every airbase in the continental United States instructing what to do in case a U-2 needed to make a forced landing.

Cunningham instructed the project officer to ask the pilot how far he could glide so they could determine which Air Force Base to alert. The pilot, now over Arkansas, radioed back that given the prevailing wind and the U-2's 21:1 glide ratio; he thought he could reach Albuquerque, New Mexico.

Within minutes, Cunningham phoned Colonel Geary in the Pentagon. Geary, the Air Force's Assistant Director of Operations, and BGen Ralph E. Koon called the commander of Kirtland AFB near Albuquerque. General Koon told the base commander concerning the sealed order and warned him to expect an unusual aircraft to make a dead stick landing within the next half hour.

The general instructed the base commander to have air police keep everyone away from the craft and quickly place it inside a hangar.

After a half-hour, the base commander called the Pentagon to ask where the crippled aircraft was. As he spoke, the officer saw the U-2 touchdown on the runway and remarked, "It's not a plane; it's a glider." It surprised the air police surrounding the craft, even more, when it came to a halt, and the pilot climbed from the cockpit in his "space" suit. One air police officer remarked the pilot looked like a man from Mars.

The pilot, Jacob Kratt, later reported to Cunningham how from the first flameout to the landing at Albuquerque, the U-2 covered over 900 miles, including more than 300 by gliding.

A few days later, Captains Setter and Meierdierck flew a recently repaired T-33 to Kirtland to pick up the U-2. They started it in the hangar, and Meierdierck taxied the plane to the nearest runway. He took off with only partial power. As not to reveal the capability of the aircraft, he kept it low to the ground. The end of the runway overlooked a deep ravine. Meierdierck flew down into the ravine where he flew a long distance from the Kirtland AFB before climbing higher and heading back to Watertown.

Aside from the Lockheed pilots, Meierdierck had become the highest flying-time U-2 pilot with 121 hours when the program deployed from Watertown, a distinction later earning him associate membership in the Society of Experimental Test Pilots.

Despite its extraordinary gliding ability, however, the U-2 remained a difficult aircraft to fly. While its light weight enabled it to achieve extreme altitude, the construction made it fragile. The pilots had to carefully keep the craft in a slightly nose-up attitude when flying at operational altitudes.

If the nose dropped only a degree or two into the nose-down position, the plane gained speed at a dramatic rate, exceeding the placarded speed limit in less than a minute, at which point the aircraft would come apart. Pilots paid close attention to the speed indicator because at 65,000 feet; they had no objects for the eye to use as a reference to give them a physical sensation of speed. Also, to avoid overheating, pilots had to critically monitor gauges for EGT (exhaust gas temperature) and EPR (engine pressure difference, comparing the ratio from air intake to exhaust).

Coffin Corner – In Their Own Words

Graham: "My pressure suit inflated, and my face piece fogged over. I began to shut down systems in the airplane to conserve battery power for an expected air start of the engine in the denser air below 40,000 feet . With a sinking feeling, I discovered that the engine was not going to start. By this time the battery was dead, and no communication with the ground was possible. I tried to get some action from the flaps and speed brakes. There was virtually no hydraulic pressure available to operate these systems." - McIlmoyle p. 76

Birkett: "The warning system also indicated I was directly over the SAM site, and the next signal was the missile launch OS light. Immediately I disconnected the autopilot and started a right max rate turn. At maximum altitude, we flew 3 to 4 knots below high-speed buffet and 5 or 6 knots above low-speed stall. In a bank or turn, this difference decreased because of increased wing loading required to turn. Above a 12-degree bank a pilot would have been in a high-speed buffet and a low-speed stall. This situation was commonly known among pilots as the 'Coffin Corner.' To recover from the coffin corner, the pilot had to descend and/or decrease bank angle. With a missile launched toward me, decreasing angle of bank or leveling out was not an option. I lowered the nose and extended the speed brakes and landing gear. Pulling the throttle back even a small amount would have risked a flameout, not where I wanted to be with a missile on the way. I lost a couple thousand feet as I turned the buffeting/stalled U-2 westbound. " - McIlmoyle p. 300

1955 aerial view, N-S runway, and SW corner of Groom Lake (at top)

Coordinating Intelligence Collection

For the U-2 program, the CIA saw the apparent need for an interagency task force or office to develop and coordinate intelligence collection workflow for the nation's covert overhead reconnaissance effort, following the lead from Eisenhower's TCP.

On 5 November 1954, Edwin Land wrote a three-page memo to DCI Dulles setting forth the idea of the Technological Capabilities Panel's Project 3 ("Intelligence"). The memorandum recommended a permanent task force that included setting up an Air Force supporting statement under suitable cover to provide guidance on intelligence collection. The memo called for gathering requests from all agencies with a legitimate interest, planning missions according to priority and feasibility, maintaining continuity of operations, and carrying out the dissemination of the resulting intelligence information according to special security arrangement.

Land's proposal went into effect when U-2 testing approached completion at the last of 1955. After a meeting with Deputy Secretary of Defense Quarles and Trevor Gardner (promoted from his special assistant post to become AF Assistant Sec'y, Research and Development), Bissell established an Ad Hoc Requirements Committee—or ARC—naming James Reber as chair of the ARC and also the intelligence requirements officer for the U-2 project.

Reber, already experienced in coordination with other intelligence agencies, headed the DI (Directorate of Intelligence) of the OCI (Office of Intelligence Coordination) for four years. The first full-scale ARC meeting took place on 1 February 1956 with representatives from the Army, Navy, and Air Force present. Members of the Directorate of Plans and OSI (Office of Scientific Intelligence) attended for the CIA, whose presence later expanded to include the Office of Current Intelligence (OCI).

In 1957, the National Security Agency (NSA) began sending a representative. The State Department followed suit in 1960, even though receiving reports from the committee all along.

The Requirements Committee's primary task was drawing up lists of collection requirements for the U-2, prioritized per the U-2's ability to meet the three top national intelligence objectives concerning the Soviet Union in the mid-1950s: long-range bombers, guided missiles, and nuclear energy.

The committee issued its list of targets for the use of the entire intelligence community, utilizing all available means of collection and not just the U-2.

ARC gave the top priority target list to the project director and the project staff's operations section to plan the flight paths for proposed U-2 missions. Although not responsible for developing flight plans, the requirements committee assisted the planners with detailed target information as required. When ready to submit a flight plan to the President, the committee drew up a detailed justification for the selection of the targets. This paper accompanied the flight plan.

In developing lists of targets, committee members considered the variables involved and the interests of their own organizations. Thus, the CIA emphasized strategic intelligence: aircraft, submarine, rocket factories; power and nuclear facilities; then geography and civilian infrastructure.

In contrast, the military services placed a heavier emphasis on order-of-battle data. The Air Force had a keen interest in gathering intelligence on the location of Soviet-bloc airfields and radars. Though the committee members kept the interests of their agencies in mind, their awareness of the vital mission kept the level of cooperation high. Although occasionally impossible, the group always attempted to reach a consensus before issuing its recommendations.

U-2 Mission Film Handling

Early in the book, the author identified the CIA's Watertown as a facility intended from day one to have the ability to project compartmentalized power and to exert influence on a global scale, with the installation in Nevada the home of such influence. One might describe the CIA operating as the Hydra, the many-headed serpentine water monster in Greek mythology.

Developing the U-2 plane at Lockheed, and flight testing it at the CIA's Watertown Flight Test Facility were parts and stages of what AQUATONE entailed. While going through these stages, the CIA was preparing to deploy the U-2 and three detachments to commence operational flights over the Soviet Union.

One might say the CIA created two Hydras, one at CIA headquarters at Langley and one at Watertown. The heads appeared wherever the CIA flew—whether it be over the United States or elsewhere around the globe—with each having a special purpose. One of the heads was Project EQUINE, the developing and analyzing of film footage obtained during the U-2 missions. Another was Project HTAUTOMAT, exploiting the film footage.

On 13 December 1954, DCI Dulles and his assistant Richard Bissell briefed Arthur C. Lundahl, the chief of the CIA's Photo-Intelligence Division (PID), on Project AQUATONE. At DCI Dulles's direction, Lundahl set in motion within his division a compartmented effort known as Project EQUINE to plan for the exploitation of overhead photography obtained from the U-2. The CIA's Photo Intelligence Division grew quickly under Lundahl.

During May 1955, the 13-member PID staff found the number of personnel too small to handle the expected flood of photographs from the U-2. Therefore, the Directorate of Support authorized expanding the PID staff to 44 persons. Soon the division moved from its room in M Building to larger quarters in Q Building.

The PID authorized a further doubling of the staff during January 1956 as HTAUTOMAT came into existence to exploit the expected U-2 photography. The change meant placing all the products from this project in a new control system.

To handle this growth, during the summer of 1956, the PID moved to even larger quarters in the Steuart Building at 5th Street and New York Avenue, NW, Washington, D.C. By then, the PID photo interpreters were already working with U-2 photography following the Unit Simulated Combat Missions (USCMs) flown by each of the detachments at the Ranch. The missions were carried out under 'combat conditions.' The U-2s photographed several US installations analogous to high-priority Soviet installations. These preparations readied the PID for the mass of photography of the Soviet Bloc that began coming as of Carl Overstreet's 20 June 1956 overflight, quickly followed by the first successful series of overflights of the USSR

Special Delivery to Langley -- by CIA Operative Herb Saunders

Herbert F. Saunders, Jr., a former Deputy Director, Office of Technical Service (OTS), CIA, was born on October 27, 1931, in Randolph, Maryland. Following graduation from the University of Massachusetts, he served in the Air Force in England from 1953 to 1955. Mr. Saunders began his career with the CIA in 1956 as a Security Officer in the Middle and the Far East. He then served as Chief of Security for a large

CIA station in Southeast Asia, responsible for the security of all CIA personnel and operations in-country. Saunders subsequently transferred to the CIA's Office of Technical Service, serving as Chief of a large OTS facility in Europe, then as Chief of Operations. He completed his CIA career as the Deputy Director, OTS, in 1984. He is the recipient of the Distinguished Intelligence Medal, the CIA's highest honor.

During November 1956, Saunders received a posting to Detachment 10-10, the U-2 unit at Incirlik Air Base in Turkey (also called Detachment B). Despite having only recently arrived, he drew the assignment to escort to Washington D.C. classified "take" from a recent U-2 overflight. The take typically consisted of a couple of rather large boxes and maybe a few smaller packages. He was to proceed to the U-2 sister Detachment A in Giebelstadt, Germany from whence he would be taken by vehicle to Rhein-Main Air Base to connect with a special flight to Dover Delaware Air Base, then to be trucked to Washington to deliver materials to the Steuart building. He carried a .38 caliber pistol in a shoulder holster, which would remain concealed except in an emergency. He carried what he described as "Mickey Mouse" orders identifying him as a USAF civilian courier, GS-12. As a relatively new GS-7, he found this morale enhancing. Carmine Vito was at the L-20 when he took off.

The first leg was uneventful. (It would prove to be the only leg that was.) Arriving at Giebelstadt, he joined some friends for dinner at the officer's club. At some point, Col Fred McCoy, the detachment commander, came to their table to introduce himself. He said he would be flying to Rhein-Main the following morning, a 45-minute flight as opposed to a five or so hour drive, and Saunders was welcome to join him if he liked. Always on the alert for a good deal, Saunders recognized one and accepted.

Very early the next day, Thanksgiving Day, Saunders was out on the airstrip to supervise the loading of his cargo onto the aircraft, an L-20 with USAG markings. An L-20 was a single-engine prop plane, a two-seater with a high wing that looks sort of like a boxy Piper Cub. Shortly after they were airborne, Saunders was in the right seat

with a top coat hiding his weapon, and Colonel McCoy was of course in the pilot's seat in full uniform and displaying rows of ribbons topped by Command Pilot wings. Saunders felt comforted.

Near an hour into the 45-minute flight, Saunders asked, "Are we there yet?" Colonel McCoy responded that he was having trouble finding Rhein-Main, that a lot of ground fog enveloped it and neither our radio nor homing device was operative. So, they bored holes in the sky for a little while, sometimes turning left and occasionally right. Suddenly, something roared over the top of their right wing and disappeared into the sky before them, leaving massive turbulence in its wake. The jets repeated buzzing the plane three or four times. Colonel McCoy speculated that they were Air Force F-102s trying to lead them into Rhein-Main, which did not seem feasible since they were slow and the F-102s were so fast. After this little caper came to an end, Colonel McCoy announced that he was returning to Giebelstadt.

About an hour later, with some trepidation, Saunders asked the same question as before. The good colonel advised him that he was unable to find Giebelstadt, that they were getting low on fuel, and Saunders should keep his eyes open for an airstrip, any airstrip. Within minutes, Saunders spotted a small strip, and they headed for it. As they taxied along, they saw a small Quonset hut with stars and bars on it and a little Army plane much like theirs next to it. They pulled in, and while the colonel was looking for a phone, Saunders rousted a GI who was napping in the hut and asked him if he had any gas. He said he could probably scrounge up 20 or 30 gallons in five-gallon cans. So, they got out a funnel and took all the gas he had. Meanwhile, Colonel McCoy advised that they were just a few miles south of Giebelstadt and all they had to do was follow the main highway right to the front gate.

After 20 minutes in the air, Saunders could see no highway. Giebelstadt was nowhere in sight, and McCoy advised that they now had a severe fuel problem. He asked if Saunders wanted to jump. Noting that they didn't seem to have much altitude, Saunders asked what his plans were. McCoy said he was going to "ride it on in." Saunders elected to do likewise. McCoy asked

Saunders if he could tell which way the wind was blowing. At that point, they were heading toward what appeared to be a small village surrounded by open pasture. Saunders could see laundry hanging in yards, and reported the wind blowing one way and then another. Absorbed in his meteorological duties, he suddenly realized that things had become very quiet—like the phenomenon of dead engines. They headed for a pasture, achieving ground contact with a huge thud, plowed through a giant pile of hay, under some high-tension wires, and finally came to a halt only feet from a high stone wall with the village cemetery on the other side. Saunders, who had yet to get on a first-name basis, congratulated Colonel McCoy on the landing. McCoy decided he would go on another phone-finding mission in the village, leaving Saunders behind to stay with his classified cargo.

As Colonel McCoy was leaving, the villagers  started to arrive. A crowd encircled the aircraft, wanting to see what was going on with the foreign plane landing in their backyard. Saunders' primary concern was the cargo. He could see some of the villager's intent on getting a souvenir. Saunders busied himself, preventing the villagers from hack-sawing a piece from the propeller.

It was a gray, misty, frigid day and his poor feet in his thin Thom McAn shoes were freezing as he circled the tundra. Some of the villagers got the bright idea of boarding the aircraft. Saunders beat them to it and decided it his best bet to stay inside the plane and hope that deterred the crowd. He had already concluded that to display his weapon could only make matter worse. So, he sat, wondering where the colonel was and eyeballing the assemblage, which seemed to think that he was lunatic as they pointed at him and shook their heads while making clucking sounds.

After a considerable time, the colonel reappeared. He had found a phone, made contact with Giebelstadt who advised they would arrive to pick them up. (It turned out that they were in a town called Freezing, which was not far from where they wanted to be.) The colonel had found a cozy little Gasthaus, had a few pops and bonded with the patrons, who now formed an admiring entourage as they followed behind him. (The plane sat for a day or so at the graveyard before Carl Overstreet came down and flew it out under the high-tension wires.)

Eventually, a security team led by Jack Harris arrived in a Volkswagen bus. The priority, however, was to rescue Harris. He had attempted to leap a rather fast running stream not far from the plane, landed right in the middle of it and washed downstream. The water was about 15 to 20 feet wide, so it would have been a feat of Olympic proportions had he made it. At any case, the security team loaded Saunders' cargo, and off they went to Giebelstadt. He had flown all over and gone nowhere. He later learned at Happy Hour that Russian radar had tracked them the entire time and their erratic movements had everybody worried since they were very close to the East Zone. The thought even occurred that they had turned off all their electronics and were trying to defect with their precious cargo. The US Air Force had ordered the F-102s to shoot them down if appearing to cross into the East. The next day, Saunders went by road to Rhein-Main and boarded a special flight to Dover. This should have been the end of his story, but it wasn't.

The special flight was a C-54, a four-engine prop plane. It was carrying a bunch of coffins, presumably occupied, Saunders, and his cargo. At some point late in the flight, the crew chief awakened Saunders from a nap and advised that they had lost the left outboard engine. Saunders was exhausted and not fully comprehending, mumbled something about waking him again if they lost anymore. He awoke again when the crew chief advised that the right outboard engine had shut down. Saunders was now paying attention. The crew chief suggested that Saunders get off the floor and buckle himself into one of the bucket seats, which he did, without a word. Saunders remained there until they landed, departing the aircraft to find it surrounded by emergency vehicles of every shape and size. As his cargo loaded on the Steuart vehicle, he noted the crew chief was eyeing him suspiciously, probably thinking that if he wasn't merely stupid, as a civilian, he had behaved well to avoid panic in a somewhat tricky situation. What he didn't know was where Saunders had been for the last few days and that he was neither stupid nor courageous, just numb.

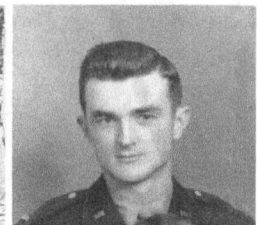

Howard Carey Wilburn "Billy" Rose Frank G. Grace

The CIA Hiring Foreign U-2 pilots

In authorizing the U-2 project, President Eisenhower told DCI Dulles he wanted non-US citizens to pilot these planes. He believed this made it easier for the US to deny any responsibility for a U-2 coming down in hostile territory. The CIA assigned the Directorate of Plans, Air/Maritime Division (AMD) the task of hiring the pilots with Lt Col Geary in charge of training recruits in mid-1955.

Using selected foreign pilots to fly the U-2 soon ran into trouble. At Luke Air Force Base, all except four of the Greek pilots flunked from the preliminary training program. None of those who flunked qualified to advance to Watertown for flight training.

During May 1956, Luke Air Force Base sent the four Greek and one Polish pilot to Watertown for pilot familiarization in the U-2. At the same time the CIA's Detachment B, the second class of 'sheep-dipped' pilots, arrived. That class included Francis G. "Frank" Powers.

SAC Instructor Pilot Louis Setter found the foreign pilots lacked the proficiency to fly operations. While Powers' class underwent training, the Greek pilots all washed out, and Setter never allowed the Polish pilot to fly the U-2. One Greek student pilot did complete the standard transition training in the T-33 (consisting of dozens of drag-in approaches to simulate the U-2 approach). He flew the U-2 only once. The flight was his first solo flight without a pressure suit.

Republic of China pilots later trained to fly the U-2 with CIA Detachment H, going on to overfly mainland China from Taoyuan Air Base in Taiwan (see Chapter 16).

CIA U-2 Pilot Marty Knutson

USAF U-2/SR-71 Pilot Tony Bevacqua

Carmine Vito   *Glen Dunaway*

Carl Overstreet

Aerial view of Lockheed Oildale facility

Preparing for U-2 Detachment A's Deployment to England

The CIA was intent on deploying the spy-planes in allied countries, and the UK was a willing partner. The Agency was keen on being able to use British crews if President Eisenhower should ever become reluctant to approve overflights flown by US pilots.

When the group of U-2 drivers for Detachment A completed their training at Watertown, the CIA proceeded to deploy them to Royal Air Force Lakenheath to see how much of what the Russians had displayed the previous month was real.

Each year, during Russia's May Day celebrations, the US armed forces went on a worldwide alert while the Soviet Union paraded its latest armament and military might. From a US newsreel at the time: "Moscow skies are filled with the first public display of Russia's newest combat aircraft, startling western observers with their quantity and quality – a shock to the complacent, a spur to the alert."

Knowing the existence of Russia's armament meant the US needed more data assessing its strength and confirming its state of advancement. Were the Soviets presenting a demonstrator, or did they have many of what they showed in their display of force?

Lockheed, for its part, felt confident in the ability of the U-2 to fly, obtain this information, avoid interception and return it, and the CIA was equally eager to receive such needed intelligence.

From January 1956, Lockheed had opened its U-2 assembly plant "Unit 80" at Oildale, California near Bakersfield and had continued delivering U-2s to Watertown.

The British government approved the U-2s deployment from Royal Air Force Lakenheath with the CIA's Detachment A still in training at Watertown. NACA announced the USAF Air Weather Service would use a Lockheed-developed aircraft to study the weather and cosmic rays at altitudes up to 55,000 feet.

The CIA U-2 Detachment A, known publicly as the 1st Weather Reconnaissance Squadron, Provisional (WSRP), operated at Royal Air Force Lakenheath as a power projection extension of the CIA's Watertown facility—still training more detachments at Watertown for deployment elsewhere. The CIA was already preparing to deploy its second class, Detachment B, to Turkey, and its third class, Detachment C, to Japan.

Setting up the deployment venues required considerable planning, negotiations, and logistics. This process was known to only a few. This top-secret activity allowed in the loop only those having a need-to-know. Advance parties deploying to the UK, Turkey, and Japan included communications personnel at not only the host base but also other countries where the U-2 might fly. The same applied to the fuel personnel, Lockheed maintenance personnel, and security. The success of the program depended on using individuals of resourcefulness, ingenuity, and a can-do attitude. Each received a top-secret classification.

As of 19 April 1956, two advance echelons of support personnel for Detachment A totaling about 18 men had begun deploying. At the same time, most other Detachment A personnel were taking pre-deployment leave, scheduled to reassemble 27

April for the move overseas. Similarly, the primary mission pilots were sent on pre-deployment leave. The pilots scheduled for five days of "Escape & Evasion" training in the first part of May. The pilots would be the last to arrive in the UK.

After the USCM for Detachment A completed, processing teams including a crew from Headquarters began packing, crating, and staging of equipment loads for the move. As of 19 April, two of Detachment A's four aircraft were disassembled, with the other two expected to be readied in the next few days. Tentatively, the number 12 aircraft from Lockheed was planned later also to be shipped overseas, to give the detachment five U-2s.

As Detachment A was preparing to leave the Ranch, Detachment B was beginning to arrive. As of 23 April, the first two pilots of Detachment B began training at Watertown; by 7 May about 50% of "B" personnel were scheduled to arrive.

Beginning 30 April 1956 for the next four days, one C-124 daily would load a U-2 aircraft at Watertown, along with security and other personnel to accompany the plane.

Another four C-124s on a similar schedule would load supplies and equipment and then depart Watertown. Records indicate that each C-124 carried approximately fifteen tons of mission equipment.

The movement of three echelons of personnel was by DC-6 or in military designation, C-118 aircraft. By the end of April all personnel processing had been accomplished, including the issuance of orders, passports, clearing of customs requirements, etc.

Thus, on 30 April 1956, the CIA airlifted two U-2A planes to Royal Air Force Lakenheath to await the arrival of all the Detachment A personnel from Watertown. The mission pilots, last to arrive, would arrive under alias via commercial airliners.

Chapter 14 - Deployment

Deployment of U-2 Detachment A

CIA Pilots Trained at Watertown for Detachment "A:"

Hervey Stockman	Marty Knutson
Jake Kratt	
Carl Overstreet	Glen Dunaway
Carmine Vito	
*Bruce Grant	**Wilburn Rose

***Howard Carey

* Grant withdrew from the program

** Rose was killed during training

*** Carey trained with "B" group but brought in to replace Rose

Stockman's account names Bruce Grant as selected:

"My introduction to Watertown started at the Brown Derby in Hollywood and a drive to a hotel on Hollywood and Vine where Glen Dunaway, Jake Kratt, Bruce Grant and I met John Raines from security. At the Lockheed Air Terminal in Burbank, we climbed aboard a C-54 and headed northeast towards Nellis AFB and Indian Springs. Mt. Charleston slid by, and we were in the restricted area. We dropped down and came into the dry lake."

"The Ops, housing, and hangars were adjacent to an airstrip. Jake, Glen, Bruce and I had a house trailer, which would be our home until we headed out for England. Carmine, Marty Knutson, and Carl Overstreet were already in place. We met Col Fred McCoy, Phil Karis and others who would participate in our checkout. Blue suiters that I can recall were Hank Meierdierck, Lou Setter, Hank Majeski, and Major Sam Cox, who was the weatherman."

Grant reportedly encountered a problem with hypoxia that prevented his participation in high-altitude flying. The pilot solo lists for the U-2 do not name him. The demands of project security most likely led to a lack of mention of him in official records. According to his nephew, Bruce Grant received a service-connected disability retirement at age 34.

For those who deployed, their travel orders identified them as Department of the Air Force civilians of the Watertown Weather Reconnaissance Squadron, Provisional. (WRSP-1)

Two advance echelons of personnel had begun to travel by 19 April 1956, and by then two of the U-2s had been disassembled and prepared for shipping. Total deployment needed 2 C-118 transports for personnel and 10 C-124s for the planes, camera supplies, spare parts, oxygen trailer, tanks, and other equipment, plus other materiel.

Detachment "A" first located to Mildenhall, England, but moved to Lakenheath shortly after that as the Mildenhall's hangar space proved inadequate. By 4 May, all of the detachment's personnel and equipment, including four aircraft had taken the long flight in a C-124 ("Old Shaky") to Lakenheath. Plans were also underway to ship the fifth U-2 to Detachment A on 14 June 1956.

After the four U-2s arrived in secrecy at Lakenheath, the USAF released a cover story with the following substance:

"On 7 May the National Advisory Committee for Aeronautics (NACA) announced that Lockheed aircraft would be flown by the USAF Air Weather Service. The flights are for the study of high-altitude air turbulence, convection, and cloud formation, wind shear, jet stream dynamics, and ozone, water vapor, and cosmic ray studies."

The first public acknowledgment of U-2 operation was providing its "cover" story. Most actual meteorological research came later.

Unfortunately, the aircraft could not fly its reconnaissance missions over hostile territory until solving the flameout problem. Detachment A began reliability testing of the old engines in the friendly skies of England until the new P-31 engines could begin arriving from Pratt & Whitney in the US

Before spy flights could begin in earnest, the agreement with Prime Minister Anthony Eden began to falter as the Suez crisis loomed on the horizon. The CIA had informed the British that they would deploy only one plane but sent the four. The CIA would soon experience two incidents that led it to transfer the U-2s to Wiesbaden in West Germany later during June.

The first incident was the notorious 'frogman' incident where MI6 sent a retired naval commander, Lionel 'Buster' Crabb, to investigate the Soviet cruiser Ordzhonikidze that took Nikita Khrushchev and Nikolai Bulganin on a diplomatic mission to Britain. Though Bulganin was Premier in 1955, Khrushchev was First Secretary of the Communist Party and therefore held a lot of the actual power. When Bulganin and Khrushchev traveled to India, Yugoslavia, and Britain, they were known in the press as "the B and K Show."

Crabb studied the Ordzhonikidze's propeller, a modern design that Naval Intelligence wanted to examine. On 19 April 1956, he dived into Portsmouth Harbor, and his MI6 controller never saw him again. Crabb's companion in the Sally Port Hotel took all his belongings and even the page of the hotel registration on which they had written their names. Ten days later British newspapers published stories about Crabb's disappearance in a mission.

MI6 tried to cover up this espionage mission. On 29 April, under instructions from Rear Admiral John Inglis, the Director of Naval Intelligence, the Admiralty announced that Crabb had vanished when he had taken part in trials of secret underwater apparatus in Stokes Bay on the Solent. The Soviets answered by releasing a statement stating that the crew of the Ordzhonikidze had seen a frogman near the cruiser on 19 April. Later, a body was snagged by fishermen. It was missing its head and both hands, which made it impossible to determine for certain who it was. Adding to the diplomatic faux-pas stacking up in Whitehall, two days later (during May 1956) a U-2 from Lakenheath inadvertently penetrated the British radar network, causing Royal Air Force fighters to scramble. The British government, embarrassed at the Russians catching the British spy in the act, asked the United States to move its assets off English soil

Watertown's First U-2 Pilot Fatality

At this time, Watertown suffered its first loss of life since the C-54 crash on Mt. Charleston. The first fatality while flying the U-2 occurred on 15 May 1956 when pilot Wilburn Rose, flying Article 345 experienced trouble dropping his pogos, the outrigger wheels keeping the wings parallel to the ground.

Rose may have tried to drop his pogos too late. With more air speed (or in a crosswind), increased air resistance could serve to jam the pogo in the socket. Once airborne, Rose made a low-level pass over the airstrip and shook the left pogo loose. He attempted a right turn to come back over the runway to shake loose the remaining pogo. Because of the bank angle and with the wings fully loaded with fuel, the right wing kept going down. The U-2 plunged to the earth, disintegrating over a wide area.

After this, a spring-load (rather than pilot release) pushed the pogos out just as soon as the wings generated lift. In case of a jammed pogo, the procedure was developed to go to higher altitude to try to shake it out - as a safety margin in case of stall. If that wouldn't work, then they would have to land to free it, and sometimes launch by hand with no pogos (ground crew would balance the wing, one crewman at the end of each wing, and run with the plane for 100 meters or so until enough lift was generated). Despite safety improvements there was another fatality due to a jammed pogo in the 1990s.

Pogos played a significant role in the lightweight design of the U-2. Note-worthy quotes follow:

Pogos – In Their Own Words

Chapman: "The last thing a U-2 ground crew does before the bird launches is to pull the pogo pins so they will fall out as soon as the wings begin to lift. This takes a couple of seconds, but when a mission is critical, seconds can seem like hours. Sometimes a pogo will fail to fall out of the wing causing a 'hung pogo' and can cost not only precious time but a hazard to the pilot. In that scenario, the pilot might have to land the plane or fly to a designated area to shake them loose." - McIlmoyle p. 222

May: "Late one day, I noticed something was causing the pogos to stay on the wings of the U-2 on takeoff. To avoid aborting a flight because of the pogo problem, I was told to hold up one wing and another airman would hold up the other wing. As the pilot started his takeoff, we were to run like hell with him until he could balance it. We only had to run about 15 or so steps before 'Kitty

Hawk' took off." - McIlmoyle p. 169

Kemp: "After preparations were made to launch, someone up the chain decided to pull the pogos and hand launch due to the rough taxiways and runways. A man was stationed on each wingtip to balance the airplane while it taxied and started the takeoff roll. With a very small amount of lift, the pilot could control the balance with the ailerons. It was quite a show for those who have never witnessed this feat. [The Base commander said]: 'Now by God, boys, I've been in the Air Force for a long time and I have seen airplanes take off every way you can think of, but that's the first time I've ever seen two guys throw one up.'" - McIlmoyle p. 242

Burrough: "After the first plane was off, a search was mounted for the pogos that were left in deep puddles of rainwater on the runway. We wasted a lot of time searching for the pogos. We decided to remove the remaining pogos on the waiting aircraft; ground crew held the wings and raced along with the plane until the wings maintained lift on their own. This worked well, and within a short time, all were airborne and became the first U-2s involved in the Cuban crisis." - McIlmoyle p. 212

Williams: (photo interpreter for Detachment B, Adana, Turkey) "Then the strange pogo sticks on wheels which supported the wings fell off—with the exception of the day one pogo stayed on and then dropped through the roof of a mosque." - Beschloss p. 141

U-2 Detachment A Leaving England for Germany

To avoid delays, on 11 June 1956, following the refusal of the British government to allow mounting U-2 operations from Britain, Detachment A moved to Wiesbaden, Germany without approval from the German government. This was while Giebelstadt prepared to become a more permanent base; crews worked on old hangars there and put up fences with mirrors.

It concerned Eisenhower that despite their considerable intelligence value, overflights of the Soviet Union might cause a war. His fears reached back to the 1955 Geneva Summit, while the U-2 was under development. Eisenhower proposed to Bulganin that the Soviet Union and the United States each grant the other country airfields for use in photographing military installations. Bulganin, who was in effect acting for Khrushchev. Rejecting the "Open Skies" proposal. Premier Bulganduring may have been receptive, but Khrushchev denounced the proposal as a plot and a way to make the Soviet Union lose 'face.' (Open Skies would have confirmed Soviet weakness in weaponry at the time)

At the time, the CIA was telling the president that the Soviets could not track high-altitude U-2 flights. The CIA based this belief on the Soviet Union still using American radar systems given them during World War II. Continental defense radar within the US could not track the U-2 practice flights. Lockheed pilot Ray Goudey recalled that he flew from Watertown to the Pentagon, "took pictures, and returned without being observed."

Although the Russians could track the overflights, they could not identify the aircraft. Often the U-2s overflying Soviet-denied territory would have an entire squadron of MiGs flying at a much lower altitude beneath the U-2s, which for the pilots was a real pain in the neck because of the MiGs blocking the U-2's view of its recon target. In the three weeks from June 20 through July 10, 1956, the CIA U-2s made eight overflights beyond the Iron Curtain, including five over the Soviet Union.

Then, a year later, the Office of Scientific Intelligence appeared more cautious, stating that detection was possible, but believing the Soviets unable to track the aircraft. Dulles further told Eisenhower (per Presidential Aide General Goodpaster) that in any aircraft loss, the pilot would not survive. Having such assurances and with the growing demand for accurate intelligence regarding the alleged "bomber gap" between the US and the Soviet Union, Eisenhower approved ten days of overflights for June 1956.

On 20 June 1956, a month after declared operational, a U-2 piloted by Carl Overstreet conducted the first "Operation Overflight" out of Wiesbaden, West Germany. This mission was the first overflight over the denied territory of the

Soviet Union. This operation was under the cover designation of the WRSP-1. Meanwhile, WRSP-2 in Incirlik, Turkey, and WRSP-3 in Atsugi, Japan were preparing to follow, as deployment continued.

Carl Overstreet -- First U-2 Pilot to Overfly Denied Territory

Carl Overstreet entered the reconnaissance business in the winter of 1954 when Gerry Johnson, his 508 SFW Wing CO at Turner AFB stopped him out on the ramp and suggested his meeting some people downtown who have something to offer. Overstreet's thoughts turn to F-86s and the Flying Tigers. Off he went to meet the recruiters for the CIA who revealed to him a photo of the Article, and he agreed to fly it.

After physical exams, fitting for a pressure suit and a scary ride in the chamber to 80K feet, he arrived at Watertown during January 1956.

Now, in the summer of 1956, the trained Weather Reconnaissance Squadron A sat ready and waiting at Wiesbaden Air Base, West Germany. With events moving rather slow from his point of view, Overstreet took a short weekend trip to Vaduz, Lichtenstein, accompanied by Max Conn, a tech rep for Westinghouse. Upon his return to Wiesbaden, he learned of the CIA scheduling him for the first operational flight of the U-2 over Eastern Europe.

On Wednesday 20 June 1956, Carl Overstreet flew Mission 2003 over Eastern Europe, flying north and west from Wiesbaden to gain altitude before looping back to the base and turning east. He entered hostile territory where the West Germany, East Germany, and Czechoslovakia borders met. He flew across northern Czechoslovakia, turned north, passing east of Dresden and into Poland, flying over every major Polish city before turning back to Wiesbaden the way he came via Prague.

The overflights of Eastern Europe continued. In another mission, Overstreet photographed the Suez from the Red Sea to the Mediterranean after Nasser shut down the canal in 1957.

On 2 July 1956, Jake Kratt flew Mission 2009 over Eastern Europe, heading southeast from Wiesbaden across Austria into Hungary. Kratt flew past Budapest and turned south along the Yugoslav border. The route extended all the way to the Black Sea back to Wiesbaden, making it a 7-hour sortie.

On the same day, 2 July, Glendon Keith "Glen" Dunaway flew Mission 2010 Eastern Europe. He headed from Wiesbaden over East Germany, southern Poland, eastern Czechoslovakia, Hungary, and Romania before turning around at the Black Sea and returning to Wiesbaden after a 7-hour sortie.

The first U-2 overflight had now occurred, using the existing authorization of Air Force overflights over Eastern Europe. On this flight, the CIA intended to test the Soviet radar to see if it could track the Angel. Eisenhower shared this worry. He nonetheless approved the first flight over the Soviet Union, Mission 2013 scheduled for 4 July 1956.

U-2 missions from Wiesbaden departed westward to gain altitude over friendly territory before turning eastward at operational altitudes. The NATO Air Defense mission in that area included No. 1 Air Division RCAF (Europe), which operated the Canadair Sabre Mark 6 from bases in northwestern France and West Germany. The service ceiling of this aircraft was around 54,000 feet. Numerous encounters between the U-2 and RCAF "ZULU" alert flights have been recorded for posterity.

Canadians Track US Spy Flights

Canada played an integral role in the defense of North America and in fulfilling the combined NATO mission in Europe. Recall that regarding the vulnerability of North America, in the early 1950s the US feared Soviet bombers coming straight over the North Polar route from air bases in Siberia.

US National Intelligence Estimate 18 in 1951 noted the extensive Soviet program to develop biological and chemical weapons: "During the period 1951-1954, the Soviet Union will be capable of delivering BW (biological warfare) agents to the United States by long-range aircraft." The US considered chemical warfare agents delivered by Soviet long-range bombers an equal threat. After the Soviet nuclear arsenal increased, the US expected BW and CW to become supplemental weapons.

In response, the US progressively developed the Pinetree, Mid-Canada, and DEW radar lines across Canada, operated largely by Canada in the integrated defense network for North America. Canada also began its own programs for the CF-100 Canuck and CF-105 Arrow interceptors.

Meanwhile, in Europe the Soviets had blockaded Berlin in 1948, the Communist Party had taken control in Czechoslovakia, and as it turned out, the Soviet Union was ready to send tanks into Hungary in 1956. Warsaw Pact forces maintained sizeable armies with potent armored divisions. Besides, air attack toward Britain would now arrive fast, because of the development of jet warplanes after WWII.

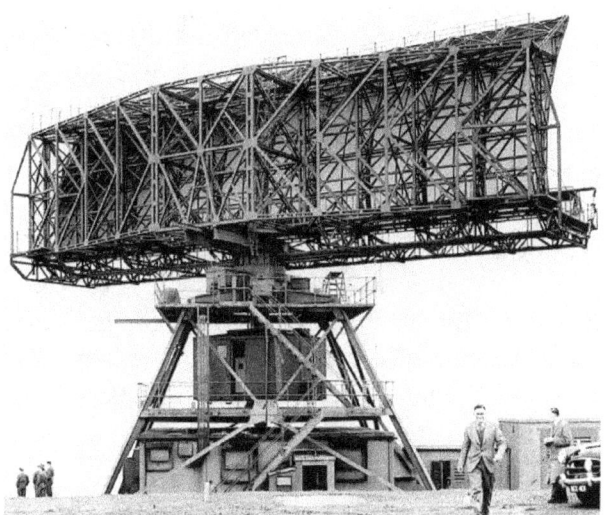

More powerful NATO radar was needed to detect all aircraft coming westward out of the Soviet Bloc. Enhanced, The mid-1950s, found the most powerful of the NATO radars located in Metz, France and operated by the Canadian 61 AC&W Squadron, RCAF. Ten or more other NATO radars in West Germany and France had direct relations with Metz, which was also headquarters of the No. 1 Air Division RCAF.

"Yellowjack" was the call sign at Metz. The Yellowjack radar was an AMES Type 80 that could normally detect a large fighter plane at ranges over 210 nautical miles. In this picture taken by an unidentified RAF photographer, notice the relative size of the human figures below the AMES Type 80 radar at the bottom left of the scene.

Enhancing the Metz radar further increased its range to give ZULU alert fighters enough time to

respond to and identify all aircraft coming out of the Eastern Bloc airspace. The ZULU alert

fighters had to be ready to go airborne in 2-3 minutes to verify any intruder aircraft in the air defense intercept zone along the East German – West German border. Canadian ZULU alert flights normally consisted of 'sections' or flights of four Sabre Mark 6s, each armed with 6 Browning machine guns.

Interception procedure provided for two fighters of the ZULU section to approach from below and behind the incoming aircraft and two to the sides. If the intruder was part of an attack, the section would engage and sound the "VICTOR" alert. VICTOR alert aircraft were other warplanes set to go on 15-minute notice, 24 hours a day, seven days a week, 365 days a year. Armed with nuclear weapons, in case of VICTOR alert their crews would receive specific targets to strike.

The very first front-line fighter equipped with nuclear weapons was the F-84 starting in 1952. This photo from *Air Force Magazine* shows an F-84 carrying the Mark 7 nuclear weapon under its wing. President Eisenhower had learned the lessons of Pearl Harbor and the invasion of South Korea: His war plans "called for plenty of nuclear-armed fighters to hold massed Soviet armies in what he called a 'bottleneck' across Europe." Other targets would be Soviet Bloc air bases, to open the way for NATO bombers to counterattack the Soviet Union.

As it worked out, the CIA U-2 pilots were often combat-experienced and selected from the ranks of F-84 wings undergoing disbandment at the time. The tactical nuclear role would pass to more advanced fighters coming into service like the F-100 and F-105.

Regarding the U-2, the National Archives of Canada record several radar tracks from the Metz "Yellowjack" installation that correspond to known missions flown by Detachment A from

Wiesbaden. According to the incidents reported in these archives, inbound U-2 flights resulted in ZULU scrambles unless the order was given that no intercept action be taken, as on 20 June 1956, for the first operational U-2 mission flown by Carl Overstreet.

For another example the 'Pinetreeline.org' website reports from the official archive for 7 September 1956: "RCAF Zulu sections were given a live scramble against an inbound track. Both sections made interceptions before 12 AF ACC identified the track as a highly classified mission." The Pinetreeline.org website comments: "It would appear that the NACA U-2, locally nicknamed 'The Beast' was airborne again."

Tracking of other US spy flights also occurred. Scant information reports that Hawker Hunter fighters intercepted in other NATO sectors USAF RB-57D Black Knight missions coming back from the Baltic. During the 1955 period, some USAF Heart Throb RB-57A missions came west out of the air space of the Eastern Bloc at 62,000 – 68,000 feet altitude on their return to Rhein-Main Air Base in West Germany. Heart Throb pilots remarked that they were not too concerned with MiG interceptors because only the Canadian Sabre Mark 6s were able to come close to the altitude of the RB-57As. Canadian archives may confirm the tracking of the Heart Throb flights.

Some connect the Canadian tracking of US spy flights over Germany with the incident involving break up of a U-2 piloted by Howard Carey on 17 September 1956, as mentioned further herein. Sources for the above subheading include Wikipedia ("1 Canadian Air Division"; "AMES Type 80"; "Distant Early Warning Line"; "Martin RB-57D Canberra"); "Victor Alert" in *Air Force Magazine*; "The U-2 and the Avro Arrow" by Patrick Bruskiewich; "National Intelligence Estimate 18"; "Heart Throb Pilot"; "61 AC&W Squadron Tracks the U-2 – National Archives of Canada" – see Bibliography

First U-2 Overflight of the Soviet Union

Carl Overstreet's first flight over Eastern Europe was significant. However, Hervey Stockman was suiting up to fly directly over the Communist homeland. The CIA was preparing to poke the Soviet bear. By the end of June 1956, the main factor preventing Stockman's flight was unfavorable weather.

On 4 July 1956, Hervey Stockman flew Article 347 marked as NACA 187 on Mission 2013, the first flight over the Soviet Union to target the Soviet submarine construction program in Leningrad, as well as counting the numbers of the new Myasishchev M-4 "Bison" bomber. From Wiesbaden, he flew over East Germany and Poland, before crossing the Soviet border near Grodno in Belarus. He flew over various bomber bases around Minsk, north to the naval shipyards and bomber bases at Leningrad, west over more bomber bases in the Baltic states and back to Wiesbaden in an 8-hour and 45-minute flight. Soviet radar detected and partially tracked this mission, and several MiG fighters attempted to intercept the U-2. *The National Air and Space Museum, Washington, DC now has this aircraft on display.

The second flight on 5 July continued searching for Russian Bison bombers. It took the only U-2 photograph of Moscow ever and investigated rocket factories at Kaliningrad and Khimki. Eisenhower realized how unrealistic his hope of no Soviet detection was and ordered the overflights to stop if the Russians managed to track the aircraft.

The CIA confirmed the Soviets were unable to completely track the U-2s. Therefore, the Russians did not know of the U-2 overflying Moscow and Leningrad. The aircraft photographs depict small images of MiG-15s and MiG-17s attempting and failing to intercept the plane elsewhere, proving the Soviets could not shoot down an operational U-2.

On 5 July 1956, Carmine Vito in Article 347 marked as NACA 187, flew this second Soviet overflight during Mission 2014, taking a similar route to Mission 2013 except further south. He continued to the royal capital city Kracow in Poland, into Ukraine over Brest and Baranovici. Vito headed toward Moscow, following the railway from Minsk to the Soviet capital. He flew over the Fili airframe plant in Moscow, northwest to Kaliningrad and the main Soviet flight test and research center at Ramenskoye, returning to Wiesbaden via the Baltic states, where Soviet

152

radars and MiG-17s again partially tracked this overflight. Vito remained the only U-2 pilot to fly over Moscow. His flight was the fifth operational flight over potentially hostile territory, or what the pilots called "hot" flights. The risk that the Russians would reveal an effective interception capability increased every day.

A sign in the Special Projects Operations Division at Edwards AFB, California states, "The man who never made a mistake is not here, the man who made too many is not here either." Pilot Carmine Vito completed his mission, but a sore throat could nearly have killed him. He had a pocketful of lemon drops when he took off headed to overfly Moscow. Accidentally mixed in with the candy was a cyanide tablet for use if his situation grew desperate. He almost popped it into his mouth.

*The Air and Space Museum in Washington, DC has the U-2 flown by Vito suspended from the ceiling. It still contains a payload from one of its earliest flights: a small lump of Tutti Frutti chewing gum under the left rail. Vito stuck it there on 5 July 1956 before taking off from Wiesbaden on his first operational mission.

On 9 July 1956, Marty Knutson flew the third Soviet overflight, Mission 2020 from Wiesbaden, north to overfly Berlin, East Germany, and the Baltic states to Riga. He headed east and south, covering targets around Kaunas, Vilnius, and Minsk before returning via Warsaw to Wiesbaden. His mission was to learn how much of a numerical gap existed between the bombers of the US and Russia. He captured photos of an airfield where all of Russia's Tupolev "Bison" bombers were based. This informed the US that the Russians did not have as many bombers as thought by the United States. Knutson's mission lasted 9-hours and 20 minutes. He landed with only 20 gallons of fuel remaining. Allen Dulles, the head of the CIA called the intelligence photos from Mission 2020 "million-dollar photos."

On 9 July 1956, Carl Overstreet flew the fourth Soviet overflight, Mission 2021 from Wiesbaden to head south into Czechoslovakia and Hungary. He flew northeast into Ukraine as far as Kiev and over various bomber bases before returning to Wiesbaden via Poland.

This same day, the NACA made another announcement about the significant research work conducted with the U-2. It informed the public of the need to carry out these types of research flights overseas. The statement was merely another cover story to explain the presence of U-2 in Germany and other locations.

On 10 July 1956, Glen Dunaway flew the fifth Soviet overflight, Mission 2024 from Wiesbaden, over East Germany, Poland, and Ukraine to Kerch on the eastern tip of the Crimean Peninsula. He headed back via Sevastopol, Simferopol, Odesa, Romania, Czechoslovakia, and Hungary to Wiesbaden with fighter aircraft radar tracking his flight near Odesa.

Bomber Gap Disproven

On 10 July 1956, the Soviets, believing it a Canberra, protested what they described as overflights by a USAF twin-engine medium bomber. The US knew by 19 July 1956 of no American "military planes" having overflown the Soviet Union. Nonetheless, the fact the Soviets' report revealed they could partially track the U-2s for extended periods caused Eisenhower to halt overflights over Eastern Europe. In truth, the president was more concerned about the Soviet protests than he was about the public's reaction to news of the US violating international law.

To avoid project cancellation, the CIA now began Project RAINBOW to make the U-2 less detectable.

The eight overflights over communist territory, however, had already proven the 'Bomber Gap' did not exist; the U-2s did not find any Myasishchev M-4 Bison bombers at the nine air bases they visited. Because the Eisenhower administration could not disclose the source of its intelligence, however, debates over the bomber gap continued.

The presidential order did not restrict U-2 flights outside Eastern Europe. During May 1956, Turkey approved the deployment of Detachment B at Incirlik Air Base, near Adana, Turkey. Before the new detachment was ready, however, Detachment A in late August used the Incirlik Air Base in Adana as a refueling base to photograph the Mediterranean. The aircraft found evidence of many British troops in Malta and Cyprus as the United Kingdom prepared for its future intervention in Suez. The US released some of the photographs to the British government. As the crisis grew in seriousness, the project converted from a source of strategic reconnaissance, which prioritized high quality over speed (its maker processed the film, and then analyzed in Washington); to a tactical reconnaissance unit that provided immediate analysis.

The Photo Intelligence Division set up a lab in Wiesbaden as Detachment B took over for Det. A. It flew over targets that remained classified. The Wiesbaden lab's rapid reporting helped the US government predict the Israeli-British-French attack on Egypt three days before it began on 29 October.

First Stealth Attempt – Project RAINBOW – the Dirty Bird

Even before the U-2 became operational during June 1956, the CIA project officials were turning their thoughts to developing a stealth plane. The CIA estimated a short life expectancy for flying over the Soviet Union of between 18 months and two years. "National Intelligence Estimate No. 11-5-55" of November 1955 noted great strides in Soviet radar development and said that by 1960, "Against a B-47 size target GCI coverage could extend to as much as 85,000 feet," also that by then the USSR might have more than 200 SAM sites.

By August 1956, the U-2's vulnerability concerned Richard Bissell to the point he despaired of its ability to avoid destruction for six months, let alone two years.

To extend the U-2's useful operational life, project officials first attempted to reduce the aircraft's vulnerability to detection by Soviet radars. During December 1956, Lockheed modified Article 341 using radar-absorbent materials for a series of radar cross-section (RCS) tests called Project RAINBOW.

Lockheed strung another U-2, Article 344, with piano wire of varying dipole lengths between the nose and wings of the aircraft to reduce the radar signature. These methods created extra drag with a resultant penalty in range and altitude. The U-2 planes modified under Project RAINBOW earned the name "dirty birds" from the planes not being aerodynamically "clean."

Anti-radar techniques relied upon radar cross-section (RCS) of an object. Radar measured how much electromagnetic (EM) energy an object reflected, expressed as an area defined in square meters. The RCS of an object became a function of the object's size, shape, and materials that varied depending upon the frequency of the EM energy. Long-distance search/acquisition radars used different frequencies than short-range fire control radars. Thus, the CIA undertook research and development of a variety of techniques to protect the U-2.

All parts of the aircraft created reflections—the fuselage, tail, wings, engine inlets and exhaust. The anti-radar technique investigation fell into two categories. It either absorbed the radar energy or created "interference" reflections.

Edward Purcell's (Harvard University) first concept placed on the U-2s fuselage an absorbent material that the BEACON HILL Study Group's MIT lab team and Lockheed developers called "Wallpaper." This second approach, tested in early 1958, involved the use of plastic material containing a printed circuit designed to absorb radar pulses in the 65-to 85-MHz range. This material was glued to parts of the U-2's fuselage, nose, and tail.

Although the "trapeze" and "wallpaper" systems protected against some Soviet radars, the systems proved ineffective against radars operating below 65-MHz or above 85-MHz. Furthermore, both degraded the U-2's performance. Weight and drag from "trapeze" reduced the aircraft's operating ceiling by at least 1,500 feet, and "wallpaper" sometimes caused the engines to overheat.

Another firm, Edgerton, Germeshausen & Grier (EG&G), which was also composed of MIT faculty members, tested the results under an Air Force contract to evaluate radars. EG&G operated a small testing facility at Watertown for this purpose.

Although closely involved with the radar deception project since the early days, Kelly Johnson cooperated only reluctantly. He disliked adding attachments that made his aircraft less airworthy. Johnson reflected his dislike of the antiradar attachments reflected in the unofficial nickname "dirty birds" for the modified aircraft.

After Lockheed mechanics had mounted the various RAINBOW devices on the prototype U-2, a Lockheed test pilot flew the plane over EG&G's installation, a little more than a series of tractor-trailers containing instrumentation. EG&G technicians could thus record and evaluate the U-2's radar returns as it traversed a specific course over their facility. This method of testing radar-deceptive modifications proved both time-consuming and dangerous

The scheme called Trapeze attempted to protect the engine inlets by running another wire diagonally from the nose to the slipper tank on each wing. To reduce low frequency (70 MHz) reflections from the leading and trailing edges of the wings, the mechanics placed a wire parallel to and ahead of each wing's leading edge and another parallel to and behind each wing's trailing edge.

A fiberglass pole attached to each wingtip to anchor the outboard end of each wire. The poles provided anchor points ahead and behind the wings. Each wire ran from the front end of each pole to the slipper tank (which projected in front of the wing) and from the slipper tank to the fuselage. Behind each wing, a wire ran from the back end of the fiberglass pole to the fuselage. The horizontal stabilizer carried wires similarly. The engineers and mechanics placed Ferrite beads on the wires to tune them to the expected frequencies, the technique called "Wires."

The disadvantage of wallpaper was it acted as a thermal insulator and trapped heat inside the fuselage. Initially, the engineers applied it to the upper and lower surface. However, after recognizing the heating problem, they applied it only to the bottom half of the fuselage.

The effect of the Wires and Trapeze installations together caused increased drag that cost the U-2 as much as 5,000 feet in altitude and 20% in range. The pilots lacked enthusiasm for flying the plane with its reduced performance. One of them likened it to being "wired like a guitar."

On 29 August 1956, two U-2s on Missions 1104 and 1105 flew from Wiesbaden during the Suez Crises, where they photographed preparations for the landings. The missions landed at Incirlik in Turkey. The following day, the two U-2s at Incirlik retraced the previous day's sortie, flying over the Suez area, and landing back at

Wiesbaden.

The CIA Pilots of U-2 Detachment B trained at Watertown

During August 1956, the second U-2 class completed their training and received transfer orders to Incirlik Air Base, Turkey as Det 10–10; Det B; WRS (P) -2.

Tom Birkhead	Bill McMurray
E. K. Jones	Buster Edens
John MacArthur	Frank Powers
J. Robbie Robinson	Bill Hall
David Dowling	Sammy Snider

Det B pilots and support officers circa 1956 – via Sue Powers
Top L-R; Powers, Snider, Birkhead, Perry, Jones, McMurray, Hall
Bottom L-R: Bohart, Perkins, Edens, Cherbonneaux, Cordes

L-R: Chet Bohart, Cy Perkins, Francis Gary Powers, Sammy Snider,
Tommy Birkhead, Ed Perry, E.K. Jones, William "Dad" McMurray,
Bill Hall, Buster Edens, Jim Cherbonneaux, Harry Cordes

Formal training at Watertown for Detachment B completed by 16 July 1956, followed by the Unit Simulated Combat Mission (USCM) performed by all the pilots, carried out under the exact conditions expected overseas. At the Ranch, the drift sight and on development testing of the B Camera continued.

Circa 1959 – L-R: Bob Ericson, Marty Knutson, Al Rand, Gary Powers
Bottom: Jim Barnes and Barry Baker
John Shinn and Glen Dunaway, not present – via Gary Powers Jr.

U-2 Detachment C's Arrival at Watertown

During August 1956, the third U-2 training class arrived.

Barry Baker	Jim Barnes
Thomas L. Crull	Lyle Rudd
Jim Cherbonneaux	John C. Shinn
Bob Ericson	Albert Rand
James A. Smiley	Harry Cordes
Frank G. Grace, Jr	Russell Kemp

From this class, Frank G. Grace, Jr. perished in a training flight, and 'Al' Smiley deployed to Detachment A after the loss of Howard Carey there.

Three months following the loss of Wilburn Rose, on 31 August 1956, a second fatal crash occurred at Watertown during a night-flying exercise. To avoid a stall, experienced U-2 pilots always cut back on the throttle as soon as the pogo sticks fell away. Grace, lacking experience, stalled Article 354 at an altitude of 50 feet when he tried to climb too steeply at takeoff. The falling craft cartwheeled on its left wing and struck a pole at the control tower.

Before the year's end, two more U-2s crashed, one of them fatal. On 17 September 1956, Article 346 lost part of its right wing while near Lindsey Air Force Base in Wiesbaden, Germany. The aircraft may have flown into the jet wash from a Canadian F-86. Whatever the cause of the mishap, the U-2 disintegrated in midair near Kaiserslautern, West Germany, killing pilot Howard Carey.

Loss of the U-2 Over West Germany

CIA historians Pedlow and Welzenbach comment on the move of Detachment A to Wiesbaden during May 1956: "The detachment commander, Col. Frederick McCoy, was disappointed in his hope that the redeployment of the U-2s from Lakenheath could be accomplished without drawing undue attention." The main reason for this disappointment was that Wiesbaden was the location for one of the busiest air bases in West Germany. The preferred location for Detachment A at Giebelstadt was still being prepared with the construction of facilities and installation of enhanced security. In addition to this issue, as the official history continues: "One of the detachment's aircraft was lost in a crash on 17 September 1956, killing pilot Howard Carey and garnering unwanted publicity."

A memorial feature on the CIA website comments further: "On Sept. 17, 1956, pilot Howard Carey took off from Lindsey Air Force Base in Wiesbaden, Germany. His U-2 mysteriously disintegrated in midair, perhaps caused by the jet wash from four fighter aircraft nearby."

Those are essentially the officially released details. Private historian Pocock comments that like all the accidents which befell CIA U-2s, the findings of the investigation remain classified; and "It is known that four Canadian F-86 interceptors were flying in the vicinity."

The official Canadian record has likewise remained classified. After 50 years, however, some memoirs and informal accounts by RCAF pilots and ground personnel who had contact with the U-2 began to emerge. One source is the 'Pinetreeline.org' website that was put together over the years by retired RCAF Captain Joseph George Reynald L'Ecuyer, to collect memoirs of Canadian Forces personnel who served in defense of North America and in carrying out the NATO mission in Europe during the Cold War. The Military Communications and Electronics Museum of Canada now host this collection of memoirs and accounts.

Among the 'Pinetreeline.org' documents is the account in the local Kaiserslautern newspaper

"*Die Rhein Platz*" of 18 September 1956 regarding the crash, and statements from eyewitnesses at the time and the report of the local fire brigade that responded. The paper states that APs from nearby Sembach Air Base used rifle butts to move local people away from the crash area.

As to 'pilot accounts' 50 years later, general agreement includes scenario of a 'section' of 4 Sabres from 421 Squadron flying ZULU alert for the 2nd Wing at Grostenquin; one RCAF plane circled the U-2 as it was going down and reported the incident location to 'Yellowjack' (1 Air Div. Headquarters at Metz); a US general came and individually interviewed the Canadian pilots at Sembach; certification had to be made that no rounds had been fired by any of the CF-86 Mark 6 Sabres; Canadians were told to "keep your Mark 6s away from the U-2"; the U-2 had come to be known informally among NATO personnel in Europe as "The Beast."

Sources for above subheading: *The Central Intelligence Agency and Overhead Reconnaissance* (Pedlow/Welzenbach); "Remembering CIA's Heroes: Agency Pilots in the U-2 Program" (CIA website); Wikipedia ("RCAF Station Grostenquin" "1 Canadian Air Division"); Pinetreeline.org ("The U-2 Incident of 17 September 1956 – Roy Gummeson", "letter from John Farnham – 17 July 2002", "Keep Your Mark 6's Away From My U-2 – Edward McKeogh" "U-2 Crash of 17 September 1956—Assorted Sources"); "The U-2 and the AVRO Arrow" by Patrick Bruskiewich – see Bibliography

Detachment A Moved to Giebelstadt during October 1956

The Detachment A move placed it within 17 miles of the Iron Curtain. In the first two months of 1956 Giebelstadt served as a launch site for the Project GENETRIX reconnaissance balloons. By July 1956, the 602 AC and W Squadron took over the Giebelstadt radar station, with added upgrades. The new height-finding radar was the AN/FPS-6 Long-Range Height Finder.

Memoirs of Canadian ground personnel from Giebelstadt at the time state that this upgraded height-finding radar was so powerful, "it would light up fluorescent tubes, up to a mile away." Also: ". we were to turn the height finder 180 degrees away from the runway whenever one of the U-2 aircraft was departing or landing. The height finder put out a lot of RF energy, and we had inadvertently exposed some film." ("Memories of the U-2 at Giebelstadt" – 'Pinetree.org')

The account by Wayne Grover goes on to relate: "The first indication we had of the coming of the U-2 group was when 'civilians' suddenly took over a hangar that the USAF had been building for a recreation facility. We were kicked out; a high-security fence was built with short watchtowers and a security gate. We received zero information. I never saw more than two U-2s at Giebelstadt. The day of the first landing, we were all told to go into our barracks until further notice if we were not at work. I was at the radar site and working when the first one landed. It came in and landed. Period. We had no Air Traffic Control at the base, so no tower or radar approach was involved. The CIA had some ATC at the base and brought the bird in. I saw it land and roll on its belly wheel, tilt and lean down on one wing runner, followed by a few vehicles. It was silver."

"The U-2s were always towed directly to the hangar and out of sight. When the U-2 flew, everyone was told to go to their rooms, don't look out the windows and take no pictures. It was like saying don't swallow. We all saw them come and go. At work as a radar operator, I often tracked the U-2 as it turned West and climbed. We tracked it as it climbed on both search and height finder, seeing it turn East and leave our maximum height capacity of around 75,000 feet. We all knew what it was doing and how. When the U-2s came back, they entered West Germany from the East, too high to see on search radar, descended and came to Giebelstadt from the West to land."

As the CIA history observes, security came to be a problem: ". a long, black Soviet-bloc limousine was parked at the end of the Giebelstadt runway whenever the U-2s took off." The memoir of Ralph Heffley further states: "The strange thing about 'security' was that when this thing flew, military personnel were not authorized to take pictures of the aircraft. However, at the end of the runway, a German magazine called the *Bundespost* was taking pictures and when their magazine came out, they gave a complete description of the U-2 along with detail such as an approximate speed and altitude."

The CIA flew nine sorties over various Middle Eastern countries involved in the Suez crisis in the Sinai Peninsula during the build-up to the 10-day Suez conflict on 29 October 1956. Detachment B flew several sorties after troops landed at the Suez on 6 November 56. The CIA flew 14 sorties during November and December 1956.

Eisenhower refused during September 1956 to reauthorize overflights of Eastern Europe. It took the Hungarian Revolution during November and his reelection that month, to persuade him to renew Eastern Bloc overflights over border areas. The Taiwan Strait crisis was also simmering during these months.

On 20 November 1956, Gary "Frank" Powers flew the sixth Soviet overflight, Mission 4016 from Incirlik, north over Syria and Iraq. He flew over Baghdad, into Iran before turning north toward the Caspian Sea. Powers crossed the Soviet border and flew over Baku before turning west to overfly Yerevan. The flight aborted heading for Tbilisi when electrical problems forced an early return to Incirlik. The flight was the first one using the B-Camera. Both radar and fighters tracked the mission.

At this time also, the Soviets protested a December overflight of Vladivostok by RB-57Ds. Eisenhower again forbade overflights of Communist-controlled countries. However, flights close to the borders continued with the first ELINT-equipped U-2s.n 10 December 1956, the CIA flew Mission 4018 over Eastern Europe from Incirlik over Albania, Bulgaria, and Yugoslavia and back to Incirlik.

The same day, 10 December 1956, Carmine Vito flew Mission 2019 over Eastern Europe from Wiesbaden over Albania, Bulgaria, and Yugoslavia and back to Wiesbaden. On this sortie, Vito, known as the Lemon Drop Kid, almost bit on the suicide L-pill, mistaking it for one of his favorite sweets. After this, the CIA put suicide pills into boxes to avoid confusion. The poison needle replaced the L-pill during January 1960.

Nine days later, Watertown suffered the loss of Article 357 on 19 December 1956 resulting from pilot hypoxia. A small leak prematurely depleted the oxygen supply and impaired Robert J. Ericson's judgment as he flew over Arizona. Because of his inability to act and keep track of his aircraft's speed, the U-2 exceeded the placarded speed of 190 knots and disintegrated when it reached 270 knots. Ericson managed to jettison the canopy before the wind sucked him from the aircraft at 28,000 feet. His chute opened at 15,000 feet, and he landed without injury, but with the aircraft a total loss.

Future Plans for Project AQUATONE/OILSTONE

In a memo of this title dated 29 July 1957, CIA Director Dulles summarized to USAF Chief of Staff General Thomas D. White the current prospects for the joint project going forward:

"As to the risk of loss, we have as yet seen no hard evidence that the Russians have developed an interception capability effective above 65,000 feet and we believe that there is a chance that electronic countermeasures may reduce the effectiveness of such an interception capability when developed. As to risk of diplomatic protest, we are still hopeful that the experience of the current season will demonstrate that at least occasional overflights of the USSR can be conducted without eliciting embarrassing diplomatic protests provided a few of the most highly sensitive areas (such as Moscow itself) are avoided and provided overflights of the USSR are not tracked by one or more of the Satellite governments to the embarrassment of the Russian military establishment. We also hope that the Russian tracking ability will be impaired by electronic countermeasures to a point where they will not have solid evidence on which to base protests.

"We conclude from the foregoing that sporadic overflight activity, at least, is quite likely to be permitted by our political authorities, but that there is little prospect for an intensive overflight program. To give effect to these conclusions, we propose to maintain only two Detachments at reduced strength instead of three as at present. This contemplated reduction in scale will render a number of aircraft and other items of equipment surplus to this program. It appears that an initial transfer of five aircraft can be made during November of this year. It is our hope that we can arrive at an agreement with you whereby equipment turned over to the Air Force by this Project can be borrowed back at a later date if a requirement for it should arise . The Air Force has been a full partner in this enterprise from the beginning, and I will, of course, be happy to discuss any of these points with you if you so desire. I will look forward to receiving your comments. (Signed) Allen W. Dulles"

and to obtain the temporary use of certain facilities at these bases when required for staging operations. I recognize the burden that the provision of this support places upon the Air Force but hope it will be appreciably reduced by the planned reduction in the scale of this activity.

The Air Force has been a full partner in this enterprise from the beginning and I will of course be happy to discuss any of these points with you if you so desire. I will look forward to receiving your comments.

(Signed)
Allen W. Dulles

Chapter 15 - The Air Force Competing

By late 1956 – early 1957 USAF 4028th SRS U-2 Pilots arrived at Watertown for training as General LeMay made his stated move for the Air Force to take the U-2 away from the CIA.

Air Force Pilots Checked Out in the U-2 at Watertown

Jack Nole (squadron commander)

Joe Jackson	Dick Leavitt
Jack (Curly) Graves	Mike Styer
Floyd Herbert	Bennie LaCombe
Ed Emerling	Richard (Steve) Heyser
Dick Nevett	Bill (Skip) Alison
Leo Smith	Joe King
Howard Cody	Tony Bevacqua
Ken Alderman	Ray Haupt
Dick (Gordo) Atkins	Richard McGraw
John Campbell	Warren (Goog) Boyd

During June 1957, the USAF's Strategic Air Command, who wanted nothing to do with the U-2 at first now needed more secrecy for its spy plane operations. SAC was concerned about its U-2 pilots assigned to what was originally the 508th Strategic Fighter Wing at Turner AFB, GA, flying the F-84G. Deactivated in 1956; the 508th became the 4028th Strategic Reconnaissance Squadron, a component of the 4080th Strategic Reconnaissance Wing. The Air Force activated the wing to fly the Lockheed U-2 spy planes out of Laughlin AFB, Texas.

A second U-2 squadron, the 4029th SRS, received assignments in expectation of the CIA Project AQUATONE ending with the aircraft turned over to the Strategic Air Command. The transfer of the U-2s from the CIA never happened, and the 4029th SRS never acquired the U-2.

SAC relocated the 4080th from Turner AFB to Laughlin Air Force Base, near Del Rio, Texas in early 1957. While the 4080th moved to the new location, the Strategic Air Command instructor pilots at Watertown were training their final group: a selected cadre of pilots to fly the US Air Force's RB-57D replacement, the U-2 Dragon Lady.

On 20 March 1957, USAF U-2 #6696 piloted by USAF pilot: Anthony Bevacqua landed at Watertown in a crosswind. Following the chase vehicle instructions led the aircraft to veer off runway, leading to damage of the landing gear and no pilot error.

Lt Col Tony Bevacqua, the son of a Sicilian immigrant, graduated from high school and enlisted in the USAF on 29 February 1952. He attended Aviation Cadet Pilot Training Program to receive his commission as an officer in the US Air Force. Bevacqua checked out in the U-2 at Watertown during March 1957, becoming the youngest U-2 pilot ever. Tony Bevacqua accumulated 1904 flight hours in the U-2 and 738 hours piloting the US Air Force's SR-71 Blackbird.

On June 11, 1957, the 4028th Strategic Reconnaissance Squadron commander, Colonel Nole led the first of two three-ship U-2 formations from Watertown to their new home at Laughlin, Texas for operational duty.

Both RB-57s and U-2s graced the West Texas skies until the RB-57s retirement during April 1960. The US Air Force's U-2 eventually became the United States' sole manned-airborne reconnaissance platform, and most of them stationed at Laughlin Air Force Base.

RB-57 and U-2

Declassified RB-57 pilot accounts reveal similarities between the two programs. One such account released in 2018 by the National Reconnaissance Office is "Heart Throb Pilot" by Gerald E. Cooke. Cooke flew the lightweight RB-57A-1 used from November 1955 to around late August 1956 for missions over the Eastern Bloc and East Asia. At the same time, the improved RB-57D and the U-2 were coming into service.

Note ways that the RB-57 and U-2 projects mirrored each other.

David Clark Company - "Between May and August 1955, we spent time at Wright Patterson Air Force Base for survival training, an altitude chamber refresher course, and flight clothing indoctrination and fitting sessions. From there, we went to the Dave Clark 'Brassiere and Girdle' factory where we were measured for our tailor made T-1 partial pressure suits."

"All heart Throb overflights required the pilot to wear the T-1 pressure suit and pre-breathe 100% pure oxygen. I believe we were the first operational Air Force unit to regularly operate with the pressure suit. It was tight mesh nylon that fit like a layer of skin from neck to ankles, with foot and glove attachments to cover the extremities."

"The helmet was a hard shell that fit over a soft helmet that contained headphones. There was a bladder that fits around the neck at the Adam's apple and held pressure upon suit activation."

"The suit had two pressure capstans that ran the length of the body. The idea was to apply pressure through the capstan and its connections, including the helmet, gloves, and boot inserts. If there was a loss of atmospheric pressure, the suit provided the external body and internal lung pressure to sustain life until you could get back to lower altitudes. This was the best that early pressure suit technology had to offer."

"The helmet tended to strangle you when the pressure suit activated. Second, the helmet also tended to pop up on the head like a champagne cork when the pressure suit activated. A pulley – cable modification was devised that allowed the pilot to pull the helmet down as pressure tried to pop it upward and off."

Compartmentation - "Hdqs, US Air Forces in Europe (USAFE) controlled the intelligence and operation of Heart Throb missions. The individual briefings I received made it clear that the Heart Throb missions would follow a special line of control between Lieutenant Colonel Holbury, acting for the USAFE Commander (Lt Gen William H. Tunner) and the individual mission pilot."

"Even our training missions, photographing military installations in NATO and other European countries, were kept secret. We always had a US uniformed security guard for our aircraft when on the ground. Mission information was not shared within the squadron."

"I also believed that BGen Andrew J. Goodpaster, then serving in the White House as Staff Secretary and Defense Liaison to President Dwight D. Eisenhower, was a point of contact for our project and that Heart Throb missions rose to that level of concern, if not direct oversight."

RB-57A-1 Performance - "The pilot lifted off at about 100 knots after a roll of 2400 feet,.the Heart Throb RB-57A had a maximum refueled mission range of about 1800 miles, so we had never conducted overflights at that limit." [Powers' last U-2 mission was to be close to 4000 miles]

"At the end of about one hour in the target area, the pilot would have been anywhere from 62,000 - 68,000 feet in altitude and near the point of exit. Heart Throb penetration altitude was above 50,000 feet, and exit altitude averaged 62,000 feet."

RB-57D - This was the modified version with enlarged wing and more powerful engines. Delivery to SAC began during April 1956. About four months later, six of the Black Knight RB-57Ds deployed to Yokota AB and Eielson AFB for missions presumably over N. Korea, China, and along the USSR periphery. Operation Sea Lion out of Yokota monitored the fallout from Soviet nuclear tests. Three RB-57Ds overflew Vladivostok during December 1956. In Europe, RB-57Ds flew out of Rhein-Main AB from 1957 on. The RB-57 did carry a heavier payload than that of the U-2. A program called *Diamond Lil* trained Nationalist Chinese pilots to fly RB-57Ds over mainland China. A Chinese Army SA-2 missile shot down one of these at 65,000 feet altitude on 7 October 1959, discussed in the next chapter.

Comparison – the RB-57A penetration altitude of 50,000 feet plus compares with around 65,000 feet for the original U-2 using its P-31 engine. The improved RB-57D and the U-2 each boosted higher in their performances using the new J-75 engine. James Cunningham noted: "You'd get contrails up to the point, depending on the day, it could be as high as 60,000 feet. What you did was to make sure that you didn't penetrate the other guy's country until you were at your penetration altitude, which was 69,500 feet [U-2 with J-75 engine], by which time you were out of the contrail area."

RB-57F – Around 1962 the RB-57 was modified with an even larger wing and more powerful engines, and the resulting aircraft was known as the RB-57F. James Cunningham observed: "BLACK KNIGHT was the D-model Canberra. They had taken the Canberra and stretched the wings on it, made them bigger, and then they went to the F-model, finally, which was supposed to be the real challenger to the U-2."

"But the F-model never really made the grade either, because the Air Force was so desperate to get all the missions they could away from the Agency, they made a lot of fabulous claims for the engine on the F-model."

"They said they could sustain a cruise indefinitely at 73,000 feet, and I sat in on briefings when they were making these claims, and then I talked to the guys at Pratt & Whitney, who were making the engine, and they said, no way, this thing hiccups and dies at 69,500, which we used as the penetration altitude."

Regardless of controversies, at the Ranch, the Air Force follow-on group of pilots was ready to train once Detachment C left. By September 1956 the first Air Force U-2s were available at Watertown, and pilot training occurred between September 1956 and June 1957.

Like Lieutenant Colonel Bevacqua, many of the US Air Force pilots who trained at the CIA's Watertown training facility went on to make notable aviation history as did the CIA pilots.

Lt Gen Lloyd R. Leavitt, Jr., a U-2 pilot at Watertown during June 1957, later became chief of standardization in the 4028th Strategic Reconnaissance Weather Squadron at Laughlin Air Force Base, Texas. He retired from the US Air Force a lieutenant general.

Medal of Honor recipient, Lt Col Joe M. Jackson was one of the first Air Force pilots to train in the U-2 at Groom Lake in 1957. Jackson, when assigned to the 4080th at Laughlin Texas, worked for Colonel Robert Holbury, chief of the Reconnaissance Branch at a SAC command post. Jackson coordinated and briefed all Idealist (CIA) overflights to the SAC key personnel, i.e., Col Holbury, General Wiseman (Chief the SAC's command post), and General Blanchard, Director of Operations. He planned and monitored all Air Force U-2 photo and air sampling operations as directed by Air Force Headquarters. He earned the Medal of Honor during the Vietnam War.

BGen Raymond "Ray" Haupt, another Watertown U-2 pilot graduate, began his association with the Strategic Air Command's reconnaissance program during May 1957 when he received an assignment to the 4028th Strategic Reconnaissance Squadron (re-designated as the 349th Strategic Reconnaissance Squadron) at Laughlin Air Force Base, Del Rio, Texas. He served there as a flight commander-in-chief of the squadron's standardization division and earned his first Distinguished Flying Cross while assigned to the 4028th Squadron. The citation accompanying the award stated that his Lockheed U-2 reconnaissance aircraft experienced severe mechanical difficulties during a routine high-altitude test flight. "After all attempts to remedy the situation failed, he contacted his home station and declared his intention to execute a flameout landing." Following prescribed procedures, he executed an emergency landing.

While at Groom Lake, Haupt handled the standardization for and was the instructor pilot for training pilots in the A-12 Titanium Goose, F-101, T-33, and U-3B. The IP job meant a tradeoff of enough flying time against an abundance of paperwork. During August 1965, he transferred to Beale Air Force Base, California as the chief of the Standardization Division, 4200th Strategic Reconnaissance Wing, and later was the commander of the 4201st Strategic Reconnaissance Squadron (re-designated as the 1st Strategic Reconnaissance Squadron). The squadron flew the US Air Force SR-71 Blackbirds: "Fred White and I did the work of converting the factory version of the A-12 aircraft to the Air Force Standard Flight Manual. Coincidentally, we did the flight manual for the SR-71. This saved lots of time when we started operations at Beale."

General Haupt earned his second Distinguished Flying Cross in his next assignment as the Chief, Operational Evaluation Division, Detachment 1, 1129th US Air Force Special Activities Squadron at Area 51. From November

1961 through May 1965, he participated in test flights in the reconnaissance aircraft. His citation stated: "The operational techniques and procedures he helped to develop, which resulted from knowledge gained during his flights, contributed to his unit reaching an operational readiness status and in providing safe operational instructions for succeeding Air Force units."

U-2 Detachment C Begins Operation

During February 1957 CIA Detachment C deployed to Atsugi, Japan. This completed the goal of AQUATONE management to locate two U-2 operating facilities to the west of the Soviet Union and one to its east. Ongoing SAC reconnaissance operations in the Far East complicated matters, along with Soviet pressures against any of its neighboring countries that hosted US reconnaissance planes. What are some of the milestones of Detachment C as it trained and formed at Watertown and awaited its deployment? What were the difficulties in obtaining its operating facility in the Far East?

At Watertown, the advance support personnel for Detachment C were arriving as Detachment B deployed to Turkey on 13 August 1956. The first U-2 for Detachment C arrived at the Ranch 10 August, and on 14 August Detachment C personnel flew for the first time. According to records:

"Training began on 6 August 1956 with the arrival of four pilots. Four more pilots reported on 20 August and the last three on 10 September. The Detachment Commander reported on 20 August. Operations staff officers arrived between this date and 5 October. The arrival of pilots in groups of three or four on pre-planned dates simplified the training staff's problems."

The pilots completed formal pilot training by 19 October 1956. This included 70 hours of ground school and an average of over 100 hours of flying for each pilot. The USCM completed 26 October 1956. Four U-2s were available for the USCM and reported 25 flameouts using the older P-37 engines. One of the engines gave repeated flameouts, also affecting camera operation. (A flameout would shut down the whole plane until re-start.)

Detachment C, now the third group of pilots to complete training, identified itself officially as the Weather Reconnaissance Squadron, Provisional-3. Thus, Detachment C was ready to deploy during November 1956 but did not yet have an overseas base. However, the CIA had been busy negotiating for locations in Asia.

On 29 May 1956, the Deputy Project Director issued a letter to Col. Shingler (Dir. of Materiel) to "expeditiously get onto the project of working with Air Force people in the Pentagon and find what Far East bases appear to be best suited to our U-2. Also, Col. Payne and his people should be contacted concerning concurrent Air Force overflight programs which will operate out of FEAF to see if we can operate on the same bases or, if not, what difficulties we may encounter in FEAF with joint operations." (FEAF = Far East Air Force)

The Air Force had by 1955 operated flights in the Far East using RB-57A Canberras, and during April-May 1956 Project Homerun from Thule, Greenland also covered the Soviet Far East using RB-47s and tankers. Now the improved RB-57D, with a maximum unrefueled mission range of about 1,800 miles, was becoming available and the Air Force wanted to use it for Black Knight operations out of Yokota Air Base, Japan.

As a result of this situation, in a 5 December meeting Air Force Chief of Staff Twining denied CIA use of Yokota and objected to any U-2 operation in the Far East. Bissell reported this to Dulles, along with the mention that as of 6 December 1956, Atsugi Naval Air Station was available for the U-2, but that "Admiral Burke told us that he did not wish to proceed against USAF opposition."

On 12 December 1956 Dulles reaffirmed in writing to Twining the desirability of deploying AQUATONE to the Far East, and that necessary facilities could probably be made available at Atsugi: "Our request for such facilities is before Admiral Burke, and we are awaiting the outcome of his consultations with you and other members of the Joint Chiefs of Staff."

Now came the sharpest Soviet protest to date, after three RB-57D Black Knights from Yokota Air Base flew over Vladivostok during midday on 11 December. According to senior photo analyst Dino Brugioni, the Air Force had been making a 'hard sell' in Washington, D.C. for using their Black Knights:

"After some strong persuasion by Gen. Nathan Twining and assurance that the high-flying and faster RB-57D would not be detected by Soviet radar, President Eisenhower approved a mission. On December 11, 1956, three RB-57Ds overflew Vladivostok, the headquarters of the Soviet Pacific Ocean Fleet. The Soviets did detect and track the planes. On December 16, 1956, they lodged a strong protest. Eisenhower, according to Allen Dulles, was furious. On December 18, still angry, President Eisenhower said he was going to 'order a complete stoppage of this entire business.' He instructed Colonel Goodpaster to call Secretary of Defense Wilson, JCS chair Radford, and DCI Dulles and inform them that, effective immediately, there were to be no flights of US reconnaissance aircraft over Iron Curtain countries." (Brugioni 2010, p. 193) As it turns out, Vladivostok had not been on the national Ad Hoc Requirements Committee's list of highest priority targets.

Political pressure was against overflights, but Soviet rocketry and the Soviet nuclear program were rapidly advancing at the time, and Bissell wanted in all due haste to have the three Detachments in place to be able to monitor and assess the situation. As it turned out, the Klyuchi area of the Kamchatka Peninsula became the designated landing area for the Soviet ICBM tested in 1957 and would go to the top of the priority target list. Therefore, Far East operation would become more important than ever.

In the face of lack of progress toward securing an Asian operating facility for the U-2, Bissell on 29 January 1957 wrote to Dulles: "We have been trying since last October 25th to obtain facilities for AQUATONE in the Far East. The Air Staff is opposing this almost wholly for reasons of jurisdictional jealousy. Specifically, I think you should ask Radford and Twining to support your request to the Navy for a hangar at Atsugi and should say to them perfectly politely that if they are unwilling to do so you are going back to the President for a specific mandate."

At length, the granting of the request for the use of Atsugi enabled the CIA to direct money for renovation of the hangar and other construction

there to meet the needs of Detachment C.

Discussion of Air Force – CIA Disunity

The RB-57D detection by the Soviets reportedly gave rise to new doubts in the minds of Admiral Radford and General Twining about the security of bases in Japan. Bissell received notice that the matter of the deployment of Detachment C to the Far East required discussion with the Joint Chief of Staff. As late as 24 January 1957, no date had been set for such a discussion nor had there been the assurance of any decision from a meeting. An important Air Force Commanders' meeting at Maxwell Air Force Base beginning on 25 January, prevented the meeting for at least another week. Bissell said that this was one of the worse cases of bureaucratic foot-dragging and executive indecision, he had witnessed in some 13 years of Government service.

While this long, drawn-out interchange had been going forward on what should have been the relatively minor and easy issue of deployment, the prospects for a favorable decision on the major issue of authority to perform overflights seemed to have deteriorated. The attitude toward overflights both in the White House and the State Department was very different than what it had been a year or even six months before. The President's Inaugural address reaffirmed a policy of peace at almost any price, and those in higher authority seemed to regard an overflight as a dangerously provocative act, a distinctly more alarmist view than that expressed in the special National Estimate on probable Soviet reaction. At a minimum, higher authority now anticipated that any detected overflight would provoke more Soviet diplomatic protest. To date, these already slowed down or stopped U-2 overflights.

Bissell pointed out that along with the discouraging developments with the Far East deployment, the attitude of the Air Force toward Project AQUATONE had undergone a marked change since mid-autumn from one of full and open support and partnership toward one of increasing jurisdictional jealousy. The most obvious manifestation of the change was the effort to prevent AQUATONE from competing with Black Knight in the Far East.

Still another concern was the time-consuming

and contra-productive insistence of processing of AQUATONE film in the field using units under Air Force command, rather than by personnel attached to the AQUATONE field detachments.

Bissell was convinced that much of the pressure behind the SAC follow-on program involving the U-2 aircraft had as its purpose not the creation of a much needed hot war reconnaissance capability, but the readying of Air Force units having the same capability as AQUATONE so as to undermine any argument for the retention of this capability by the CIA. This whole attitude of increasing competitiveness, suspicion, and unconcealed eagerness in some quarters to have AQUATONE terminated was not only unpleasant, but was beginning to interfere with the CIA's activities and necessary security arrangements.

Rightly or wrongly, these circumstances—the inability to obtain any decision in the Far East, the growing fear that overflights were not resuming, and the increasingly evident Air Force disfavor—was having a major effect on the morale of the personnel assigned to the Project. The fact that a definitive meeting on the Far East problem looked for each week since before Christmas, and never scheduled appeared as evidence that this Project no longer had a higher priority claim on the time and attention of senior officials. The failure to use the capability for so many months after alerting the Russians to its existence was deeply discouraging to everyone, especially to the pilots who knew well that the Russians were working hard to develop the means of interception. The increasing Air Force disfavor was particularly hard on the Air Force personnel assigned to the Project. There were several senior officers already feeling their Air Force careers prejudiced by their loyalty to this Project, which had aroused the criticism of Generals Lewis and Everest (and quite possibly of General LeMay).

Bissell reminded the Director that many of the personnel concerned felt that their involvement in this Project brought real discomforts and disabilities, including separation from their parent organizations (unwelcome even to many Agency employees who had been pulled out of their regular offices where they had their careers), a long period of duty overseas or deployment to

locations remote from recreational facilities, and long separations from their families. By and large, they had freely accepted in the belief that this was an urgent and enormously important enterprise, and that hardships accepted with cheerfully if this belief maintained. But, when compounded by long inactivity, the loss of a sense of urgency, and the beginning of organizational jealousies, the effect on morale was extremely serious. Bissell felt it was all very well to argue that in their businesspeople must cultivate patience, but it was difficult to make this demand of people if they felt that delay and uncertainty were the products of indecision rather than of unavoidable circumstance.

Bissell recommended to the Director that should they not obtain a decision to deploy Detachment C, that the Director approach the President within the next three weeks to lay out the problem to him. Bissell was convinced that further obstruction of their deployment would be evidence, in part, of the Air Force's jealousy and in part of general discouragement in all the Services concerning the possibility of overflight activities during the next year. This was a state of mind that he could not live with. He felt the Agency could not wait for another six weeks to obtain clarification of their position. In that light, he recommended not advancing the Rainbow program until they received a decision. In his opinion, 10 February 1957 was the cutoff date.

In the aftermath of all these tribulations, the final decision was favorable to the CIA, and an advance party for Detachment C deployed on 19 February to Atsugi, Japan. By 21 February 1957 Detachment C personnel flew their 1,000th hour in the U-2, starting from their training.

In Germany, Detachment A moved from Wiesbaden to Giebelstadt during February 1957 with all three CIA units now deployed, and the Air Force group in training at the Ranch.

Work on Rainbow

On 6 May 1957, Bissell reported to the president about the progress of Project RAINBOW, saying that in operational missions, most flights went untracked. President Eisenhower again authorized overflights of the Soviet Union with the CIA promising Soviet tracking of the U-2 as unlikely.

On 2 April 1957, during a Project RAINBOW test flight at Watertown, Article 341 suffered a flameout at 72,000 feet due to airframe heat build-up caused by the "wallpaper" modification acting as insulation around the engine. Lockheed Test Pilot Robert Sieker's pressure suit inflated, but his helmet faceplate failed, and he lost consciousness. The plane stalled at 65,000 feet and entered a flat spin.

At a low altitude, Sieker recovered enough to bail out. Without an ejection seat, Sieker died for lack of enough altitude for safe manual egress. The U-2's tail struck and killed him in midair. The aircraft crashed in an area, so remote search team needed four days to locate the wreckage. The extensive search attracted the attention of the press. An April 1957 article in the *Chicago Daily Tribune* headline read, "Secrecy Veils High-Altitude Research Jet; Lockheed U-2 called *Super Snooper*." The rescuers found Sieker's body near the wreck with his parachute partially deployed. Instructor pilot Louis Setter returned his body from the crash site to Watertown in the L-20.

Because of its wide, glider-like wingspan, an out-of-control U-2 tended to enter a classical flat spin before ground contact — this slow descent lessened the impact. Having no fire to occur after impact often made the remains of crashed U-2s salvageable, as was the case with the wreckage of Article 341. Kelly Johnson's crew at the Skunk Works used the recovered pieces, along with spare and salvaged parts of other crashed U-2s, to produce another flyable airframe.

Development Projects Staff noted the U-2's ability to survive a crash in fair condition. The survivability after a crash became a consideration in its contingency plan for the loss of a plane over hostile territory. The thought of a plane surviving a crash meant an easy compromise of the weather research cover story because of the equipment on board the aircraft.

The loss of one of Lockheed's best test pilots, as well as the prototype "dirty bird" U-2, led Kelly Johnson to suggest that Lockheed install a large boom at the radar test facility. Using the boom, which could lift entire airframes 50 feet in the air, technicians could change the airframe's attitude and run radar tests almost continuously without having to fuel and fly the plane. Meanwhile, other means of making the planes stealthier to enemy radar continued. Lockheed conducted an extensive program of designing and building configurations of the aircraft, measuring radar returns from as many as 40 of these models.

Passport Visa - Project 6268, Radar Camouflage Study

For years, the Wave Propagation Section of the Aerial Reconnaissance Laboratory in Washington had been developing a radar absorber material suitable for structural application to aircraft.

ARDC assigned a T-33 aircraft to the Passport Visa project. They proposed completely covering the T-33 with the absorber (except for a part of the canopy, the engine intake, and exhaust ducts) in a way that allowed removal and replacement by other material. They anticipated a test continuing up to 2 years after completion of initial installation. The test called for flying an uncoated T-33 beside the test vehicle and comparing the reflection levels from a wide variety of S, X, and C-band radars.

In summer of 1957, boom-testing of the 'dirty bird' was complete and on 21 July 1957, Detachment B flew the first overseas operational mission of a U-2 using this radar deception system. The CIA flew nine flights of the 'dirty bird.' It became apparent to the team that the system offered only marginal improvement, so its use ended during May 1958. Bissell and his Air Force assistant, Colonel Jack Gibbs decided the need for a follow-on aircraft, one with a stealth design.

Bissell and Gibbs consulted with various laboratories and experts about what materials and designs might succeed. During December 1957, Bissell conducted a meeting presenting various techniques:

Robert Sieker

Engines and other metal structures inside the aircraft required shielding by reflection.

Some structural members were impossible to shield and required transparency by using plastic and eliminating metal components inside.

Protecting against S-band and X-band radars required shaping the exterior of the aircraft to reflect the energy away from the radar unit.

To reduce reflections, exposed edges would require "softening" to have a gradual change in the impedance of the structure.

During May 1957, Eisenhower had again authorized overflights over certain significant areas, based on promising improvements with RAINBOW. The President continued to authorize each flight plan. At that time, Detachment A was in Germany, Detachment B in Turkey, and Detachment C in Japan.

Detachment C staged one U-2 and supporting operation to Eielson Air Force Base in Alaska during the summer of 1957. On 8 June 1957, the U-2 took off from Eielson to conduct the first intentional overflight of the Soviet Union since December 1956. This mission broke new ground in two respects: it was the first overflight conducted from US soil and the first by Detachment C.

James Barnes flew this mission. Inclement weather over the Soviet ICBM impact area near Klyuchi on the Kamchatka Peninsula spoiled the mission. The aircraft stayed offshore and landed at Eielson.

Part of the reason for approval of renewed overflights was flight planning to enter from the Far East, where Soviet radar defense was less effective. On 18 June 1957, Al Rand flew the seventh Soviet overflight, from Eielson over the Kamchatka Peninsula and back to Eielson. Following this, the CIA flew "Soft Touch" missions over Russia and China during 4–27 August 1957.

Operation Soft Touch took advantage of gaps in Soviet early warning radar coverage in the eastern, remote republics of Turkmen, Tadzhik, and Kirgiz. The CIA had gained permission to launch U-2 overflights from neighboring Pakistan, which President Eisenhower finally approved. During five weeks starting 5 August 1957, Detachment B flew deep into the Soviet Union nine times on "Soft Touch," successfully photographing many high-priority targets,

including Kapustin Yar; the newly discovered ICBM test launch site at Tyura-Tam (Tyuratam); and nuclear weapons development facilities as far north as Tomsk.

Other flights were originating from Lahore, Pakistan. A Lahore flight on 5 August piloted by Buster Edens provided the first photographs of the Baikonur Cosmodrome near Tyuratam; the CIA unaware of its existence until then.

On other flights, the CIA investigated the Semipalatinsk nuclear test site and the Sary Shagan missile test site. After a few more overflights that year, only five more would occur before the May 1960 incident because of Eisenhower's cautiousness. The president sought to avoid angering the Soviets as he worked to achieve a nuclear test ban; meanwhile, the Soviets began trying to shoot down even U-2 flights that never entered Soviet airspace, and the details in their diplomatic protests showed that Soviet radar operators could effectively track the aircraft.

The Soviets had developed their own overflight aircraft, variants of the Yak-25, which in addition to photographing various parts of the world through the early 1960s, acted as a target for the new MiG-19 and MiG-21 interceptors to practice for the U-2.

To counter the new threat, Lockheed attempted to hide the aircraft by painting the plane in a blue-black color to blend it in against the darkness of space. The CIA also countered by powering the aircraft with the more powerful J75 engine that increased its mission cruising altitude to 74,600 feet.

Proposed Advanced Reconnaissance System

On 19 November 1957, the President's Board of Consultants on Foreign Intelligence Activities stated that it was aware of two proposed reconnaissance systems, the Department of Defense reconnaissance satellite and the Central Intelligence Agency's manned reconnaissance aircraft design study for greatly reduced radar cross-section. The study in question originated with the RAINBOW project, the purpose of which was to develop radar camouflage applicable to the U-2 aircraft without serious impairment of performance and sufficiently effective to permit a percentage of the reconnaissance missions to go undetected and greatly to reduce the accuracy and extent of radar tracking of reconnaissance missions even when detected.

Although achieving considerable success toward this objective, it became apparent by mid-summer 1957 that only limited and temporary success possible through the application of passive camouflage to an aircraft of a conventional structure due to either being too bulky or too heavy.

At the same time, the Russian radar system, already characterized by the use of a very considerable degree of frequency diversification had rapidly improved. Any feasible combination of narrowly banded camouflage solutions could cover only two or three regions of the whole spectrum and therefore could give only limited protection.

As a proposed course of action, the program of studies, measurement, and experimentation would proceed with all possible speed, in conjunction with further work on the RAINBOW camouflage, looking toward the choice of a design approach for a possible new aircraft within three months. The work would be under the technical direction of the scientific staff with actual systems responsibility remaining in the AQUATONE Project Headquarters in Washington, DC.

Bissell had hoped that the RAINBOW project would be the U-2's salvation. During flight tests in the first half of 1957, U-2s coated with radar-absorbing materials or fitted with radar-deflecting wires had fooled some US radars enough so that the CIA deployed the modified aircraft to Detachment B and the new Detachment C based in Japan.

However, operational test flights along the Soviet border encountered technical difficulties in defeating both the S-band Tokens and the lower, VHF-band Soviet radars nicknamed Dumbo and Knife Rest. In addition to this, the added weight of 'stealth' modifications had reduced the U-2's maximum altitude by up to 5,000 feet, negating advantages of the new J75 engine.

The stealth issue added to the increasing frustration in the US intelligence community over the political constraints. Bissell noted, "the result of keeping the car in the garage until better times will merely ensure that it becomes obsolete before it is ever used at all." In plain words, the U-2 was a "dying resource," and its follow-on programs now needed more than ever.

The CIA let study contracts with Convair, Boeing, Hughes, Marquardt, and Goodyear, followed and reviewed by a special panel under the chairmanship of Dr. Edwin Land. Meanwhile, Bissell was already looking at launching Corona satellites to take photographs during their orbits over the Soviet Bloc. The satellites would drop their film in a capsule in a preselected ocean impact area for recovery.

The CIA's Watertown Facility and its Nuclear Neighbor

Because of impending nuclear tests, Watertown was now readying to shut down. During June 1957, the entire facility evacuated with all remaining CIA and USAF personnel, materiel, and aircraft transferring to Edwards AFB, California, as Detachment G.

An AEC information booklet called *Background Information on Nevada Nuclear Tests* published in 1957 gave a cover story for the Watertown operation. It stated that NACA was operating U-2 aircraft at the Groom Lake site "with logistical and technical support from the Air Weather Service of the US Air Force to make weather observations at heights unattainable by most aircraft." The U-2 aircraft had remained unpainted except for NACA markings in the event of losing one.

No one paid much attention to the booklet. What the CIA was doing was merely a footnote to other more newsworthy activities occurring in the area.

During May 1957, AEC Radiological Safety Officer Charles Weaver, with Oliver R. Placak and Melvin W. Carter, participated in two meetings at Watertown, where he revealed and discussed the film, *Atomic Tests in Nevada*. The AEC briefed Watertown personnel on nuclear testing activities, radiation safety, and the possibility of radiation hazards from the upcoming Operation Plumbbob test series. Before leaving Watertown, the AEC men met with two Air Force officers, Colonel Jack Nole, and a Colonel Schilling, and CIA commander Richard Newton to discuss arrangements for radiation monitoring personnel to visit the flight test facility whenever anticipating the fallout in the Watertown area.

The atomic bomb testing did not deter Kelly Johnson from operating at Watertown. He realized the nearby Atomic Proving Grounds was already providing a blanket of security for the facility. For the next few years, Watertown residents would learn to live with their atomic neighbor, evacuating the facility during nuclear tests and returning to repair the damage caused by the atomic detonations. At Watertown, the workers wrapped chicken wire around the fluorescent lights in the mess hall and hangars to catch any bulbs shaken out by the underground tests or the sonic booms from the planes.

On one occasion, Watertown received notification of an underground atomic test in the range to the west. About nine in the morning, they felt the earthquake-like shake of the test. A brief time later, they learned of the underground test venting and the prevailing winds blowing the radiation toward the CIA's Groom Lake facility

The "Connies" evacuated the non-critical people, and everyone else assembled in the metal mess hall for possible evacuation. The authorities informed them of the planes returning from Burbank to evacuate them, and for them to remain inside the mess hall that remained open 24 hours with free food.

No planes came, and after more than an hour, someone told them of two AEC cars driving around the facility, monitoring the radiation levels. One person in a gray, four-door Chevy with the AEC name on the doors stopped nearby while he fooled around with some instruments in the rear of his vehicle. He did not want the Watertown residents around and refused to answer questions concerning the radiation levels he collected. He said, "It's OK," and drove a couple of blocks to the other car. They stopped side-by-side for a few minutes before driving off in different directions.

During June 1957, a month following the US Air Force U-2 pilot class moving to Texas, that the CIA's U-2 test operation moved to North Base at Edwards Air Force Base, California. The Plumbbob series of nuclear tests was to last several months, so there was no choice but to evacuate Watertown.

After the completion of U-2 student training, Setter was still stationed at Groom Lake, acting as CIA's Mr. Bissell's test engineer. Bissell had called him to Washington DC, and on the flight back to the Ranch in a T-33, Setter was cruising solo at about 35,000 feet over Texas and was probably 80 miles west of El Paso, when his engine began to flame out. The fuel pressure had dropped to near zero, and the RPM was dropping steadily. Setter had no idea what the problem was, so he set up a glide from high altitude headed west toward Biggs AFB, El Paso. There were a few clouds under him, and he could see the Interstate 10 highway below, and no alternate airports. There were low clouds over El Paso. Setter had never bailed out of an airplane and did not trust his ejection seat after some of his friends were killed or injured using ejection seats. He seriously considered landing on the highway if all else failed. But he continued straight ahead. Biggs radar control picked him up and vectored him to the runway. The problem was—for the last few miles, Setter was IFR—gliding in the clouds, and utterly dependent on radar coverage to keep him out of trouble. Finally, at about a mile from the runway, he broke out and landed safely. Setter had enough speed, so he turned off the runway and parked, very lucky to be alive. Suddenly he noticed that his engine had come back to life—by itself! It was at idle RPM. The Aerodrome Officer drove up, along with the fire trucks, and asked: "what's going on?" "Why did you declare an emergency?". To him, everything looked normal. He shut down the engine, climbed out of his cockpit, jumped off the wing, and said to the officer, "come look under my left wing." He took out his big screwdriver, loosened the main wing fuel drain cap, and out comes a steady flow of fluid—water! The water in his fuel lines had frozen at high altitude, then, while Setter was gliding, the ice in the lines had melted, allowing the engine to keep running at idle RPM, without his knowledge! Setter had them drain all his fuel lines, refuel the airplane, then took off again for the Ranch. AND THEN, THE FUEL LINES FROZE AGAIN! He pulled back the throttle, let down to about 10,000 feet, below the freezing level, and continued on to the Ranch (Setter was now over Phoenix, Arizona). (Note: In those days, the Air Force had not yet added a solvent to jet fuel, so his problem had happened before. Later, when they added a solvent to jet fuel, the problem was solved.)

Watertown a Nuclear Fallout Monitoring Station

The Atomic Energy Commission was using the Watertown facility as a monitored area for its nuclear tests, even before the departure of the U-2s. This was especially so when Watertown became a virtual ghost town in caretaker status with a site manager, security, and minimal complement of personnel.

During April 1957, a mere five miles northwest of Groom Lake in Area 13, the AEC experimented with an XW-25 warhead. This Project 57 preceded Plumbbob. The device, with a design yield of one to two kilotons, simulated an accident without a nuclear detonation with mock-ups of sidewalks, city surfaces, and objects placed to see how contamination would affect them. The test spread plutonium over 895 acres.

Interestingly, nine years later, over Palomares, Spain, a mid-air accident led to a similar release of plutonium when two H-bombs hit the ground without parachutes. A collision at 30,000 feet killed all the crew of a KC-135 tanker and three of the crew of a B-52. The high explosives of two of the H-bombs with failed parachutes detonated on impact, spewing plutonium contamination over about a square mile. The site still undergoes cleanup.

The firing of the 12-kiloton tower shot BOLTZMANN on 28 May 1957 at Yucca Flat was followed in early June by two minor blasts, FRANKLIN and LASSEN, all at about the time the U-2 pilot classes finished training at Watertown and the U-2 operation moved to North Edwards AFB.

The entire series of Plumbbob occurred between 28 May and 7 October 1957, following Project 57 to become the biggest, longest, and most controversial test series in the continental United States. The operation involved 29 explosions, with 21 laboratories and government agencies participating.

Operation Plumbbob nuclear testing continued with five additional safety experiments, and 18 more full-scale detonations conducted during August 1957 with several shots dropping significant fallout on Watertown. These were DIABLO, DOPPLER, SMOKEY, and WHITNEY. SMOKEY, with a yield of 44 kilotons, fired on top of a 700-foot tower in Area 8, 14 miles southwest of Groom Lake. The dirty mushroom cloud spread radioactive debris over the Groom Lake area.

Most Operation Plumbbob tests contributed to the development of warheads for intercontinental and intermediate-range missiles; they also tested air defense and anti-submarine warheads with smaller yields. They included forty-three military effects tests on civil and military structures, radiation and bio-medical studies, and aircraft structural tests.

One nuclear test involved the largest troop maneuver ever associated with US nuclear testing. In all, approximately 18,000 members of the Air Force, Army, Navy, and Marines received testing physically and psychologically to the rigors of the tactical nuclear battlefield.

At shot HOOD, the Marine Corps conducted a maneuver involving the use of a helicopter airlift and tactical air support. At shot SMOKEY, Army troops conducted an airlift assault, and at shot GALILEO, Army troops tested to determine their psychological reactions to witnessing a nuclear detonation. (Around 45,000 Soviet armed forces personnel performed similar tactical exercises after a 40-kiloton nuclear detonation at the Totskoye proving ground in 1954)

The Atomic Energy Commission subjected almost 1,200 pigs to bio-medical experiments and blast-effects studies by placing them in elevated cages. The pigs wore suits made of varied materials to test which materials provided the best protection from thermal radiation.

The AEC detonated PASCAL-A, the first underground shaft nuclear test to study the possibility of a nuclear weapon detonation following a plane crash.

The JOHN shot on July 19, 1957, tested the US Air Force's AIR-2 Genie missile with a nuclear warhead. The Air Force fired it from an F-89 Scorpion fighter over Yucca Flats with five Air Force officers and a photographer standing beneath ground zero of the blast detonated at

19,000 feet altitude.

Outside the Atomic Proving Grounds, the AEC used the CIA facility as a monitoring area for an atomic test code-named WILSON, which deposited fallout on Watertown that the AEC recorded closely. The AEC measured radiation inside the buildings and vehicles at the facility to study the ability of various materials to withstand the fallout. In effect, Watertown was a laboratory to determine the shielding qualities of typical building materials found in any average American small town.

The nuclear fallout and damage to Watertown continued with the 37-kiloton PRISCILLA shot detonated at Frenchman Flat. HOOD, the sixth nuclear shot of Operation Plumbbob followed on 5 July 1957 14 miles southwest of Watertown.

A balloon lifted HOOD to a height of 1,500 feet over Yucca Flat, where it exploded with a yield of 74 kilotons. The blast caused substantial damage to the Watertown facility, its shockwave shattering the windows on two buildings at Watertown and breaking a ventilator panel on one of the dormitories. The west and east doors buckled on a maintenance building on the west side of the Groom Lake facility and buckled the south door of the supply warehouse west of the hangars.

During its era at Area 51, CIA Project OXCART would subject the workers to five underground tests at Yucca Flats. These were FADE on 25 June 1964, DUB on 30 June 1964, CENTAUR on 27 August 1965, DERRINGER on 12 September 1966, and TORCH on 21 February 1968.

Today, some wonder about the radiation levels at Area 51 lying downwind of the atomic test site where it often took two or more hours for the AEC to give the facility an "All Clear."

Did the test expose them to atomic radiation without notice as it did the people of St. George and Cedar City, Utah during the aboveground tests of the 1950s?

Department of Energy, in their "Openness Report" stated 114 of the 723 Underground Tests, since 1963, released radioactive material into the atmosphere. The most common radioactive effluent was Iodine-131, and they mentioned Krypton-85. They say, in part, that they contained the leaked radiation within the border of Nevada Test Site and the adjacent Government controlled Nellis Air Force Range, which included Area 51. However, those working at Area 51 never experienced any health anomalies from working there during the atomic testing. Interestingly, at the time of this writing, most of Area 51 veterans of the CIA era are now in their mid-80s, and an unusual number of the men have outlived their wives who experienced no exposure to nuclear fallout.

Unauthorized Visitors

During July 1957, security at the CIA facility detained a civilian pilot when he made an emergency landing at the Watertown Airstrip. Edward K. Current, Jr., a Douglas Aircraft Company employee flying a cross-country training flight became lost. Running low on fuel, he decided to land at Groom Lake. Here, security officers held him overnight for questioning. The Nevada Test Organization (NTO) security officials reported the incident occurring at the "Watertown landing strip existing in the Groom Lake area at the northeast corner of what would later become Area 51."

In another incident, a flight of three F-105 Thunderchiefs, led by British exchange pilot Anthony "Bugs" Bendell, was on practice nuclear weapon delivery sortie 80 miles north of Nellis Air Force Base when one aircraft experienced an oil pressure malfunction. One F-105 returned to Nellis while Bendell led the stricken craft to the airfield at Groom Lake. After making a pass over the field with no response to distress calls, Bendell advised the student pilot to land. At this point, two F-101 Voodoos intercepted Bendell and forced him to land.

As the US Air Force RED FLAG exercises increased in attendance, participating pilots began declaring emergencies in a guise to obtain permission to land in the restricted airspace at Groom Lake.

The emergency landings ceased after the CIA retained one of the pilots for a prolonged period for debriefings and held the plane even longer to send a message to the US Air Force that the CIA would no longer tolerate such security breaches.

PCS and Temporary Duty Basis for Field Assignment

From the start, the CIA went to great extremes to hide the agency's affiliation with the U-2 and to hide the operating facility in Watertown. In 1955, the permanent cadre had received change of station (PCS) orders to Los Angeles with instructions to settle their families there. They returned to Watertown on temporary duty orders. Other personnel assigned to Watertown for training before going overseas received a PCS to Washington and temporary duty orders to Watertown. (Not Area 51)

The CIA tried to equalize per diem rates among all categories of personnel. Any employee, civilian, or military, reporting to the agency's flight test site on or after January 1956 received $12 per diem for the first 30 days and $10 a day after that. The same per diem applied to the CIA's flight test site at Edwards Air Force Base North Base established during July 1957.

Mr. Robert Macy of the Bureau of the Budget questioned this policy during a visit to Watertown during February 1956. It concerned him that individuals drew that much per diem but paid only $4.25 for room and board at the facility. After an explanation of the philosophy behind the policy, Mr. Macy decided not to bring the matter up in his report.

When Detachment A deployed to England on a PCS basis, it was without dependents or household effects, but in anticipation of a full tour in England. The unforeseen events that necessitated the hurried move to Germany turned the deployment into an 18-month overseas tour when, before the year was out, the detachment moved to another German base before returning to the ZI.

This experience led to the decision of the detachments deploying on temporary duty rather than PCS given the inability to predict the length of stay at a given base. General Cabell approved this change of policy during August 1956 when Detachment B deployed on temporary duty to the Incirlik Air Base in Adana, Turkey without dependents or household effects. During March 1957, Detachment C deployed to Japan on the same basis.

The CIA's U-2 program spent much of 1956 and the spring of 1957 with only a minimum of mission tasking for Detachment B; a few Soviet penetrations, several peripheral SIGINT missions, and proficiency flying. The detachment flew one test mission with the weather sensing package aboard, about four hours around Turkey. The entire operation became boring with only periodic breaks to go into Adana. At times, Adana itself was off-limits.

There was only limited recreational activity on the base. For this reason, the CIA furnished a boat, motor, trailer, and water skis for the detachment to use in a nearby reservoir. Colonel Perry felt the detachment was greatly over-staffed and sent Bohart, Perkins, Don Curtis, and others home. There was much drinking and poker playing, particularly by the contract pilots.

In anticipation of future tasking, Colonel Perry ordered the detachment to undergo a simulated deployment exercise in which everyone packed up and cataloged all supplies and equipment they would need at a bare based operation in Iran, Pakistan or somewhere else. His foresight was remarkable when during May-June 1957, the detachment moved a mobile unit to Lahore, Pakistan to fly penetrations of the Soviet Union.

The preferred base was Peshawar, but the runway was under repair. Thus, Colonel Harry Cordes, along with weather, maintenance, communications, and flight planners, plus the inevitable CIA security moved to a Pakistani Air Base for staging. They moved mostly by C-54 through Karachi but ferried the U-2 aircraft and a T-33. They found the Pakistani air force people extremely helpful and cooperative. They lived in a Lahore motel-like hotel that had excellent food and service. The menu offered a choice of English or Pakistani, the latter being hotter with curry.

On one of the flights, Gary Powers made a dead stick landing on the 6,000-foot runway at Lahore. All the U-2 pilots had made many simulated flameout (SFO) landings; however, Powers landing with no damage to the U-2 was a further demonstration of his flying skill.

Colonel Perry sent Cordes back to Incirlik because the chairman of the Joint Chiefs of Staff, Air Force General Twining, and other high-ranking officers, including General William Blanchard, would be visiting Incirlik. General

Twining had been personally involved in the fight with the CIA over who would control the U-2 program. (Air Force reconnaissance pilots also looked askance at the U-2 pilots with mixed feelings: According to RB-57 pilot Gerald E. Cooke, U-2 pilots boiled down to "demilitarized Air Force fighter pilots receiving salaries almost seven times our own.")

On 15 August 1957, Colonel Perry returned to Incirlik and resumed his arguments with Richard Bissell over several issues with overstaffing and lack of meaningful activity; CIA reluctance to turn over Soviet Targets Photography to SAC for use by crews who were dependent on *National Geographic* and German 1942 photography; project future plans being too open-ended; and the apparent prospect of dead ends for military personnel careers.

Colonel Perry left Incirlik and the U-2 program shortly, after that, leaving Maj Harry Cordes in command. Bissell did not trust a major to command Detachment B and sent a headquarters officer, Major Joe Richmond to assist him. Two months later, Col Stan Beerli arrived from Detachment C in Japan to take over Detachment B. Major Cordes requested his return to the Air Force.

On 24 September 1957, the project director wrote to the Deputy Director, Support to advise him of the desired change in policy: with the prospect of continuing AQUATONE operations overseas at least through the calendar year 1958, he suggested they make plans to have the dependents of project personnel to join them at their locations. He pointed out how the concept to date had centered on maintaining a high degree of mobility for staff and equipment.

The events of the past eighteen months showed the political impact of having an AQUATONE unit within the border of a friendly country less time than anticipated, and this consequently shifting them to a fixed base concept with a forward staging capability. A fixed base operation made them consistent with the secrecy concerns of including dependents for unit personnel. The concept included the CIA's contract pilots, some of them· married with dependents wanting to join them overseas.

The DD/S approved their establishing this policy and initiating a crash program to prepare dependent housing. They accomplished this at Adana by rental and renovation of local economy houses and by using trailers shipped from the US.

In Atsugi, they remodeled existing Agency billets and constructed more units through a local builder. This program cost several hundred thousand dollars in each case. The CIA could not recoup this cost when the two detachments returned to the ZI.

Soviet Reaction and Adaptation to U-2 Overflights

The USSR's counterintelligence information on the U-2 started accumulating with the first overflights. According to Soviet Colonel Alexander Orlov, the U-2's route showed what the Americans wanted to know. In 1955-1956, it was a reconnaissance of Soviet air defenses. In 1956-57, the main zeal was Soviet strategic bomber bases. In time the Soviets came to discern the U-2 could carry out missions of 6,500 kilometers (4000 miles) without refueling and of flying for 8-10 hours.

The USSR located airfields used to base the U-2 and for their forward staging in the FRG (Federal Republic of Germany, West Germany), Turkey, Japan, Pakistan, and in Alaska. By the end of the 1950s, it became clear to the Soviets how the U-2 prepared for a mission. Usually, a few days before a combat mission, a U-2 accompanied by a C-130 departed from the Incirlik Air Base in Adana, Turkey for Peshawar, Pakistan. The U-2 pilot would stay there for 1-3 days waiting for favorable meteorological conditions. After that, the pilot would carry out the mission.

The CIA did not select Peshawar airfield as a base at random. First, the Air Defense radar network in the Turkestan military district was weak, which made it possible to penetrate without detection. Secondly, the route was the shortest for a spy plane to reach the most crucial testing ranges in Tyuratam (Baikonur), Sary Shagan, Semipalatinsk, Kapustin Yar, and other vital installations.

It had become clear to the Air Force and CIA that in the field of strategic aviation and air defense that the USSR was far behind. However, the situation with Soviet nuclear missiles was not so clear. The flights of the U-2 had a sporadic character about them, and Washington did not know the whole picture in this area. For the United States, it was imperative to know the genuine answer of how the USSR military compared with that of the US. By 1960, crucially, the USSR had established its Strategic Missile Forces and identified the regions already surveyed by the U-2. The USSR knew to expect new high-altitude spy missions.

Additionally, in late 1957, mobile S-75 AA missile complexes appeared in service at the Soviet Air Defense units. The S-75 (also called SA-2, Surface to Air-2, 'Guideline' by the US) could hit air targets flying up to 1,500 km/hr. at 30-40 kilometers. By 1960, the S-75 missiles were in service in many air defense units, tightening the grip or noose on the U-2 overflights.

Nonetheless, on 9 April 1960, U-2 pilot Bob Ericson flew a 6-hour mission over the most critical installations, including Semipalatinsk, Sary Shagan, and Tyuratam. He got away with it and landed in Iran. That day put the USSR Air Defense to shame because of several chances to shoot Ericson down. General Mikhail was furious and gave the 'bitter lesson learned' on 9 April every consideration in every Air Defense unit, and not in vain. Russia would shoot down the next visitor, Francis Gary Powers.

The CIA's U-2 overflights of the USSR did not answer questions about the Soviet rocket and space programs, which surprised even Wernher von Braun when the USSR launched the Sputnik I & II satellites on 4 October and 3 November 1957.

Former Soviet Colonel Orlov's Comments

Soviet Colonel Alexander Orlov, Retired, who became a member of the Russian Academy of Natural Sciences, spent U-2 overflight years developing Soviet strategies to counter them and helped prepare the questions used in interrogating Francis Gary Powers, shot down over Russia on 1 May 1960.

According to Colonel Orlov, the CIA carried out its U-2 missions at the time when Soviet air defense was relatively weak as it was undergoing rearmament, putting into operation modern air defense warfare that included new radar stations, high altitude interceptors, AA missiles, and surveillance equipment. According to Col Orlov, formerly with the USSR Air Defense Forces, in 1955 Russia had around Moscow the stationary anti-aircraft system S-75 Berkut (SA-2 "Guideline") equipped with 60 missiles. It could engage 20 targets flying up to 1,250 kilometers per hours at altitudes up to 20 kilometers. This system protected only Moscow.

There were two sides to the overflight issue. During this period, the Soviet Union was developing and testing different missile weapons, as well as bombers and nuclear-propelled missile submarines. All of this occurred in the remotest regions of the country far away from anyone's eyes and strictest secrecy.

Col. Alexander Orlov

According to Colonel Orlov, in the early 1950s the United States might have been tempted by the idea of massive surprise nuclear attacks on vital objects of the Soviet Union deliverable by strategic aviation forces, a theory grounded on the assumption that the USSR had no adequate means to attack targets on the territory of the United States, and that the air missions the United States and other western spy planes were carrying out on a regular basis since the late 1940s near borders of the USSR (sometimes intruding into the airspace of the USSR), proved that Soviet defense was weak.

However, by the middle 1950s, the situation started to change. In 1954 and 1955, air parades in Moscow demonstrated Soviet M-3 and Tu-95 Bear Strategic Intercontinental long-range bombers and supersonic MiG-19 fighter planes. The USSR showcased its considerable success in the field of aircraft construction. Also, this sent the message of the Soviet Tu-95 bomber being capable of hitting installations on US territory, at least on a one-way strike mission.

The USSR May Day parades raised the

question of how strong is the bomber fleet? How strong will the Soviet air force be in the immediate future? At this time Khrushchev believed that so long as the Soviet Union was the weaker power, it had to practice brinkmanship to keep its adversary off-balance: "On January 8, 1962, in a speech that remained secret for over forty years, Nikita S. Khrushchev announced to his colleagues in the Kremlin that the Soviet position in the superpower struggle was so weak that Moscow had no choice but to try to set the pace of international politics. 'We should increase the pressure, and we must not doze off and, while growing, we should let the opponent feel this growth.' He adopted the metaphor of a wine glass filled to the rim, forming a meniscus, to describe a world where political tensions everywhere brought to the edge of military confrontation. 'Because if we don't have a meniscus,' he said, 'we let the enemy live peacefully.'" – Fursenko & Naftali p. 5

As remembered by Colonel Orlov, "Soviet successes were accompanied by a noisy propaganda campaign on the part of the Khrushchev regime. Khrushchev declared publicly that the USSR 'was making missiles like sausages.' Soviet influence on Third World nations also was on the rise in the late 1950s, as were rates of economic development in many Communist-ruled countries. All this was received painfully by American public opinion."

"The Soviets were deploying strategic missiles, building nuclear-powered submarines, adding new surface-to-air missiles to their inventory, and equipping the Air Defense Forces with advanced radar and other equipment. This meant U-2 flights were among the most important sources of information on the USSR's arms programs. Even so, CIA leaders recognized that the constant upgrading of Soviet air defenses was making deep U-2 penetrations increasingly risky."

At the time, Col Orlov knew that the Soviet Air Force and Air Defense intelligence and radio technical surveillance troops followed the U-2 missions right from the very first flight on 4 July 1956. Initially, there were strong doubts regarding the capability of the U-2 aircraft to fly at such high altitudes. This was due to the many radars in service at that time (P-8, P-10, and others) equipped with altimeters and capable of locating targets at an altitude of 20 kilometers and even more, but there were few of them, and therefore it was not always possible to lock on to the target without gaps in the coverage. In total, during the four years, the USSR detected and locked onto eighteen U-2 aircraft. However, the United States reported approximately 30 missions.

In the beginning, the Soviet knowledge about the aircraft was very obscure. In the 1950s, the reaction of the Soviet leadership to the U-2 flights was very nervous. The plane obtained credible information on the deployment of strategic weapons in the USSR in the way of photos taken that undermined Khrushchev propaganda campaign praising the numbers and delivery accuracy of the Soviet missile. The U-2 helped the Americans collect a great deal of valuable information regarding the Soviet nuclear programs and other weapons.

The U-2 established that while Soviet air defense was strong around Moscow and the western USSR, it was rather weak and ineffective in the critical northern and eastern directions.

The U-2 found that Soviet strategic aviation was far behind the US Strategic Air Command and not usable as the main force of a decisive blow on American territory. The CIA flights did establish that the primary strategic weapon of the USSR would be ICBMs. However, the flights failed to give the CIA a distinctive answer about the USSR's military capabilities, especially that of the Soviet Strategic Rocket Forces and the power of their atomic warheads. The CIA estimated the USSR as having 500 missiles by 1960.

According to Orlov, who personally participated in the USSR's Air Defense Forces, the USSR's first ICBM, the R-7 did not become operational until September 1960. In 1960-62, the Soviets had only six operational ICBMs (now re-named SS-6), and about 20 test launches accomplished to date.

The highest-ranking Soviet military officials believed that an aircraft, even of special design, was not capable of flying at an altitude of 20,000 meters for 7-8 hours. Only Gen Tupolev acknowledged such a plane, and he drew a sketch which showed a strong resemblance to the actual U-2. Others thought that the U-2 was able to fly at that altitude only during part of its mission and that it carried out the rest at the lower altitudes within striking distance of Soviet fighters. All available fighter planes, mainly MiG-19s, received orders to intercept U-2s flying over their area. Russia developed and tested the 'zoom climb' method and later used it against the mission flown by Francis Gary Powers.

U-2 Flights Continue as Area 51 Develops a Replacement

On 5 August 1957, Eugene 'Buster' Edens flew the eighth Soviet overflight, Mission 4035 from Detachment B deployment at Lahore in Pakistan, where he found and photographed the Soviet missile test facility at Tyuratam at a distance.

The president was seeking to avoid angering the Soviets as he worked to achieve a nuclear test ban; meanwhile, the Soviets were trying to shoot down even U-2 flights occurring outside Soviet airspace. The details contained in their diplomatic protests revealed that the Soviet radar operators were tracking the aircraft. Because of Eisenhower's increasing cautiousness, only five more flights occurred during the year before the May 1960 incident.

On 12 August 1957, a flight from Lahore on the ninth Soviet overflight returned with no details. That same day, the CIA flew the 10th and 11th Soviet overflights from Lahore over Semipalatinsk, Novokuznetsk, Tomsk, Berezovskiy, and back.

Ten days later, on 22 August 1957, Jim Cherbonneaux flew the 13th Soviet overflights from Lahore and returning. Mission 4050 discovered the nuclear weapons testing facility in Semipalatinsk. The photographs revealed many of the ground zeros from previous nuclear tests. The other sortie discovered Sary Shagan (used to test radars against missiles fired from Kapustin Yar and later a center for Soviet ABM development). Bill Hall flew another sortie to map the whole of Tibet.

On 28 August 1957, E. K. Jones flew the 14th Soviet Overflight, Mission 4058 from Lahore where he overflew and photographed Tyuratam.

On 10 September 1957, Bill Hall flew the 15th Soviet Overflight, Mission 4059, from Incirlik. He overflew and photographed the Soviet missile test center at Kapustin Yar, photographing an R-12 missile on the launch pad. Hall intended to continue his flight north toward Moscow. However, when he saw several MiG planes trying to intercept him, he turned south to cross Ukraine instead. Near Kiev, the Soviets fired a barrage of anti-aircraft artillery at the aircraft without success.

On 16 September 1957, Barry Baker flew the 15th Soviet overflight, Mission 6008, from Eielson over Klyuchi on the Kamchatka Peninsula and back to Eielson using the radar-evading 'Dirty-Bird.' Nonetheless, five MiG planes managed to detect and trail it. The extra equipment limited the aircraft to only 59,000 feet, where, looking down through the drift sight, Baker could make out the 'bone dome' of one Soviet fighter pilot only a few thousand feet below the U-2.

Jacob Kratt's Mission 2039, ELINT monitoring of Soviet fleet exercises, was the next-to-last mission for Detachment A. On 13 October 1957, Hervey Stockman flew the 17th Soviet Overflight, Mission 2040, from Giebelstadt, north to Norway, east parallel to the Soviet coast, then south toward the Kola Fjord. He flew over Polyarny, Severomorsk, and Murmansk. Stockman flew south as far as Monechegorsk before leaving Soviet territory in northern Norway. He landed back at Giebelstadt after more than 9 hours in the air, the last mission for Detachment A.

During November 1957, the CIA disbanded Detachment A, returning its planes and personnel to the USA. From this point on, the CIA used Giebelstadt only to refuel U-2s en-route to and from Detachment B.

On 2 March 1958, Tom Crull flew the 18th Soviet overflight, Mission 6011, from Eielson. He flew over the Soviet Far East naval aviation bases at Komsomolsk and Khabarovsk. He flew south following the Trans-Siberian Railroad to the Chinese border and back to Eielson. Again, despite the use of a 'Dirty Bird,' radar and interceptors tracked this flight.

Late 1958 to early 1959, Lockheed engaged in an engine change program for the remaining 13 Agency U-2s. The upgrade outfitted the aircraft with the more powerful Pratt & Whitney J75-P13 engine, making the aircraft known as the U-2C. Upgrades also to come included in-flight refueling capability and a model of the U-2 capable of operating from aircraft carriers at sea.

With the failure of Project RAINBOW to reduce the radar cross-section by using radar-absorbing materials and techniques, the CIA adopted the blue-black paint called 'Sea Blue.'

Op Congo Maiden

Toward the end of March 1957, seven U-2s staged from Eielson and returned. Flown by 'Buzz' Curry, Rudy Anderson, Bobby G. "BG" Gardiner, and 'Snake' Bedford, these photo missions along the Soviet Northern Siberian coastline in 1959 determined the status of upgraded World War II Soviet airfields in this frozen region. They enabled the CIA to evaluate Soviet air defenses in case of a pre-emptive US strike on the Arctic airfields became necessary along the extreme coastlines of Siberia. In later years, Gardiner provided the rare footage for the video "***The Inquisitive Angel.***"

During April 1958, the CIA source Pyotr Sehisonovich Popov told his handler George Kisevalter of a senior KGB official boasting of having "full technical details" of the U-2. Bissell concluded the project must have a leak. However, the CIA never identified the source of any leak. Many speculated Lee Harvey Oswald, a radar operator at the U-2 base in Atsugi, Japan, might have leaked the information to the Russians.

Col. Orlov relates, "Richard Helms, at that time a senior manager of CIA's Plans Directorate and of the high-altitude reconnaissance operations against the USSR, recalled that when the CIA learned from Pyotr Popov, one of its prize moles in Soviet military intelligence, that the Russians had amassed much information about the U-2, 'it brought me right out of my seat.'"

Unknown to Bissell at the time, the clouds of war were gathering much closer to home with Cuban President Fulgencio Batista fleeing Cuba for the Dominican Republic ahead of the Cuban revolution. His departure made Fidel Castro, the leader of Cuba. Soon, Castro declared the country was going Communist. Cuban-inspired guerrilla movements sprang up across Latin America. In Vietnam, North Vietnam formed the National Front for the Liberation of Vietnam with the intent to overthrow the government of South Vietnam.

These developments helped set the stage for future use and service by the U-2 in Southeast Asia and over Cuba, as the Cold War conflict levels increased and heated up in those areas too.

Hypoxia – In Their Own Words

Espinoza: "Other risks were less benign, as I found when I was the ground officer for a pilot who radioed, 'My skin feels like it's crawling.' He had the bends so badly from changes in pressure that when he landed huge welts covered his body. Had the weather not cleared in time for him to land, these bubbles of nitrogen might have lodged in his brain or optical nerve — as they had in other U-2 pilots." - Cholene Espinoza, from 2010 NY Times op-ed

Henry: "The first thing that happened was the pain in my knees.' Over the next five hours, Henry developed an intense headache, nausea, and extreme fatigue. The pain got worse. At one point, he hallucinated that the airplane had rolled 30 degrees to the left. At one point, he snapped awake, not realizing he'd dozed off." - "Killer at 70,000 Feet," *Air & Space Magazine*, May 2012

Halloran: "[At 63,000'] All was going well when I became aware of an increase in my rate of breathing to the point of obvious hyperventilation. I suddenly felt the breathing bladder of my pressure suit helmet collapse around my head. I knew that I was in serious trouble. OUT OF OXYGEN. My vision was seriously deteriorating, and my physical actions getting a bit spastic. With all the concentration I could bring to bear, and with an awareness that this was going to be my last shot, I managed to twist around, and with both hands, reconnected the breathing hose. The relief was immediate as my helmet filled with oxygen and my sight and breathing returned to normal."- McIlmoyle p. 77

Chapter 16 – The Black Cats

Detachment H - The Black Cat Squadron

In 1949, the government of the Republic of China (ROC) left mainland China for Taiwan. Following the Korean war ceasefire, the Chinese Communists threatened the ROC in Taiwan and the offshore islands. Former President Chiang Kai-shek, hoping to return to the mainland to liberate his compatriots, conducted photo reconnaissance flights over the mainland coastal areas opposite the Taiwan Strait.

Note for clarification, PRC is the People's Republic of China, occupying mainland China and founded by the Communists after they won the civil war in 1949. The Republic of China (ROC) government retreated to the island of Taiwan upon losing the war, declaring Taipei their provisional capital city. Republic of China Air Force (ROCAF) was Taiwan's air defense.

After the Soviet Union exploded their A-bomb in 1949, the keen interest of the Chinese Communist Party increased: "The leadership may have felt even during this early period that it would be necessary to develop nuclear weapons for reasons of national prestige, to attain and preserve great power status." - Ryan p. 22

By the early 1950s, it was beginning to become apparent that Mao Zedong wished for the People's Republic of China to play the role of 'the center of the world revolution.' For this, China would have to develop its own nuclear capability. "This perception is well captured by Mao's statement to senior Chinese officials in 1958 that, without nuclear weapons, 'others don't think what we say carries weight.'" - Lewis & Xue p. 36

In 1958 the People's Republic of China Central Military Commission stated flatly: "We must concentrate human, material, and financial resources. Any other projects for our country's reconstruction will have to take second place to the development of nuclear weapons." - Lewis & Xue p. 70

China had been "compelled" to develop nuclear weapons (Premier Zhou EnLai quote) - Lewis & Xue p. 1

"Marshal Chen Yi boasted that the Chinese must have their atom bomb even if they had to go without pants." - Perkovich p. 79

The Soviet Union also took notice, and the Sino-Soviet split began in 1958 over who would lead the Communist world. Naturally, not just Taiwan but the US and the rest of the world were concerned over the activity and ambitions of the PRC.

Accordingly, in the spring of 1960, an advance party of Americans came to Taipei to establish an agreement with Chiang Kai-shek for a joint U-2 operation. ROCAF officer 'Jude" BK Pao was selected to help secure a suitable airfield: "Because I was stationed in Taoyuan for years, I knew the base well. I had no difficulty to make arrangement and to bring them. The initial party recommendation was in favor of Taoyuan Airfield."

"In the protocol American side provides two vehicles [U-2], operation funding, maintenance, and logistic support while the Chinese side provides combat crew and base facility. The project was given a code-name RAZOR and the highest priority."

ROCAF General Hua relates: "During March 1959 Major Joe Jackson of 4028th SRS came to Taiwan and brought six of us: Shihchu 'Gimo' Yang, Tsiyu 'Tiger' Wang, Yaohua Chih, Huai Chen, Chungkuei Hsu and me, to Laughlin AFB. The following week ground school started. A few weeks later, we received pressure suits at Carswell AFB and then went through a pressure chamber at an altitude of 80,000 feet. After we got back to Laughlin, instructors began to introduce us to the glider-like U-2."

"We five ROCAF pilots completed the training during September 1959. Almost a year later, in the fall of 1960, we returned to Laughlin for further proficiency training. Two U-2Cs arrived in Taoyuan during January 1961." [Chungkuei Hsu did not finish training]

"To avoid being confused with other air force organizations stationed in the Taoyuan AB, the section became the 35th Squadron with the Black Cat as its insignia. On the American side, there was Detachment H. All US personnel were ostensibly employees of the Lockheed Aircraft Company. The Chinese participants consisted of five pilots, a few support people, and the squadron commander, a ROCAF colonel. ROCAF and US representatives signed a joint agreement, codenamed 'Razor.'"

The Agency rotated another U-2 in from the Agency inventory to replace a U-2 lost was for any reason, so that there would always be two on hand. Once the pilot detachment completed a certain number of combat missions the Agency relieved the pilots of duty and brought in a new crew to carry out the flight missions of the Detachment H / 35th Black Cat Squadron.

To create misdirection typical of the time, the CIA created the unit under cover of high-altitude weather research missions for ROCAF. All documents identified the unit as Detachment H to the US government. However, instead of being under normal USAF control, the CIA directly ran Project Razor, with USAF assistance.

From 1959 to 1971, 30 ROCAF pilots went to the US to receive training in the U-2. The pilots went in nine small groups, usually with two to four pilots in a group. Twenty-eight of the Black Cat pilots completed their training.

Their American counterparts gave all of them Western handles: Pete, Jack, Charlie, Sonny, Spike, Terry, Mickey, Mike. Hsichun Hua, the retired ROCAF 4-star general who lived in Maryland with an aeronautical engineering doctorate from Purdue, retained "Mike" as part of his name ever since. Hua, who graduated from the Republic of China Air Force Academy, was a dedicated fighter jet pilot and U-2 high-altitude reconnaissance aircraft pilot in his early years. He studied in the United States in the 1960s and obtained his Ph.D. from Purdue University's School of Aeronautics and Astronautics. After returning to Taiwan, he led the research and development program for the AIDC F-ChingKuo-1 jet, named after late ROC President Chiang Ching-kuo and more commonly known as the Indigenous Defense Fighter jet.

On the night of 3 August 1959, a U-2 on a training mission, out of Laughlin AFB, Texas piloted by then-Maj Hsichun "Mike" Hua, made a successful unassisted nighttime emergency landing at Cortez, Colorado. The incident later gained fame as the "Miracle at Cortez." The US Air Force awarded the Distinguished Flying Cross to Major Hua for saving the top-secret aircraft.

Col Yang "Gimo" Shichu, a former pilot and commander of The Black Cat Squadron in Taiwan, was another of the first six Chinese pilots chosen for training at Del Rio to fly the U-2. Before that Yang conducted RF-84 photo recon missions at low-level over the mainland from Taiwan. In Taiwan, he then continued to serve in Project Tackle, the joint CIA U-2 operation the ROC referred to as Project Razor. Yang frequently traveled to Washington with Chinese Air Force Chief of Staff General Yang Shao Lien to discuss U-2 business with OSA.

'Jude' BK Pao became Operations Chief for Project Razor on the ROC side and retired as Major General in the ROCAF.

Meanwhile, during August 1958, the Chinese Communists (ChiComs) had unleashed massive bombardments of the offshore island of Quemoy. Jet fighters from both Taiwan and mainland China engaged in air battles at a cost of 31 MiG-17s and one F-86. Curtis Peebles comments on some of the politics involved:

"Mao held a meeting with the Central Military Commission and senior air force and navy officers; he said that the 'Arab people's anti-imperialist struggle' needed more than moral support and China should take real action to restrain the United States in the Middle East. He believed that the shelling of Quemoy would shock the United States as well as Europeans and Asians, delight the Arab world, and cause African and Asian peoples to take the Chinese side. Mao also saw the shelling as a means of testing US intentions and resolve toward China and to ensnare it in the Far East. Orders were issued to move PLA units to the Taiwan Strait, then begin the shelling. By July 27, 1958, forty-eight MiG-17s were at Liancheng and Shantou" – Peebles p. 218

The US 7th Fleet dispatched ships to provide logistics support to the ROC. The ROCAF pilots were still training at Laughlin AFB, requiring the US to use American pilots to fly any high-altitude reconnaissance over the PRC mainland from Okinawa. Thus, came the need to recruit ROC pilots to fly the U-2.

The USAF first sought to furnish the U-2 to Taiwan in 1958. However, the CIA opposed exporting the U-2 and risking its cover story. In 1958 the U-2 was still covert and carrying out overflights of the Soviet Union. However, after the May 1960 incident that revealed the U-2's existence, Eisenhower approved U-2 transfer to Taiwan without formally affiliating the US with the Republic of China pilots flying the aircraft.

In the cover story, the US sold Taiwan two U-2s for weather research, and Detachment H was the US weather support component of personnel. Colonel Hugh Slater was the commander of Detachment H. (Slater later took command of the 1129th SAS in support of the CIA Project OXCART at Area 51.) Detachment H personnel deployed on a temporary duty basis without dependents or household effects.

The personnel TO&E on the Chinese side according to Jude BK Pao: "A detachment under the name of Weather Research & Analysis Group was activated in the Air Force layout. However, the detachment should carry a mask of the 35th Squadron in the field. The personnel strength shown in the organization charts comprised of a commander, a deputy, a chief of operations, six pilot officers, a flight planner, a flight surgeon, two personal equipment (PE) technicians, a medic, a weather officer, a supply officer, a supply technician, six drivers, a security officer, ten uniformed guards, a cook, and three orderlies."

Colonel Michael Lu, a former RB-57 pilot, supervised the hangar remodeling and acquired an old barrack from the Photo Technical Squadron for the combat crew. He rounded up qualified personnel to fill the slots while General Huang worked out the personnel and support, a Herculean task dealing with hawkish superiors and negotiating with hostile colleagues on a secret project that excluded them under the CIA's need-to-know protocol.

Lt Col Danny Perling (real last name: Poston) showed up with security as the first American designated manager. The following day, Colonel Poston and Maj Pao selected the Taoyuan Airbase Hostel, as living quarters for his men. No sooner than when they arrived at the hostel, the hostel boy ran on a double to report to the American major in charge of the hostel. The major appeared, wearing a long face, and loudly shouted to Colonel Poston, "Anyone coming to this hostel, I should be first notified." With a poker face, Colonel Poston told him quietly, "Major. You'll be notified." A

coordinator had sent word ahead, but the message had failed to get through.

Within a week, Colonel Poston was remodeling the hostel, expanding it to 50 bedrooms, one large mess hall, a giant bar counter linking to the hall, one tiny theater, one library reading room, a game room with a pool table, a small swimming pool, and gym equipment for recreation.

Individually and in groups, the CIA personnel started arriving in Taipei, the first being Chief of Security, Bill Charapa, a gray-headed man whose real name was William Coogan. He deployed Ed Cahill in Taipei for liaison with other American agencies such as NACC, TDC, ATF 13, MAAG, and most importantly, the Commissary. Cahill rented a house for the lodging and to serve as an office in Jenai Road, section 3, east of the Chinese Air Force headquarters. Cahill reported only to Coogan and no one in Taoyuan.

Jude BK Pao relates, " Chief of Operations Fred Wingback (Lt Col Frederick Webster) and the supply chief, Pete, arrived a few days later. Colonel Webster was a very serious, dedicated hard worker. Losing no time, he and Pete talked to people in USAF Air Task Force 13 Provisional to ensure the transfer of ground support equipment on time and smoothly. The equipment comprised of a flatbed, two forklifts, a crane, three heavy-duty trucks, a PE van, three weapons carriers, two shuttle coaches, eight jeeps, four Ford Motor sedans, two gasoline tankers, etc." Continuous supply of fuel was also to have no delay in availability.

By November 1960, the CIA had the post fully manned with 28-30 individuals in Taoyuan. Additional arrivals included crew chiefs Jim Baker and Ralph Mason along with some mechanics from Lockheed; engine representative James Travis from Pratt & Whitney; Ed Helsburg and his cameramen from Kodak and Hycon; and electronic comm technicians including Vernon Hayward. Others included a flight planner, an autopilot specialist, a Perkin Elmer representative, an administration/financial clerk, and five security personnel, including John Raines who would 22 years later bailout 'Chubby' C. T. Yeh and 'Jack' L. Y. Chang imprisoned in China.

Up to now, all personnel arrived through the

regular customs channel in Taiwan. One such arrival, however, required special arrangements. This was Russ Egjelon, (alias for Bob Ericson) who arrived on a moonless night during November. ROC Lt Col Pao received instructions to pick him up. Due to secrecy, Pao took a jeep without a driver to the meeting point at 2300 hours. The pick-up was "snappy." Pao took Ericson in his jeep to the Taoyuan Air Base hostel. The first of seven U-2 instructor pilots had arrived in Taiwan. It is unknown how Ericson got into the country. However, it is known that the Keelung Customs Office detained his flying gear, the customs inspector had never seen a partial pressure suit, a helmet, or the seat-pack before.

The American personnel in Taoyuan Air Base listed as foreign technical employees of the Chinese Air Force. Each carried an issued identification bearing their pseudo name for safe passage. The CIA messengers carrying the secret document pouch carried letters that allowed them to bear arms. However, the passes and letter were invalid unless informing the security organizations (police) in Taipei City and Taiwan Province.

Getting the CIA couriers authorized to carry arms required arrangement through General Ching-kuo Chiang's office who arranged it with the Provincial Police Headquarters and Mr. Wang in the Taipei City Police Bureau.

During January 1961, the CIA provided the Republic of China with its first two U-2Cs. Lieutenant Colonel Pao received an urgent early morning call from Colonel Webster and Coogan, asking him to meet them at the Songshan Airport American Military Transport Terminal immediately. A Douglas C-133 Cargomaster, an American large turboprop cargo aircraft carrying the first U-2 plane was supposed to land at Taoyuan but had to shift to Songshan due to the gross aircraft weight. The Americans needed Pao to contact Gen W.C. Huang to arrange with the Air Force Installation and Facility for security guards.

According to Col Teh-sui Lu, Chief of Installation in the Air Force Logistic Department the Taoyuan runway tolerance could handle an aircraft tandem-wheel weight up to 100,000 pounds. Landing a heavily loaded C-133 was out of the question. However, to move the long crate containing the U-2 plane through the twisting, narrow highways to Taoyuan via either the Taipei or Taoyuan city complex was impossible. Security also presented a grave concern. Pao reached a compromise by explaining the RF-101 plane had made repeated high-speed landings every day with high-pressure, narrow tires. He argued that this high touchdown speed probably caused more impact on the 8,000-foot runway than the C-133 reducing its fuel load and making a slow landing.

Pao observed: "As soon as the Black Lady got assembled in one piece, ROC twelve-point star national insignias were painted on each wing, and the ROCAF serial number was also painted on the tail to cover the US national code 'N.' Test hop was flown by the 'Hamburger,' nickname for Russ Egjelon. Preparation was made to give Chinese jockeys transition and training flights." This involved ground school and transition because the U-2C used a different power plant than the U-2A planes the Chinese pilots had trained in at Laughlin AFB in Texas. Most of the Chinese pilots spoke English linguistically as well as technically from their assignments, and training with the SAC 4080th Wing in Texas and the six-month solid photoreconnaissance training in Rapid City, South Dakota.

Long before the CIA partnered with the ROCAF for flying the U-2, many of the ROC pilots already knew the English language and had served in the CONUS and elsewhere with the US military forces. Lt Col Jude BK Pao was one of those pilots. Born Pao Bing-Kwang in Kwangtung Province, the southern part of China on 29 June 1924, he had studied in the primary and middle schools of the famous Dr. Sun Yat-sen's University in Canton. The outbreak of World War II interrupted his education.

As Pao related, "Japanese invasion of China had torn the country apart, and the constant bombing had destroyed many schools. Seeing many of my young friends get killed by the Japanese bombs, I decided to join the military. In the spring of 1944, I applied and was accepted in the air force flying school in Kunming. The school later became Chinese Air Force Academy."

Owing to the Japanese blockade, constant harassment and incessant bombardment on mainland China, the Black Cats confined all flying to combat missions only. The lack of aviation fuel and aircraft parts made it impossible for pilots to engage in training flights. Commissioned a second lieutenant in the fall of 1946, Pao served with the CAF 12th Photo Recon Squadron at Nanking, the only reconnaissance unit in the Chinese air force. He flew P-38 combat missions that extended to the Franco-China Islands in the South China Sea. He accumulated more than 5,000 hours in flights that covered 75 percent of the entire China mainland.

Pao and his fellow students served with the 13th Detachment in Lahore, Punjab, Pakistan for their preliminary training. (At the time, Pakistan was still a part of India.) They received US Army Air Corps basic and advanced training at Douglas Field in Arizona. Pao graduated from US Army Air Corps Class 45-F and received his American pilot wings ten days before the surrender of Imperial Japan. At the same time, the Chinese Aviation Commission (the forerunner of the Chinese Air Force) authorized him and the others to wear ROCAF pilot wings as well. From Douglas, Pao and five other new graduates rode a train to Oklahoma City, where they received extensive B-25 instrument training at the Will Rogers Field. From there, the Army Air Corps sent Pao and 11 others to the Rapid City Airbase, South Dakota for P-38 combat training. Pao earned his Senior Pilot wings in 1960. (He earned his USAF Command Pilot Wings in 1978.)

The CIA, US Air Force, and Chinese personnel worked and played harmoniously as one big family. The Chinese pilots soon had American nicknames such as Gimo, Tiger, Bill, Phil, Windy, etc. Bill Coogan reproduced several musical soundtracks for the Chinese listening pleasure. CIA's John Raines formed an English class to teach the non-comms (enlisted non-commissioned) personnel. Ed Helsburg, the kitchen supervisor, introduced many delicious recipes to feed the group. Vern Hayward gathered electronic gadgets from time to time to satisfy the Chinese learning pleasure, and George Muiling clowned around to entertain everyone. The CIA provided recreation equipment such as basketballs, baseballs, tennis rackets, golf club, and go-carts.

Everyone participated in selecting the squadron insignia. Maj H.S. Chen designed two drawings, a Fierce Tiger and a Cunning Wolf.

Neither met approval. Carl Duckett, Deputy Director for Scientific and Technology, later told Pao how the Black Cat portrayed the U-2 high-altitude equipment:

"The cocked ears like a pair of air sampling bottles, two eyes for Gamma-ray detection/reception, the downward long neck for versatile camera systems and the six whiskers resembled three pairs of electronic (F, S, & X bands) reception antennae. Everyone in the hangar enjoyed the emblem and took the honor to wear it on the cap, the jacket and (for pilot) on the flight shirt."

Each of the 35th Squadron's coming operational missions would require approval by both the US and the Republic of China presidents beforehand. All US military and CIA/government personnel stationed in Taoyuan enforced a further layer of security and secrecy. The CIA issued official documents and IDs with false names and cover titles depicting them as Lockheed employees, all assigned to Detachment H. The ROCAF personnel did not know their US counterparts' real names and rank/titles, or with which US government agencies they were dealing.

During 1961 Detachment H conducted training missions, but the Kennedy administration was not yet ready to approve over-flights. Marty Knutson flew in a replacement U-2, using the alias Marty Knight to replace one pilot killed in training on 19 March 1961. By fall 1961, the President's Foreign Intelligence Advisory Board suspecting significant progress in PRC nuclear and missile development recommended overflights.

The US Intelligence Board began setting up requirements for Detachment H missions over mainland China. The approval process also began on the ROC side with film from the missions processed in the US, and duplicate positives returned to Nationalist China within ten days.

Overflights began in early 1962. The first attempt was on 12 January 1962, planned to cover the missile testing range at Shuangchengzi, but obtained only oblique photography—not the preferred stereo vertical coverage.

The second mission was 23 February 1962, over the nuclear weapons facility at Lanzhou. The third and fourth missions took place on 13 and 16 March 1962, covering airfield and industrial targets in Kunming and central China. Satellite images had uncovered new facilities, and now the U-2 coverage would produce better detail.

The US Intelligence Board, therefore, recommended further missions to cover similar targets in northeast China. Also needed was coverage of missile test ranges in north China. Accordingly, Detachment H flew three more missions during June 1962.

An apparent buildup of PRC troops was occurring across the Taiwan Strait from the island of Quemoy interrupted this strategic reconnaissance later the same month. Recall that in 1958 the PRC shelled Quemoy and threatened to invade there. In 1958, the PRC staged nearly 100 MiGs, but lost many of these to ROCAF fighters firing new Sidewinder missiles. During the 1958 crisis Detachment C flew U-2s over the PRC mainland to verify whether carrying out invasion plans at that time.

Now, four years later, a similar scenario was developing. Defense Secretary McNamara immediately ordered Detachment H U-2 coverage, to verify the extent of any PRC buildup. Imagery from expedited photo processing indicated no invasion preparations, so U-2 coverage, there ended in late July 1962.

During August 1962, a mission flown by Maj. Mike Hua covered Peking and Manchuria. Once before, an RB-57D had tried to fly past Peking, but it never returned. Ordinarily, Peking was the most heavily defended city in China, yet the U-2 photos showed that the SA-2 missile sites there were empty. The PRC had begun the tactic of making their surface-to-air missiles mobile and moving them to where they thought the U-2s would come next.

During September 1962 the Black Cat Squadron launched two missions, one over south China on the 8th and another over Giangsu Province on the 9th. The squadron lost a second one near Lu-Shan, where a flameout caused the U-2 to come down lower where PRC fighters downed it with an air-to-air missile. The PRC captured the pilot after parachuting out. At this point, the PRC accused the US of masterminding the reconnaissance operation.

President Kennedy ordered a suspension of the missions, but publicly denied US involvement.

The previous president, Eisenhower, had been the one who sold two U-2s to Taiwan. Either way, this spurred China to focus on defense in response to spy flights, like the Soviets just a few years previously.

General Hua observed: "The ChiComs did not ignore the intrusions. They always sent MiGs to intercept and follow the U-2s, although the MiGs could not reach the altitude at which the U-2 flew. The MiGs waited for the opportunity when an equipment malfunction would force the U-2 to descend."

On the PRC side, one of the People's Liberation Army Air Force pilots named Han DeCai went on to become a Lt General and vice commander of the Nanjing Air Command which faced Taiwan. In the interview, DeCai has discussed what it was like to try to counter the U-2. The PLAAF in DeCai's Nanjing Command equipped with MiG-17s; made the U-2 their main target.

Han: "The U-2s conducted their reconnaissance missions in the daytime. We did our best to attack them, but the problem was we could not reach them. They usually entered the mainland from the northeast of Shanghai."

"Chasing a U-2 made for a pretty dull flight. Every time, we sent up two aircraft to track it. All we could do was to try to reach the U-2 with a zoom climb . In the end, there was nothing we could do with our aircraft against the U-2. We had to leave the job to our surface-to-air missiles."

Soon the story would be the cat-and-mouse game played between U-2 missions and mobile SAM batteries disguised as geological equipment. In all, a total of over 100 missions crossed the Bamboo Curtain, with five U-2s shot down.

General Hua sums up early stages of the program: "Because the number of photos from each U-2 mission was enormous, it took time between missions to analyze them. Thus, the rate of dispatched missions was not very high. Huai Chen flew the first mission on January 13, 1962, Gimo Yang flew the second on February 23, and I flew the third on March 17. Tiger Wang flew the fourth on March 26."

"In addition to the camera, wideband ELINT receivers recorded a large amount of new electromagnetic emissions in and above VHF frequencies, including radar signals excited by U-2. The intelligence collected by these U-2 flights was tremendous."

The 35th Squadron eventually covered other countries such as North Korea, North Vietnam, and Laos, but conducting reconnaissance missions to assess PRC nuclear capabilities would remain the primary objective of the Black Cat Squadron.

For this purpose, the Republic of China pilots flew as far as Gansu and to other remote regions in northwest China. Some missions—because of mission requirements and range, plus to add some element of surprise—flew from or recovered at other US air bases in Southeast Asia and Eastern Asia, such as Kunsan Air Base in South Korea, or Takhli in Thailand.

Initially, the Black Cat Squadron flew all film taken to Okinawa or Guam for processing. The US forces refused to share any mission negatives with Taiwan until the late 1960s when the USAF agreed to share complete sets of mission photos and help set up a photo development and interpretation unit at Taoyuan.

ROCAF personnel morale was good because most of them lived nearby. Colonel Lu and Gimo Yang had dependent quarters in Kwei-Than. They went home to their families after work. Major Hsichun "Mike" Hua was a fond reader and a family man. He returned to Taipei each evening to spend with his wife, who worked in a translation branch outside the NACC compound. Hua obtained a doctorate and wore a pair of glasses under his pressurized helmet when he went on a flight. The Lockheed mechanics enjoyed his friendly manner. They called him "Sugar Hua-hua." Maj Tiger Wang was a quiet man who shared the housework with his wife, who worked in Linkou. Maj H. Chen ran a Bible class in his BOQ or in the 6th Group for the enlisted men. Lt Col Jude Pao was married, but rode a scooter to the base where he spent most of the week. He flew early morning flights in the T-33 to check weather conditions over the Strait where a rapid, low cloud buildup often caused nightmares for the flyers in Taoyuan.

Ranking officials from Washington, DC, BGen Jack Ledford, Colonel Songer, and Richard Burton in Dr. Bud Wheelon's DD/S&T office at CIA paid routine visits. The famed retired Surgeon General Don Flickinger of the USAF came often and strongly recommended changing the U-2 pilot's preflight diet from a bowl of beef noodle to a four-ounce steak, toast, milk, and orange juice. Dr. Ed Green from Kodak also came frequently to see his people and his baby; the 36-inch focal length seven-position-rocking gyro-stabilized camera. Generals Chiang and Fu-en I, together with Dr. Ray Cline, frequently dropped by to share lunch with the crew.

Mr. Bob Tildock (Lt Col Bob Tomlinson) replaced Poston as the manager when Poston ended a one-year tour of duty. Tomlinson was a playboy-type manager, taking things easy with a no-sweat attitude. Just the opposite was Fred Webster, a workaholic who often worked his maintenance personnel overtime to midnight. His poker face and quick temper earned him the nickname "Panic Button."

For mission preparation, the American Target Acquisition Branch selected the targets based on a satellite photo, agent reports, and documentary intelligence. The mission pilot didn't need to know the nature of the targets. All he had to do was stay on course and turn the equipment switches on and off.

A mission began with a briefing before launch. In the briefing room, Colonel Lu, the American manager, and the operations officer, sat on the bench with the others in a row. On behalf of Chinese authority, Lt Col Jude Pao or Colonel Lu commenced the session by announcing the mission number and top-secret classification before proceeding to outline the general target area, requested mission goals, probable enemy hideout, and an interception. Fuel at takeoff was 1,345 gallons, with payload usually a B camera and certain electronic systems. Survival kits were in the seat pack with evade/escape items.

The weatherman briefed everyone on the terminal weather for takeoff and landing. The American chief operations person briefed on weather en-route. The flight planner illustrated flight lines, major turning points, and the equipment (camera) mode and time to switch on and off according to the flight plan studied the night before. Forty-five minutes before launch, the PE hooked the pilot to 100% oxygen to ensure no nitrogen remained in his body.

On takeoff, the pilot climbed away from PRC radar coverage and reached 62,500 feet in minutes. He then turned northwest to enter China, the aircraft picking up altitude on a cruise climb as fuel gradually burned out until reaching 70,000 feet. The Command Post monitored the pilot's position by SSB radio and a radio reception facility in Ta-hsi.

On 9 September 1962, the first loss occurred when the Chinese downed the U-2 and captured pilot Chen Hay-sheng. He died in a hospital. The State Department denied involvement in ROCAF flights over China, but it was then that people realized the use of U-2s flown by Taiwan pilots.

The United States later announced that they would not send American pilots to fly to or over the "Bamboo Curtain." However, from the American strategic perspective, the CIA wanted to collect information on China's nuclear weapons and long-range missiles. It used the status of Taiwan being technically at war with mainland China to ask Taiwan air force pilots to carry out the tough mission.

Washington, DC did the mission planning based on the priority of the targets and the weather forecast, sending an encrypted mission plan directly to the manager of the 35th squadron. The commander of the squadron then submitted it through Gen Fu-en I to Ching-kuo Chiang, the vice-chairman of the National Security Council in the president's office.

The ChiComs' early warning radar had good coverage all over Mainland China. They could monitor the U-2 all the time. The COMINT station located between Taipei and Taoyuan could intercept the ChiCom Air Defense communication. By "listening in," Detachment H would know the aircraft's exact location with only a few minutes delay.

At the time when Taiwan and the United States had a Mutual Defense Treaty, the existence of the U-2s indicated the alliance between Taipei and Washington. The U-2 pilots became regarded as heroes of the age.

The CIA maintained Detachment H's U-2s and replaced them as necessary. American Detachment G pilots began using one unmarked Taiwanese U-2 for North Vietnam overflights during February 1962, but as tactical intelligence became more important after the Gulf of Tonkin Resolution of August 1964, the Strategic Air Command took over flying all U-2 missions in Indochina.

In late November 1962, Indian Prime Minister Jawaharlal Nehru requested military aid following the Sino-Indian War during October–November 1962. The CIA responded by deploying Detachment G to Takhli Royal Thai Air Force Base, Thailand, to carry out overflights of the Chinese-Indian border area.

In 1963, India agreed to an American request for a permanent U-2 base for Soviet and Chinese targets, offering Charbatia. The CIA used it only briefly though, and Takhli remained Department G's main Asian base. After the Vietnamese ceasefire during January 1973 prohibited American military flights, CIA pilots would again use the unmarked Detachment H's U-2s over North Vietnam in 1973 and 1974.

The demand for intelligence on the Chinese nuclear program grew. The number of SAM sites and use of the Fan Song radar increased as well. The People's Republic of China overflights became more dangerous with two more ROCAF U-2s shot down, one on 1 November 1963, and one on 7 July 1964. Tracking of others occurred with none of them shot down.

Following the 7 July shoot-down, Lin Pao, Minister of Defense, issued an order commending the PLAAF unit which "heroically shot down a US-made high-altitude reconnaissance plane of the Chiang Kai-shek brigands." His widely publicized order called for further combat training, improved strategy, tactics, and diligence.

The Detachment H's U-2s possessed the System XII radar detector, so, understandably, after the U-2 losses, the Taiwanese demanded improved ECM (electronic countermeasure) equipment. The United States Department of Defense feared the loss of the new System XIII radar jammer to the Chinese. So, at this time, neither the CIA planes nor those flown by the Republic of China carried the sophisticated "System XIII" ECM equipment.

However, the need for intelligence on the Chinese nuclear program grew so great the Defense Department finally agreed to install the improved ECM equipment, only under the condition that pilots did not turn System XIII on until System XII detected FAN SONG. After losing another ROC U-2 in the circumstances remaining classified as of July 2013, Taiwan refused to conduct further overflights unless its pilots could use System XIII when over mainland China. The PRC had improved defensive tactics and strategy—beginning to play a "shell game" of making SA-2 units mobile and disguising them as mining/drilling rigs.

Han DeCai describes PLAAF missile tactics: "The way we did this was just like guerrilla warfare. With our launchers fixed on trucks, we could move them around. We fired at the U-2 when it was within a range of 15 kilometers, and we used certain tactics to bring the U-2 into that range. When it started getting within range, we would suddenly turn on the radars, and it was too late for the U-2."

Ben Rich mentions that Lockheed developed ECM "calculated to confuse Chinese radar operators working their SA-2 missile systems. On radar screens, the U-2 would present a false display so that the missile would be launched in the wrong piece of sky." – Rich p. 181. But keeping radars turned off to the last second helped the PRC overcome the U-2's ECM equipment.

After the People's Republic of China had conducted its third nuclear test on 9 May 1966, the US was eager to obtain information on the Chinese capabilities. To this end, the CIA initiated a program, code name TABASCO, to develop a sensor pad to drop into the Taklamakan Desert, near the Chinese nuclear test site. The CIA designed the pod to deploy an antenna after landing and radio back data to the US SIGINT station in Shulinkou, Taiwan.

After a year of testing in the US, the pod was ready. Two pilots of the 35th squadron trained for the dropping of the pod. On 7 May 1967, a ROCAF U-2 (article 383) flown by Spike Chuang took off from Takhli Royal Thai Air Base with a sensor pod under each wing. The aircraft successfully released the pods on target, near the Lop Nor (also spelled Lop Nur) Nuclear Weapons Test Base but received no data from the pods.

The failure to obtain any desired data was unfortunate as the People's Republic of China conducted a test of its first thermonuclear device in Test No. 6 on 17 June 1967. A Black Cat

Squadron U-2 flown by Bill Chang on 31 August 1967 flew a second U-2 mission in the area to plant operating sensor pods near the Lop Nor Nuclear Weapons Test Base. This U-2 carried a recorder and an interrogator to contact the pods. This mission was unsuccessful as it heard nothing from the pods. The failure set the stage for the Black Bat Squadron conducting the epic Operation HEAVY TEA, a CIA plan to deploy two battery-powered sensor pallets near the base.

Among notable highlights, U-2s carried infrared cameras to evaluate the technology used by the Lanzhou nuclear processing plant. "Charlie" Wu experienced a MiG-21 in a zoom climb 300 feet to his left, as three missiles exploded ahead of him. "Spike" Chuang had eight SAMs fired against him on another mission.

In 1968, the newer U-2R replaced the ROC U-2C/F/G fleet. However, following the Sino-Soviet split and a rapprochement between the US and the PRC, the ROC U-2s stopped entering Chinese airspace. They only conducted ELINT missions or photo missions using Long-Range Oblique Reconnaissance (LOROP) cameras on the U-2R from above international waters. The LOROp cameras were still able to obtain useful imagery to add to the electronic signals intelligence.

The last U-2 mission over mainland China took place on 16 March 1968. After that, all missions had the U-2 fly outside a buffer zone at least 20 nautical miles (37 km) around China. Still, PRC MiGs and missile launches attacked some of these.

During his visit to China in 1972, President Nixon would promise the PRC to cease all reconnaissance missions near and over China. By then, US satellites were providing images without risking losing pilots and provoking incidents. Sungchou "Mike" Chiu flew the last Black Cat mission on 24 May 1974.

By the end of ROC's U-2 operations, the CIA had lost 13 U-2 aircraft in the US and Taiwan training and missions from 1959 to 1974. The squadron flew some 220 missions, with about half over mainland China, resulting in five aircraft shot down, with three fatalities and two pilots captured, and another eight U-2s lost in training or mission, with six killed. On 29 July 1974, the two remaining U-2R aircraft flew from Taoyuan to

Edwards AFB, where the CIA turned the planes over to the USAF.

Sources for the above include:

Lewis & Xue = ***China Builds the Bomb***, by John W. Lewis and Xue Litai, Stanford University Press, 1988

Perkovich = ***India's Nuclear Bomb***, by George Perkovich, University of California 2001

Ryan = ***Chinese Attitudes Toward Nuclear Weapons***, by Mark A. Ryan, M. E. Sharpe, NY, 1989

Hua & Pao = http://roadrunnersinternationale.com/u-2_cia.html for full narratives

(Shown on above picture from left to right Lt Col Mike Hua, Maj. Zulu [Operations Off], Lt Col Bob Tomlinson [Det H Commander] Don Flickinger [Ret. Surgeon General], General Fuen I [Razor Project Director], Maj. Gen W K Huang, Lt Col Gimo Yang, Lt Col Tiger Wang, Colonel Michael Lu and Lt Col Jade HK Pao)

Project Razor leadership

35 隊 飛 行 官 於 U — 2 型 機 前 合 影

35th Black Cat Squadron pilots

Maj Hsichun "Mike" Hua's nighttime dead stick landing at Cortez, Colorado

The author TD Barnes and General Hsichun "Mike" Hua at the 50th anniversary of the U-2 reunion in 2005

Factory 5704, Jilin, China
U-2 Mission GRC-112 June 16, 1962

Factory 5704, Jilin, China – U-2 mission 16 June 1962

"Jude" BK Pao   *Hsichun "Mike" Hua*

Taiwan Pilot "Taoyuan Bandit" Losses

The 9 July 1964 editorial of the *People's Daily* newspaper about a U-2 shoot-down, titled "Another Reprimand to the US-Chiang Kai-shek Gang" gave the PRC view of Taoyuan "brigands:"

"All the people hail the great victory. It was the third time that we have shot down a US-Chiang U-2 spy plane. In the past several years, we have downed nine US-Chiang spy planes of different types, and the shooting down of the U-2 plane makes the total around ten, including one RB-57A, one RB-57D, one B-17, three P2V-7s, one RF-101, and three U-2s."

General Hua notes that RB-57A overflight missions from Taoyuan began 6 December 1957, "but were stopped two months later when an RB-57A piloted by Kunhun 'Charlie' Chao, was shot down over the Shantung province by MiG-19s." The US turned two RB-57As formerly used in Heart Throb missions by the US out of Yokota, Japan over to the ROCAF for this program.

After the loss of Chao, in 1958, the program *Diamond Lil* renamed *Fair Treat* trained ROCAF pilots to fly the newer RB-57D (Hopkins p. 203). During November 1958, the ROCAF received two RB-57Ds. The ROCAF flew deep penetration missions successfully with them until a loss on 7 October 1959.

That 7 October RB-57D mission piloted by Capt Ying-Chin Wang flew near heavily defended Peking but never came back. According to General Hua, "No one in the free world knew how the PRC shot down the RB-57D near Peking until they officially in 1989 attributed it to a SA-2 missile."

This was the first destruction of an aircraft by a surface-to-air missile. One source stated the RB-57 was hit by a salvo of three missiles at 65,600.' The success was credited to PRC fighter planes, to keep the SAM program secret. – Zaloga p. 8. The RB-57D program out of Taoyuan canceled that month.

Delivery of the first two U-2s to the ROCAF occurred during July 1960. More rotated in with equipment upgrades or to cover losses, or as new models, using U-2 models A, C, F, G, and R. During December 1964, Article 342 replaced 359 until 359 returned to Taoyuan with dorsal canoe modification in late spring, 1965. Article 348 (U-2G), fitted with a specially instrumented engine by Lockheed, ferried to Taoyuan around June 1964.

Lockheed test pilot Bob Schumacher, and the Chinese and CIA pilots on hand flew Article 348 to diagnose and find a cure for the flameout problem that had developed after installation of a new fuel control in all the aircraft a few months earlier.

This sort of shuffling of aircraft continued throughout the program. Detachment H lost 13 U-2s: 5 shot down and another 8 suffered crashes. The CIA inventory replaced destroyed U-2s so that usually two were on hand for the Black Cats. Article 372 was the first aircraft deployed to Taoyuan in "Black Velvet" finish. All the earlier planes were "Midnight" blue.

The first U-2 fatality for Detachment H occurred on 19 March 1961 during training. Pilot Yao-Hua Chih had returned from a four-hour night flight and was practicing touch-and-go before landing. Bob Ericson and Tiger Yang in the "mobile" car saw Chih's U-2 veer and crash. Rescue crews rushed to the scene, but could not save him.

On 9 September 1962, Huai-Sheng Chen took off from Taoyuan at 0600 for his fourth mission to military deployments in the Jiangxi region in U-2C "378." A SA-2 missile fired by PLAAF 2nd Surface-to-Air Guided Missile Battalion shot him down over Nanchang. Chen died in a PRC hospital.

On 1 November 1963, "Chubby" Chang-Di Yeh was flying over a nuclear weapons facility in northwestern China. Yeh was a newlywed with family in Hsinchu. He went home only on weekends. Off duty, he enjoyed playing the trumpet. Now, a SA-2 missile fired by the PLAAF

2nd Battalion over Jiangxi shot him down. The ChiComs captured him alive and did not release him to Hong Kong until November 1982.

23 March 1964, Teh-Pei "Sonny" Liang flew a training sortie over the Strait, where his U-2 experienced excessive motion during a turn at high altitude, according to telemetry. A fishing boat recovered Sonny's body and parachute near the mainland coast. Part of the plane may have hit Sonny after he ejected.

On 7 July 1964, "Terry" Nan-Ping Lee took off from NAS Cubi Point in the Philippines on a mission on China's supply lines to North Vietnam. The PLAAF 2nd Surface-to-Air Guided Missile Battalion fired three SA-2 missiles and shot him down over Fujian. The PRC recovered the wreckage, finding the pilot dead in the cockpit. They reportedly found the ejection seat unarmed.

On 14 August 1964, Shih-Li Sheng survived the crash of his U-2 aircraft on a training mission near Boise, Idaho, USA. On 19 December 1964,

John Raines

Shih-Li Sheng was on another training mission from Davis-Monthan AFB when he lost control while climbing through a thunderhead storm. The aircraft crashed, and though Sheng survived, it forced him from the U-2 program.

On 10 January 1965 Li-Yi "Jack" Chang took off in a night mission around 6:30 pm, flying past Shanghai. Three hours into the flight, a SA-2 missile shot him down over a nuclear facility in Mongolia by a SA-2 missile. Chang suffered serious injuries from missile fragments and had to crawl in the snow the next day to some yurts for help. The Chinese arrested and imprisoned him like "Chubby" Yeh in 1963.

CIA officer Joe Donoghue was in the command post at Taoyuan the night the Chinese shot down Chang. The CIA knew within minutes of the Chicoms canceling their air defense alert. Donoghue spent the next several hours monitoring the HF and UHF radios, hoping that Chang might somehow be alive.

The next morning, Donoghue tuned in Radio Peking on an R-390A radio receiver and heard the announcement of the PLAAF 1st Surface-to-Air Guided Missile Battalion shooting down a U-2. The announcer made no mention of the fate of its pilot.

Therefore, the ROC government did not know about their survival and regarded them as killed in action. Taiwan placed two symbolic tombs for them in the military cemetery in Taipei.

After Beijing and Washington inked the 17 August 1982 Shanghai Communiqué, PRC authorities issued Hong Kong passes to Yeh and Chang and let them go, almost twenty years after their arrests. This move signaled Beijing's intention to lower animosity between Taipei and the PRC. Neither pilot knew of the other's existence, much less anyone in the free world until this move. This sudden appearance of the two embarrassed the Taipei government since former president Chiang Kai-shek had publicly eulogized them. The military rejected their return to Taiwan because their release was a Communist trick or that they had been 'turned.'

Yang Shih-chu, another U-2 pilot, later went to Hong Kong to help the duo. He gained the consent of the US Consul General for them to settle in the US on the condition that the pair obtain Taiwan passports. But Taiwan authorities still refused to give the two officers passports as Republic of China citizens.

The CIA and Chinese in the Black Cat Squadron assumed that Chang and Yeh had not survived and did not know differently until 1983 when another CIA officer, John Raines, told of his bringing them back home to the United States.

PRC, by their standards, did not physically mistreat the two captured U-2 pilots. After five years in solitary confinement (although not in a prison environment), the Chinese released both to labor camps for "re-education." The hard labor and poor rations that they endured were the norms for the average peasants in the communes. The cruel action of the ChiComs was the failure to admit that Yeh and Chang were still alive.

Thus, of their government's declaring the pilots KIA, both their wives eventually remarried and US and Republic of China intelligence circles, as well as their families, were amazed when Robin and Jack turned up alive in Hong Kong.

Yeh and Chang obtained US green cards, and the CIA gave them each $220,000. They lived on the interest from the deposits. Yeh moved to Houston where he still flew small planes. Chang became a doorman in a New York apartment. Neither remarried.

Years before the release, the PRC had attempted to send the two pilots back to Taiwan. However, their government refused to accept them back, fearing them turned by their captors.

The ROC government refused to allow the officers' return and even threatened court-martial if they returned to the Island. John Raines and others at the CIA and a network of former ROCAF pilots and former Chinese and American U-2 squadron mates arranged their release and brought them to the United States.

The CIA was under no legal obligation to become involved but saw a moral duty to these brave men. They were eventually able to have a homecoming to Taipei in 1990, but never regained the lost years or their families who thought they were dead and had moved on with their lives. The two pilots remained in the US.

CIA Detachment H suffered other losses. On 22 October 1965, Cheng-Wen "Pete" Wang went missing while flying a proficiency flight. He had already completed six overflights of the PRC and numerous training flights. Telemetry signals to Taoyuan indicated excessive G-forces, then silence. People on the southern tip of Taiwan and a boat saw a plane come straight down into the ocean. Searchers found nothing in the deep water there.

On 17 February 1966, Tsai-Shi "Charlie" Wu undertook a high-altitude photo training flight. After receiving an over-temperature warning, he shut down the engine. Trying to land on the alternate CCK Air Base with its long runway, Wu ran into problems with the windshield and canopy frosting over completely. Wu partly made out a shorter airstrip nearby, but could not see the windsock through the frost. Landing downwind by

mistake, he overshot the runway during the landing and crashed into housing. He and five civilians perished.

The missions of the 35th Squadron changed during February 1965 when the US needed to know whether the PRC was preparing to become involved in the Vietnam War as they had the Korean war. Codenamed Project Trojan Horse, the 35th Squadron flew over North Vietnam and the southwest border of the PRC to monitor the military movements in the region during the Rolling Thunder B-52 bombing of North Vietnam.

On February 29, 1965, Pete Wang flew from Taoyuan via the South China Sea toward Yunnan province. His course covered Haiphong and Hanoi in North Vietnam and the China-Laos-Burma border. South of Kunming, a SA-2 missile site triggered his radar warning receiver and jammer. He landed at Tahkli AB, Thailand, where the detachment moved to fly many missions over the Sino-Vietnam area for two months. The missions flew over the coastal provinces from the Sino-North Korea border to the Sino-Vietnam border and photographed targets in Korea, Vietnam, and Laos.

Initially, three U-2s deployed to Vietnam on 11 February 1964, flying to Hickam Air Force Base, Hawaii and then on to Clark Air Base in the Philippines on 12 February. The U-2s were at Clark only for a short time, as Philippine President Ferdinand Marcos had not been told in advance of the U-2 coming to the Philippines and was so disturbed that he gave the USAF 72 hours to get the "Spy Planes" out of the country. The CIA quickly moved the aircraft to Andersen AFB in Guam, and finally after some weeks of inactivity to Bien Hoa on 5 March 1965 with operations starting almost immediately.

These flights were to start with restrictions in the areas of operations, as there was still a CIA U-2 in the theatre of activity. The CIA was covering the flights over Communist China and North Vietnam from Takhli AB, Thailand and these could still classify as a civilian flight if downed. The Americans were still at the time only "Advisors" in the Vietnam War. This changed, and SAC took on the role left by the CIA, as Detachment G at Takhli withdrew their operations on 24 April.

The U-2s began missions to gather intelligence on North Vietnam. Initially known as "Lucky Dragon," this project became "Trojan Horse," then "Olympic Torch," "Senior Book," and finally "Giant Dragon." The sorties involved flying along North Vietnam and Chinese borders, generally gathering SIGINT. The U-2 flights also monitored the roads and trails from North Vietnam still used to send both weapons and personnel into South Vietnam and the surrounding states of Laos and Cambodia. They also supplied the target data for the forthcoming deployment of the B-57 Canberra tactical bombers to South Vietnam.

SAC flew three types of missions in Southeast Asia. SAC U-2 missions identified by the unclassified designator LUCKY DRAGON had two sub-categories. One was missions flown against OPLAN 34A clandestine targets, while the other was missions for hamlet coverage identified by feminine names, both with LUCKY DRAGON as the prefix.

The CIA classified U-2 missions identified by the designator LAZY DAISY covering other areas of Southeast Asia not accessible to SAC.

LAZY DAISY missions flown by the CIA covered road networks in Laos, North Vietnam, and Cambodia located outside the area assigned to SAC. The daily coverage of roads and logical routes from North Vietnam to South Vietnam through Laos and Cambodia required continuing analysis by MACV and by the Washington intelligence community.

A resource compiled by scholar Wei-bin Chang carries information about Black Cat Squadron missions. - See http://www.taiwanairpower.org/u2/index.html Excerpts:

"26 November 1964 – Johnny Wang entered the mainland from Fujian Province and then flew to Lanchou. At about 0510, PLAAF 2nd Surface-to-Air Guided Missile Battalion fired three SA-2 missiles. Fortunately, they all missed."

"16 March 1968 C058C - 4-hour-and-32-minute mission. Oscar Sierra activity noticed [missile launch warning] and one bogey MiG-21 sighted with no hostile action taken. The mission covered a total of 88 targets in China, Vietnam, and Laos."

While flying south of Kunming on his second mission on 14 March 1965, Charlie Wu's System 13 began jamming. System 12 showed no indication, making it difficult for him to know which way to evade the threat. A MiG-21 Fishbed suddenly zoomed to within 300 feet from his left wing tip and then dropped down to disappear. At the same time, three heat-seeking air-to-air missiles detonated in front of Charlie's aircraft. He brought back an excellent photo of the silver MiG-21 beneath his U-2.

On 22 March 1966, Huang-Di "Andy" Fan lost his U-2 when he ejected from his first training flight out of Davis-Monthan AFB. The aircraft went into a diving spiral from 20,000 feet. An investigation showed that a flap actuator had malfunctioned.

On 21 June 1966, Ching-Chang "Mickey" Yu was on a long-range training flight when the engine flamed-out at cruising altitude due to a ruptured fuel supply line. Without fuel, the U-2 was not able to restart at a lower altitude. Wu then attempted an unsuccessful forced landing on one of the islands near Okinawa. He bailed out, but his parachute did not inflate. Wu, unfortunately, died of his injuries.

On 8 September 1967, a Chinese-made Red Flag 2 by PLAAF 14th Surface-to-Air Guided Missile Battalion over Jiaxing shot down and killed Jung-Bei Huang while he was flying over the Jiangsu Province, Shanghai, and Hangzhou.

"Penetrating" ROCAF flights ended in 1968, but operations continued until the eventual withdrawal of all U-2s to the US. On 16 May 1969, Hsieh Chang died when he lost control of his plane approximately 100nm south of Cheju (Jeju) Island, ROK during a mission along Hebei seacoast. On 24 November 1970, Chi-Hsien "Denny" Huang crashed and perished during take-off and landing training in Taoyuan. He was flying the newer, larger U-2R.

Physiological Aspects of Punching Out of a U-2

Captain Kittinger bailed out of a balloon at 100,000 feet, but did so under very controlled conditions. The conditions necessitating bailout in a U-2 were not so controlled. Robin Yeh managed to get out and pull his parachute cord after a SA-2 missile hit his U-2, but the ChiCom surgeon removed fifty-nine pieces of missile fragments from his legs. Yeh remained imprisoned for ten years.

Terry Lee died in his cockpit after a SA-2 fire hit his U-2. At this date before 1965, the pilot had manually to open the canopy. Later, the U-2 provided a power jettison for the canopy.

The pilot's next problems started the moment he ejected from the cockpit. At cruising altitude, if the parachute opened immediately, he died. If his faceplate shattered, he died. If his suit ripped, he died. If his oxygen wasn't flowing, he died.

Assuming none of this happened, the pilot would be free-falling through the thin air. A spin could send him unconscious, as centrifugal force pressured a large weight of blood into his head. A spin could rip his parachute when it opened at a safe altitude. USAF U-2 pilots Nole and Stratton had to go 'over the side' in high altitude mishaps but survived the above perils.

Twenty-six of the twenty-eight ROCAF pilots who completed training in the US between the years 1959 and 1973 went on to fly operational missions. During 15 years of Project Razor, Republic of China pilots flew 102 missions that penetrated the Bamboo Curtain. Surface-to-air missiles shot down five U-2s over mainland China, killing three ROC pilots, and resulting in the capture of two others. Detachment H also lost seven Black Cat pilots during training flights and coastal patrol missions. These losses were greater than those for Detachments A, B, and C, but the Black Cats flew more missions over their main denied area.

The intelligence gathered by them, which included evidence of a military build-up on the Sino-Soviet border, may have contributed to the US opening to China during the Nixon administration by revealing the escalating tensions between the two communist nations. Shortly after Nixon's 1974 visit to Beijing, all reconnaissance flights over the PRC ceased, and the Black Cats officially disbanded.

Shortly after Nixon's visit to Beijing, all reconnaissance flights over the PRC ceased, and the Black Cat Squadron officially disbanded in 1974.

The knowledge deemed critical at the time, gained by the Black Cats about the U-235 processing plant at Lanzhou (Lanchou) and the nuclear test site at Lop Nor provided intelligence almost impossible to acquire any other way.

Senior analyst Brugioni agreed that overhead photography would have to provide the bulk of intelligence on Chinese nuclear and missile targets, "because so little information was available from other sources. In some respects, China represented a more challenging intelligence problem than the Soviet Union because we had so little information." – Brugioni 2010 p. 306

Unquestionably U-2 flights were necessary and productive. Ironically, securing intelligence this way spurred China to develop an integrated air defense system and to establish an effective air force. Now the PLAAF is one of the most capable in the world.

General Hua summarized, "It was impossible to collect intelligence in the PRC through traditional espionage methods. Without the U-2's ELINT, Dr. Henry A. Kissinger would not have known that Marshall Lin Piao had died after the failure of an attempted coup. Lin Piao was the PRC defense minister and the officially designated heir of Chairman Mao."

"Without the U-2 overflight intelligence, President Nixon might not have been able to implement his strategy of rapprochement toward the PRC."

"The pilots of the 35th Squadron did not expect to contribute towards easing the tension of the Cold War. They endured long, hard hours, facing the danger of enemy missiles, and their aircraft malfunctioning deep inside the enemy territory. They were executing the missions. Nevertheless, they had performed the most courageous operations in U-2 history."

General Pao's view is similar: "Project Razor was put into the right track at the very beginning. Mutual understanding, cordial cooperation, and whole-hearted support of both Sino-American governments set the project in motion. Of course, it was a show of those brave young combat pilots without whom nothing could have been achieved."

"From 1962 onward everything went smoothly on a rising trend and gradually ascended to the zenith. 1963, 64, 65, and 66 were the most fruitful and harvest years. Target acquisition coverage stretched all over China. It made the technicians, photo interpreters, and electronic analyzers busy day and night. They had a handful. No target on Mainland China was too far for the Black Lady."

Chapter 17 - The Black Bats

Extracts from Chapter 7 of *The Black Bats: CIA Spy Flights Over China From Taiwan 1951-1969*, by Chris Pocock with Clarence Fu. Atglen, PA: Schiffer Publishing, 2010:

The CIA had not given up on the idea of air-dropping sensors into northwest China. Starting in early 1968, the CIA and the Sandia Labs devised a much larger collection package that communicated the data that it collected in burst transmissions to a satellite passing overhead.

Bob Kleyla, while still an air operations officer in the SOD's Air Branch, suggested airdropping the new sensor package into the remote area using a C-130 Hercules transport. The aircraft had to be specially equipped to fly low over China, to avoid detection by Chinese radars. Who could fly a large airplane from Thailand into NW China, on a 'deniable' basis to the US? Of course, it was the 34th Squadron of the Chinese Air Force.

Despite the cooling of relations between the US and the ROC, the government in Taipei remained willing to help. General Yang, the chief of intelligence, summoned Squadron Commander Lu De Qi to a meeting in the CAF headquarters. General Yang asked him to select two crews, pilots, navigators, flight engineers, radio, and electronic warfare operators, and loadmasters. Most of them were from the C-123 operation in Vietnam. The mission became nicknamed Qi Long (Magic Dragon) by the Chinese side and Heavy Tea by the American side. The group of 24 airmen left for training on the C-130 in the US in late September 1968, led by Col Sun Pei Zhen (孫培震)

The Chinese pilots learned to fly the C-130E version of the Hercules during a short course at Stewart AFB, TN. Then they joined their colleagues who had already moved to Groom Lake, the CIA's top-secret airbase in Nevada, where the CIA test flew the U-2 and A-12 spy planes. The U-2s were long gone, and the A-12s had recently retired, but the remote base still housed secret projects, notably the MiG fighters covertly obtained by US intelligence. The Republic of China aircrews were the first foreign nationals to train at Groom Lake. It amazed

Americans with high-level security clearances to see the Chinese group. However, Sun and his team lacked having day-to-day contact with other workers at the base.

At Groom Lake, Colonel Sun and his men found a C-130E specially modified for covert insertions by the Skunk Works for the project. It had a high-speed, low-level aerial delivery system. For navigation, in addition to the standard APN-115 radar, a new inertial system promised only a 2-degree deviation per hour. A Forward-Looking Infrared (FLIR) system and a much-improved terrain-following radar aided low-level night flight. For self-protection, there was a radar warning and jamming system. The Chinese airmen were very impressed with the Herk's avionics; some of them had come straight from the ROC's elderly and basic C-46 and C-119 transports.

The crews began the first phase of operational conversion by flying daylight missions at various altitudes, all the way to the Mexican border in the south, and the Canadian border in the north. The USAF launched fighters to intercept them so that they could become familiar with the C-130's defensive systems.

Then came the phase two conversion, which was flying at night and low level, and the airdropping of pallets. The FLIR was a key aid here, although when the training started, it could only display still frames of scenery. Later, it received an upgrade to provide motion imagery. Over the high ground of Oregon and Washington State, the CAF crews flew over terrain that matched what they might encounter flying north across the eastern Himalayas. Around Groom Lake itself, the desert terrain matched that of northwest China.

The survival training provided might have occurred as a morale booster since it is hard to imagine how the airmen from Taiwan could have made their escape if their aircraft went down so far from home. In any event, the crews spent two cold and uncomfortable weeks on a mountain in Oregon.

The airmen from the 34th Squadron stayed at Groom Lake for over six months to complete training. Before they left, leaving the US during April 1969, Colonel Sun selected members for the primary and backup crews. They flew from San

Francisco to Hawaii for a one-week holiday. Then it was onwards via Midway to Kadena Airbase, Okinawa, where they practiced some more night flying on the C-130 flown there by an American crew.

The Magic Dragon Mission by Lt Col Feng Hai Tao (馮海濤)

"At the secret desert base, our movements were strictly controlled. We could go to the bathrooms, the gym, the laundry, and the cinema. But we were very comfortable. All meals and drinks were free, and the Americans met our needs, flying in rice and other Chinese food. They even provided preserved Tofu! The base commander joined some of our leisure activities. An American flight surgeon lived with us full-time. And I remember four other Americans who were always around. I guess the CIA sent them to watch over us."

"At first, our English was not very good. And there was no interpreter at the ground school. But we got better. We were specially selected and realized that we were representing our country. So, we studied hard, so as not to lose face. The training we got in the US was crucial to my future career. I got good grades and grew in confidence."

"In class one day, I heard a strange-sounding jet taking off. Looking through the window, I saw a MiG fighter. The instructor immediately pulled down the blinds and said: 'Don't look outside!' He also told us not to talk to other people, when we went to lunch, or to the laundry. The Americans didn't want other people on that base to know we were there. Even the Republic of China Embassy didn't know exactly where we were."

-Lt Col Feng Hai Tao, C-130 navigator, from ***Black Bat Squadron Oral History***, p. 218-222

*(Note, the author, TD Barnes, and Calvin Dawson with EG&G Special Projects, while inside

the RATSCAT working on a Nike radar during the early evening, witnessed the C-130 land. Instead of taxiing to the usual area for security check, the plane taxied out onto the lakebed near the RATSCAT where the CIA security officers met them. A few minutes later, the Chinese marched past without noticing Barnes or Dawson. They never saw them again, nor did they ever mention this to anyone as this was something that none of them possessed a need-to-know until recently declassified by the CIA.)

The Outcome of Magic Dragon

In review, by the 1960s, it was clear that the PRC focused on pursuing nuclear weapons capability and was making considerable progress without Soviet help.

The US considered it a high priority to get information on all Chinese nuclear development and capabilities. After the failure of the sensor pods planted by the Black Cat Squadron near the Lop Nor Nuclear Weapons Test Base using Detachment H U-2s, the CIA developed a plan named *Heavy Tea* aka **Magic Dragon** to deploy two battery-powered sensor pallets near the base. The training for the operation is outlined in the previous excerpts.

. The crew of 12, led by Col Sun Pei Zhen, took off from Takhli Royal Thai Air Force Base in the unmarked C-130 on 17 May 1969. Flying for six and a half hours at low altitude in the dark, they arrived over the target and dropped the sensor pallets by parachute near Anxi in Gansu province. After another six and a half hours of low altitude flight, they arrived back at Takhli. The sensors worked and uploaded data to a US intelligence satellite for six months before their batteries failed. The Chinese conducted two nuclear tests, on 22 September 1969 and 29 September 1969, during the operating life of the sensor pallets. A proposed follow-up mission was planned, but called off in 1970. (Please see Pocock 2010 for further details)

'Missile Gap'

During March, April 1963, CIA satellites noted the Russians were building a massive launch pad at the Baikonur Cosmodrome in Kazakhstan. The size of the assembly building had all the earmarks for a very large launch vehicle. With the

CIA, it was, "Hello. What is going on here?"

A very large causeway led up to two launch pads side by side. In the analysis, back in Washington, the launch pads were first-hand evidence of the preparations of a huge rocket. The activity could mean only one thing. The United States was in a space race to land a man on the Moon.

At the beginning of the 1960s, the Russians were ahead of the US in the space effort. They had better rocket engines capable of doing things that the US could not do. The first US satellite to be prepared was the Navy's Vanguard, a 3 lb. sphere about the size of a grapefruit. Its first launch attempt was in early December 1957, about two months after Sputnik I orbited. It rose ever so slightly then fell back onto the launch pad, exploding spectacularly. The second Vanguard attempt also failed. The Explorer I successfully orbited by the US Army in late January 1958 weighed about 30 lbs. In the meantime, Sputnik I weighed 180 lbs., Sputnik II 1100 lbs. and Sputnik III (May 1958) 2900 lbs. by comparison. The period of space race superiority was a wonderful one for the Russians.

Even before that, during August 1957, while the CIA's U-2 planes were overflying the Soviet Union, the US watched as Russia launched the first rocket with potential as an ICBM. During October that year, the CIA woke up one morning to the beeping sound of the Sputnik 1 in orbit. It grew worse. The United States watched Russia place the first man in space. Then, they launched the first woman into space, and then the first three cosmonauts into space. The first spacecraft to hit the Moon followed.

The US knows now that this all happened thanks to the anonymous chief designer Sergei Pavlovich Korolev, head of the OKB-1 (Experimental Design Bureau #1) and Russia's most senior rocket scientist. Russia recalled Korolev, tortured and exiled in Stalin's purges of the 1930s, to Cold War service. The US had no counterpart in its space exploration program.

In its effort to beat the US to place a man on the Moon, the Russians realized the lack of infrastructure to design a gargantuan engine. Where the US planned to use five rocket engines of 1.5 million pounds of thrust each to power its launch vehicle, the Russians designed their N-1 rocket to launch using 30 lesser-powered engines. These engines used a full-cycle design that the US considered too risky. The death of Korolev and four failures of the N-1 ended the Soviet manned lunar program.

The launch of Sputnik on 4 October 1957 gave credence to Soviet claims concerning an ICBM. The soviet success prompted a crisis in the US, but Eisenhower stated on 9 October that the launch did "not raise his apprehensions, not one iota." He refused to disclose the U-2's intelligence. During December 1958, Khrushchev boasted of a missile delivering a warhead 8,000 miles. Pointedly: "The next war will be fought on the American continent, which can be reached by our rockets. " To the UN: "Production of rockets is now a matter of mass delivery—like sausages that come out of an automatic machine." This was all bluff because reliable, operational Soviet ICBMs were more than ten years away.

Rudd, Edens, Stratton, Haupt, Overstreet, Powers, and Meierdierck

went on to fly the U-2 after serving in the 508th at Turner AFB

Top L-R: Wilson, Horne, Billy Edens, Wynn, Rudd
Middle L-R: Carson, Kruk, Bahman, Buster Edens, Trice, Herbert,
Stone, Stratton, Cooper, Haupt, Inkster, Overstreet
Front L-R: Powers, Nunrley, Graves, Schuler, Meierdierck, Kondracki, Maggert

On 15 May 1959, Lyle Rudd flew the 19th Soviet overflight from NAS Cubi Point in the Philippines. He flew north over Mongolia and crossed into the USSR, flying as far as Lake Baikal. Rudd crossed the Chinese steppes, overflew Lhasa, and landed in Dhaka after 9 hours 40 minutes in the air, the longest operational U-2 mission to date covering 4,200 miles. Slipper tanks and flight profile enabled this long-range.

On the 9 and 18 June 1959, the CIA pilots flew Operation Hot Shop, where U-2 and EB-47TT Tell-Two flew a border flight to obtain the first telemetry ever of a Soviet R-7 launch vehicle / SS-6 ICBM during the first stage burn–80 seconds after launch.

Although the Soviets' SS-6 operational missile program stalled because of technical failures, their loud boasts and the US Secretary of Defense Neil McElroy's statement during February 1959 caused Congress to believe the Soviets had a three-to-one temporary advantage in ICBMs. The suggestion of a 'missile gap' caused widespread concern in the US.

The American intelligence community divided in opinion on where the Soviet Union stood with their missile technology. The CIA suspected technical delays whereas the USAF believed the SS-6 ready for deployment. Khrushchev stroked the US concerns by continuing to exaggerate the Soviet program's success. President Eisenhower was regularly receiving CIA and State Department support concerning the missile gap. Having this support prompted him, after 16 months, to reauthorize one Communist territory U-2 overflight for July 1959. He authorized several ELINT flights along the Soviet border.

The CIA U-2 program now had a significant interest in the intercontinental ballistic missile (ICBM) situation. The US wanted to know the differences, capabilities, and the number of missiles possessed by the two countries. The Russians had located their missile sites in the eastern part of the country, requiring the U-2 to fly out of Pakistan to reach their destination and return to base. The U-2 coverage so far had not found operational Soviet ICBMs

These missions flew all over the USSR and Eastern bloc countries, plus the Middle East and the Mediterranean countries. Over one 17-month period, the CIA flew 23 missions, six over the USSR, five over European Eastern Bloc countries with most of the remainder over the Mediterranean.

On 9 July 1959, Marty Knutson flew the 20th Soviet overflight; mission 4125, Operation Touchdown from Peshawar in Pakistan. He flew north over Sary Shagan missile range and the Semipalatinsk nuclear test site followed by the nearby Dolon Airfield. He flew over the Urals to Sverdlovsk and Tyuratam before landing at Zahedan in Iran. The sortie lasted 9 hours, 10 minutes with only 20 gallons of fuel remaining when he landed.

During August 1959, Detachment C deployed U-2s to the new Thai airbase of Takhli and mounted three flights over China and Tibet in early September. Detachment B overflights of Israel discovered the Dimona nuclear reactor and processing facility under construction.

On 6 December 1959, Squadron Leader Robbie Robinson flew the 21st Soviet Overflight, Mission 8005. The flight was the first mission flown by the RAF out of Peshawar. He flew north over Tyuratam, Kyshtym, Engels airfield near Saratov, Kapustin Yar and the bomb factory at Kuybyshev. The U-2 exited Soviet airspace over the Black Sea and recovered to Incirlik.

When the CIA deployed the third detachment from Watertown, it realized a need to amalgamate its air operations at Langley under one division (Development Projects Division, DDP). Its T/O, Table of Operations, had grown to 600 Agency personnel at the end of 1956. The T/O fell to 412 in early 1958, and to 371 during March 1959 where the amalgamation went into effect. Further reductions continued through 1959, dropping the T/0 to 362 at the end of the year.

On 5 February 1960, Flight Lieutenant John MacArthur flew the 22nd Soviet overflight, Mission 8009, out of Peshawar. This second mission flown by the RAF headed northwest over the Aral Sea, looking for missile sites. Instead, MacArthur discovered a new Soviet Bomber at Kazan. He captured eight Tu-22 BLINDER aircraft on film. From there, he headed south down the Volga over the missile factory at Dnepropetrovsk. After leaving Soviet airspace at Sevastopol, MacArthur landed at Incirlik.

On 9 April 1960, Bob Ericson flew the 23rd Soviet Overflight, Mission 4155, Operation Square Deal out of Peshawar, and north over Sary Shagan, the strategic bomber base Dolon, Semipalatinsk, and Tyuratam before landing at Zahedan in Iran. The Soviet Air Defense organization manually tracked this flight the whole time, and several MiG-19s made unsuccessful attempts to shoot down the U-2 aircraft. (A MiG

lost; pilot killed)

Royal Air Force Participation

Bissell had suggested bringing the British into the program to increase the number of overflights. Prime Minister Harold Macmillan agreed and sent four RAF officers to Laughlin Air Force Base in Texas for training during May 1958.

On 8 July 1958, the senior British pilot, Squadron Leader Christopher H. Walker died when his U-2 malfunctioned and crashed near Wayside, Texas.

The CIA did not disclose the circumstances of the British death involving the U-2 for over 50 years. The CIA selected and sent another pilot to replace Walker.

After training, the group of RAF U-2 pilots arrived in Turkey during November 1958. The US and the United Kingdom remained jointly involved in the CIA's Detachment B to Adana, providing valuable intelligence during the 1958 Lebanon crisis.

Detachment B flew its first mission tasking from the CIA in Washington on 11 September 1956, flying over friendly British and French fleets, bases, and units in the Mediterranean. Shortly after that, the detachment began daily missions overflying the battle and sea areas of the ongoing Israeli-Egyptian War. The U-2 pilots observed actual fighting between tanks, units, and aircraft through the U-2 drift sight. Since the September 1956 disclosure of Mediterranean photographs, the United Kingdom had received U-2 intelligence, except during the Suez Crisis.

The CIA and Eisenhower viewed using British pilots as a way of increasing plausible deniability for the flights. The CIA saw British participation as a way of obtaining additional Soviet overflights the president would not authorize. The UK gained the ability to target flights toward areas of the world of less interest to the US and to avoid another interruption of U-2 photographs.

Although the RAF unit operated as part of Detachment B, the UK received title to the U-2s their pilots flew, and Eisenhower wrote about the nations conducting two complementary programs rather than a joint one because of the separate lines of authority.

A secret MI6 bank account paid the RAF pilots using employment with the Met Office as their cover. While most British flights occurred over the Middle East during the two years the UK program existed, the Brits flew two successful missions over Soviet missile test sites. Like Eisenhower, Macmillan approved the Soviet overflights. The British's direct involvement in overflights ended after the May 1960 U-2 downing incident. Although four pilots remained stationed in California until 1974, the CIA's official history of the program stated no Royal Air Force pilots ever conducted another overflight in an Agency U-2. Even though their U-2 experience remained secret, between 1960 and 1961, the first four pilots received the US Air Force Cross.

The Four British pilots who trained at Laughlin AFB in 1958. L-R: Sqd Ldr Chris Walker and Flt Lts John McArthur, Mike Bradley, and Dave Dowling. Walker died in a training accident, but the other three graduated and moved on to the CIA's Det B in Turkey (via Pat Halloran)

RAF U-2 pilots Chris Walker, John McArthur, Mike Bradley, Dave Dowling

Final U-2 Flights Over the Soviet Union

One British U-2 overflight of the Soviet Union occurred during December 1959, and another during February 1960. Neither proved nor disproved the missile gap. Nonetheless, the British flights' success contributed to Eisenhower's authorization of the 23rd Soviet flight on 9 April 1960.

Khrushchev claimed in his memoir a new Soviet surface-to-air missile should have shot down the April flight had the missile crews not reacted too slowly. CIA pilot Ericson flew over the Soviet Union on 9 April 1960, the last successful overflight. By this time, the CIA concluded the Soviet SAMs had "a high probability of a successful intercept at 70,000 feet, provided they detected the plane in sufficient time to alert the site."

Despite the now much higher risk, the CIA failed to stop the overflights because of hopefulness and some overconfidence from the years of successful missions, and because of the strong demand for missile site photos. The US suspected ICBMs to be at Plesetsk, the last target of Powers' planned mission.

By this time, the U-2 had photographed 15% of the country, resulting in 5,500 intelligence reports. But what about the other 85%--were ICBMs there? Eisenhower authorized one more mission to occur no later than 1 May because of the desired sun angle and weather, and because of the Paris Summit on 16 May.

On 24 September 1959, while conducting a test flight in Article 360 from Detachment C in Atsugi, Japan, Tom Crull had encountered problems on a test flight and eventually ran out of fuel. With great skill, Crull managed to dead stick the aircraft onto a small civilian airfield in Fujisawa, where curious Japanese civilians promptly surrounded and photographed it.

The CIA shipped the damaged U-2C aircraft back to Lockheed in the USA for repairs. Article 360 returned to Detachment B at the Incirlik Air Base in Adana in Turkey, where it gained a reputation as a 'Hangar Queen' for a variety of reasons. As a matter of fate, Gary Powers draw Article 360 to fly on Mission 4154, Operation GRAND SLAM.

When the U-2 became operational in 1956, an official had predicted a useful lifetime over the USSR of two years. The U-2's first flight over Soviet territory revealed that their defense system detected and tracked it. Nonetheless, the U-2 remained an invaluable intelligence source on the Soviet Union for four years.

1960 - The "Air War" Heats Up

Adding to US concern in 1960 was confirmed that the Soviet air force and Air Defense units now had the P-30, the FAN SONG radar system that could detect aircraft above 66,000 feet. The Soviet air regiments now had the SU-9 FISHPOT high-altitude interceptor aircraft as well, with supersonic speed and an altitude ceiling above 20 kilometers. Additionally, the Soviet SAM units now had the highly regarded S-75, the SA-2 GUIDELINE missile system with an 82,000 feet engagement altitude against targets flying at up to 930 miles per hour.

On 9 April 1960, Soviet radar in the Turkestan Military District had acquired U-2 pilot Bob Ericson's 6-hour mission over the Semipalatinsk nuclear test site. The Soviets tracked Ericson as it flew over or near the SAM forces at Sary Shaghan, and from there to the Tyuratam strategic missile testing range, and over the city of Mary, from which Ericsson departed into Iran.

Soviet leadership appointed a commission to investigate the reasons for the failure of the Air Defense forces to attack the aircraft that had violated Soviet airspace for so many hours. Many omissions occurred in the operation of advanced radio equipment. Information related to Ericson's activity acquired by Soviet communications interception facilities in the TransCaucasus confirmed it happened, but went unreported to the command element because of chance happenings. Khrushchev was indignant and penalized many generals and other officers.

Everything changed when the Soviets' conclusions drawn from the 9 April failure led to dramatic success only hours before the beginning of the annual May Day parade of 1960. Soviet Air Defense forces detected a high-altitude target flying at 60,000 feet over the Tajik SSR in Central Asia. Near Sverdlovsk on May Day, 1 May 1960, and while the command element of the Air Defense Forces was demonstrating Soviet military prowess, Russian missiles shot down Francis Gary Powers in Article 360, the Hangar Queen as it flew over the heart of the country. Out of 365 days in a year, the CIA could not have picked a worse day to fly over Russia, picking May Day when each year, the US and its allies went on military alert because of this being an important Soviet holiday where the Soviet Union held massive parades to show off their war equipment. Also, this being a Soviet holiday meant much less air traffic than usual.

All eyes turned on Watertown, and the U-2's proposed replacement, the Mach 3 A-12. The U-2 continued flying, but not over the Soviet Union.

The CIA had chosen for the mission—the 24th deep-penetration Soviet overflight—Operation GRAND SLAM, an ambitious flight plan for the first crossing of the Soviet Union from Peshawar, Pakistan to Bodo, Norway; previous flights always exited in the direction from which they entered.

The route permitted visits to Tyuratam, Sverdlovsk, Plesetsk (suspected ICBM location), Kotlas, Severodvinsk, and Murmansk. The CIA chose Francis Gary Powers, the most experienced pilot for the flight. Powers had flown 27 missions at this point. After several delays, the flight, Mission 4154, in the U-2C known as the Hangar Queen finally occurred on May Day, 1 May 1960.

Frank Power's mission that day was to fly over Russia and land at Bodo, Norway, where Marty Knutson was waiting to take the aircraft on its return flight. Knutson was sitting in Norway breathing oxygen and preparing to don his pressurized space suit to take over the aircraft that Powers was flying. At Bodo, they knew something was amiss when the aircraft failed to arrive within the required timeframe.

The Soviet radar systems had detected the Hangar Queen while still flying 15 miles outside the Soviet border and tracked it over Sverdlovsk. Four and a half hours into the flight, a SA-2 missile detonated behind the aircraft at 70,500 feet. Another missile hit a Soviet interceptor attempting to reach the American plane. Powers survived the near-miss that brought down his plane.

The CIA did not know the Russians had captured Powers. Nor did the CIA know that the crash failed to destroy the U-2 and that it was in Soviet hands. The government planned to say, if necessary that the NASA aircraft drifted. The conspired cover story went into effect with NASA issuing a press release.

NASA used the CIA's cover story about a U-2 conducting weather research that may have strayed off course after the pilot "reported difficulties with his oxygen equipment."

On 3 May, NASA announced one of its aircraft missing while making a high-altitude research flight in Turkey. NASA director Dr. Hugh L. Dryden's press release stated the U-2 plane was conducting weather research with Air Force support and had gone missing and presumed lost while operating overseas.

Foreseeing an event such as the Powers shoot-down, U-2 pilots received the following instructions: First, it was their duty to ensure the destruction of the aircraft and its equipment to the greatest extent possible. This was best accomplished by bailing out and actuating the destructor so that the aircraft would suffer the effects of the detonation in the equipment bay. If for any reason the pilot could not carry out this maneuver, he was to make a forced landing and to actuate the destructor as he left the aircraft.

If the U-2 program became compromised, it would be better to stand revealed as an intelligence collection activity under civilian, not military, control, and having no harmful or aggressive purposes. The revelation that this Agency conducted illegal intelligence collection activities was less damaging than the admission of such a role by the military establishment.

Bissell and other project officials believed it impossible to survive a U-2 accident from above 70,000 feet. Consequently, they used the pre-existing cover story.

****

Watertown at Area 51 came back into play when on 6 May 1960 NASA rushed to display one of its U-2s with a fictitious serial number and NASA markings. These played into the guise of the NASA weather pilot cover used by the CIA for its pilot training in the U-2 at Watertown.

Khrushchev learned of America's NASA cover story and developed a political trap for Eisenhower. By remaining silent, Khrushchev lured the Americans into reinforcing the cover story. As the saying goes, "The Russians gave the Americans enough slack rope to hang themselves." Politically, the United States did exactly that. At this point, Soviet Premier Nikita Khrushchev exposed the cover-up by revealing on 7 May that Powers was alive and confessing to spying on the Soviet Union.

Eisenhower turned down Dulles' offer to resign and publicly took full responsibility for the incident on 11 May. By then, the CIA had canceled all overflights. The Paris Summit collapsed after Khrushchev, as the first speaker, demanded an apology from the US, which Eisenhower refused to extend.

Powers had received little instruction on what to do during an interrogation. Although he said that he could reveal everything since the Soviets could learn what they wanted from the aircraft, Powers did his best to conceal classified information while appearing to cooperate.

The trial of Francis Gary Powers began on 17 August 1960. Powers—who apologized on the advice of his Soviet defense counsel—received a three-year sentence in prison. However, on 10 February 1962, the USSR exchanged him and American student Frederic Pryor for Rudolf Abel at the Glienicke Bridge between West Berlin and Potsdam, Germany.

Two CIA investigations found that Powers performed well during the interrogation and "complied with his obligations as an American citizen during this period." Nonetheless, the government was reluctant to reinstate him to the USAF because of its statements that the U-2 program was civilian. The CIA promised to do so after his CIA employment ended. Powers resolved the dilemma by choosing to work for Lockheed as a pilot.

NASA, concerned about the damage to its reputation in the wake of the Powers U-2 affair, disengaged from the CIA and no longer provided them the cover story support needed for their covert U-2 Operations.

Four months later, the May Day incident resulted in a cessation of overflight operations. The CIA reduced the number of pilots in Detachments B and C and then returned them to the ZI. Other air activities, however, increased, including the U-2 successor program that would reopen the CIA's Area 51 station. Satellite activity (Corona) and clandestine air operations in various areas of the world and the Far East increased as well. So did the staffing of cadres for the detachments at Eglin, Kadena, and the new detachment on Taiwan.

The Russians used the debris of Powers' aircraft to design a copy under the name Berijev S-13. They discarded the Berijev S-13 for the MiG-25R and reconnaissance satellites.

From 1956–60 U-2 aircraft flew 24 missions over the USSR. Detachment A flew six; Detachment C four; and Detachment B flew 14. It would now be up to Corona to confirm no 'missile gap' with its surveillance ability.

Following the Powers shoot-down, Col Stan Beerli ran the CIA's reduced U-2 operations for the next two years. As the Director of Special Activities, he was also in charge of the developing Corona satellite and A-12 OXCART aircraft

reconnaissance programs. (Richard Bissell, head of the U-2 project, named the Corona satellite after the favorite cigar of one of his staffers.) It was Corona that resolved the missile gap in the fall of 1960. Bissell was an actor in the unfortunate Bay of Pigs affair, which took him out of the Area 51 activities that he had started.

Colonel Beerli returned to the US Air Force in 1962, where he attended college before assignments to the Strategic Air Command's 7th Air Division in the United Kingdom and the Strategic Reconnaissance Center (SRC) at Offutt AFB. At the SRC from 1964–67, Beerli was a key figure in the introduction of the 'Lightning Bug' air-launched reconnaissance UAVs. His final duty before retiring in 1971 was at HQ, US Air Force as Chief of Staff in the Reconnaissance Division.

Looking back at the cover story should the CIA lose a U-2, the reality was that at the time Dryden made NACA's early proclamations during May 1956, the first U-2 aircraft deployed overseas. By 20 June 1956, the CIA had conducted its first operational overflight, undertaken over Poland and East Germany. By 4 July 1956, the CIA had already flown three more overflights of Eastern Europe, including the first clandestine overflight of the Soviet Union. Making the CIA's cover story even more vulnerable was that the Air Force also had operated the U-2 for three years.

The cover story about the U-2's peaceful scientific research purposes might have worked for the US Air Force. After all, Air Force U-2s were conducting peaceful, high-altitude atmospheric and meteorological research flights throughout the world and would continue to do so until 1968. The CIA's cover story could not stand because NASA did not receive their own two U-2 aircraft until June 1971, fifteen years and one month after the first deployments from Area 51.

U-2 Incident – Life Magazine Coverage May 23, 1960

Soviets walk out of the Paris Summit meeting, cancel the invitation for President Eisenhower to visit the Soviet Union, and denounce the United States at the U.N.

Scrambling to rebut, U.N. Ambassador Lodge shows a state gift from the Soviet Union to the US government, a Seal of the United States fitted with listening devices inside.

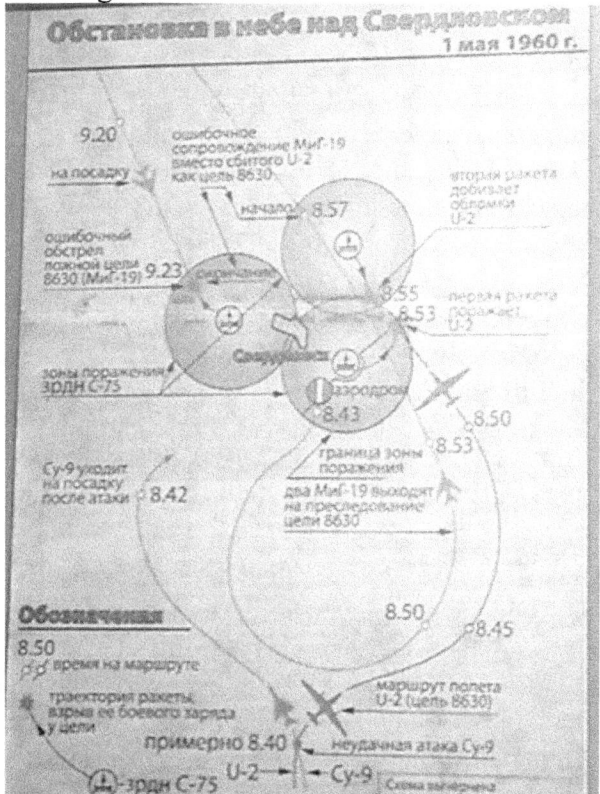

The Soviet Plot of the Powers Flight

Dr. Sergei Khruschev, son of Premier Khrushchev, was 25 at the time and with his father in Red Square and at the Air Defense Command Center as an eyewitness.

See Khrushchev's account "The Day We Shot Down the U-2" in American Heritage online: www.americanheritage.com/
day-we-shot-down-u-2

Changes to the CIA program

With the shoot-down of Gary Powers and the cessation of Detachment B and C coverage of the Soviet Union, change and consolidation sped up—now on a forced basis.

Detachments A, B, and C had already adapted somewhat throughout their deployment. One change was to allow dependents to deploy abroad. The CIA considered lengthening contract term beyond 18 months, in the interests of a stable detachment and in retaining experienced personnel.

The year 1958 saw a new proposal for creation of another Detachment in addition to and independent of A, B, and C. Obstacles to this would include logistics: a base needed with hangar, proper fuel and fueling equipment, tugs, flatbed, forklifts, jeep, pickup, bus, 1 ½ ton truck, ambulance, oxygen trailer and so on. The amount of personnel required for another detachment with 2 U-2s flying four sorties a week would come to be about 85 persons. The expense would also be a challenge: cost at that time was about $150,000 per month to run an overseas detachment.

Detachments B and C also adjusted by staging from remote locations. From Atsugi, Japan, Detachment C staged overflights of the Soviet Union from Alaska and flew from the Philippines to cover Indonesia during a revolt against Sukarno in 1958.

The phase-out of Detachment A resulted in normal changes and consolidation of personnel during 1957.

Now in 1960 after the Soviets announced that Powers was alive, the CIA evacuated the British pilots from Detachment B as Turkey did not know of their presence in the country. The end of Soviet overflights meant that Detachment B would soon leave Turkey, and during July Detachment C would leave Japan following a request by the Japanese government.

The two detachments merged into Detachment G at Edwards Air Force Base, California, where the CIA had relocated the U-2 program after nuclear testing forced it to move from Groom Lake in 1957.

By the next U-2 flight, during October 1960 over Cuba, the National Security Council Special Group had replaced the previous informal procedure in which the president personally approved or disapproved each flight after discussion with advisors. The expansion of satellite intelligence partly compensated for the overflights' end, but because U-2 photographs remained superior to satellite imagery, future administrations considered resumption at times, such as during the Berlin Crisis of 1961.

In early 1955, Richard M. "Dick" Bissell, Jr., then SAPC (Special Assistant for Policy Coordination), Office of the DCI (Director Central Intelligence)—(SAPC/DCI)—had grafted a small project group into the staff of the OSA to begin its life organizationally.

The reason was for the covert development and operation of the U-2 aircraft in conjunction with the USAF with the cryptonym AQUATONE. Mr. Dulles and General Twining had signed a formal agreement with the Air Force during June of that year, delineating areas of responsibility for both parties to the pact.

The SAPC/DCI relationship continued as the growing project emerged and gave signs of surviving beyond the 13-17 months operational life forecast for it at the beginning. Each time a field detachment deployed from Watertown, the number of personnel in the project grew until it peaked in mid-1957.

Mr. Dulles learned of the number involved and was horrified. He directed that something needed doing to reduce the obvious swelling in his staff. The AQUATONE personnel separated from Mr. Bissell's staff and became the Development Projects Staff, retaining their organizational affiliation with the director only in the term DPS/DCI.

Hence, Mr. Bissell wore two hats as SAPC/DCI and as chief, DPS. The group which became later known as COMOR attached to SAFC/DCI as the ARC (Ad Hoc Requirements Committee), the first unified effort to codify strategic intelligence requirements and to differentiate them from tactical intelligence requirements.

This condition sustained until February 1959, when Mr. Bissell became DD/P. At that point, the group became the Development Projects Division headed for the first time by an Air Force officer.

At the same time, DPD absorbed the old Air-Maritime Division of DD/P, which was by then all air and no maritime, and, functioned as a staff supervising clandestine air operations around the world.

By the time DPD (the Development Projects Division) blended into DD/P (Deputy Director of Plans), it was no longer single project-oriented, even leaving aside the inclusion of AMD assets. In 1958, preliminary work had begun on a successor system to AQUATONE (by then named CHALICE) under the cryptonym GUSTO.

Active work was underway in the development of Project CORONA, a satellite photosystem which emerged from the Air Force System 117L (Pied Piper). The cancellation of System 117L occurred during March 1958 as a ploy to permit the covert development of CORONA ostensibly under Advanced Research Projects Agency (ARPA).

During September 1959, OXCART became a project succeeding GUSTO, so on 1 January 1960, DPD was on the verge of launching with Air Force the first CORONA mission (the first twelve Corona attempts were unsuccessful). DPD operated two major CHALICE U-2 detachments overseas and was starting to work cutting titanium for the first OXCART Mach 3 aircraft.

CHALICE, Project AQUATONE renamed, was comprised of personnel and support for the two Detachments overseas, the ZI base at Edwards Air Force Base, and almost all the Development Projects Staff. The operation and maintenance of the U-2 aircraft remaining in possession of the Agency numbered thirteen. Any remaining development work on U-2 aircraft and other sub-systems employed in CHALICE, notably a new ECM device and considerable production flight testing of items to be delivered to the Strategic Air Command.

The GUSTO project was nearly a year old and consisted of feasibility studies looking toward a successor aircraft to the U-2. The Lockheed Aircraft Corporation had conducted an extensive program involving at least preliminary design of no less than 30 to 40 configurations of aircraft. It had also carried out an extensive program of model building and of measuring radar reflectivity of models. Lockheed had also built a partial full-scale mock-up of a possible GUSTO aircraft. These activities would involve substantially larger funds than presently proposed in the operating budget for FY-1959.

CORONA covered all aspects of the program for the launching of 12 reconnaissance satellites to take photography during overflights of the Soviet Bloc and would contain provisions for the storage of the exposed film in a capsule which would re-enter, drop in a preselected ocean impact area and recovered.

If CHALICE was to continue through the full fiscal year and CORONA carried through, there was little room for maneuver in the reduction of their costs. Concerning feasibility studies, the philosophy of this office had been that the objective in view was so important, and the cost of exploring technical possibilities was so small a part of the cost of a whole new reconnaissance vehicle, that all promising technical opportunities should be explored with urgency.

During November 1960, the Deputy Director for Plans, Dick Bissell, contacted the chief, Development Project Division (Col William Burke), notifying him of his intention to take advantage of the reduction of Detachment B. He hoped to achieve a decrease in the authorized strength of the division, thus reflecting the gradual shift of resources away from the U-2 into newer programs. The sizable build-up at Watertown (renamed "Area 51") for the A-12 program provided the needed evidence to support his intention. At the time, the T/O had increased to 656. However, the continuing overall personnel review within the DD/P complex cut 60 slots before the end of 1960. CORONA was now returning film.

James Cunningham noted about the Development Project Division workload in the 1960 period: "We were playing all of these keyboards: the CORONA over here, and the OXCART here, and U-2, here. And we also had to do all the planning and staffing for the Bay of Pigs. . We had to share the business of making on-orbit decisions for cameras for the CORONA missions"

The staff remained static until February 1962, when Bissell left the CIA and a six-month period of reorganization ensued. On 28 February 1962, with the resignation of Mr. Bissell as DD/P (Deputy Director of Plans), DPD transferred to the newly formed DD/R (Deputy Director Requirements). The move coincided with its move during March of 1962 to the new CIA building at the same time renamed OSA, Office of Special Activities.

During May 1965, the CIA separated its satellite operations from the other activities within OSA under the Special Projects Staff (SPS). Effective 15 September 1965, the CIA established the Office of Special Projects within the DD/S&T to carry on these operations. Twelve positions

transferred from the OSA Table of Organization to help staff this new office. The transfer involved the DPD's special projects staff to the new Deputy Director for Research while the air support function remained within the DD/P.

The OSA, Office of Special Activities would accomplish:

1. Developed and operated the U-2 program for ten years (1956-1966)

2. Developed the first operational satellite reconnaissance system in the world.

3. Developed first operational stealth aircraft in the world.

4. Developed TALENT and TKH systems, plus ? system. Established CIA worldwide security courier system and first CIA industrial security system

5. Furnished entire air support to Cuban Bay of Pigs operation.

7. Located first SA-2 site in Cuba during the 1962 missile crisis. Flew the last U-2 flight over Cuba (50 flown in all).

8. Established first combined worldwide communications and avionics capability in the CIA.

Development of CIA Detachment G

The mission of Detachment G came to be the maintenance of capability for U-2 flights over denied territory whenever and wherever needed— from either "bare strip" forward areas or established U-2 sites like Adana, Turkey. For this, Detachment G also proved integral to the development and testing of improved cameras and sensors.

As it worked out, Detachment G became a virtual "twin brother" in support of Detachment H, the Black Cats. ECM and sensor packages used by H first tested in rehearsal missions carried out by Detachment G, where the weather was favorable for extensive technical testing over the continental US.

Security personnel of Detachment G delivered the new gear personally to Detachment H, and often acted as couriers to bring back security items from Detachment H to project headquarters.

In chronological order, here are some of the developments for CIA Detachment G:

In 1960 due to shutdown of Detachments B and C Soviet Union coverage, eight pilots from B and three from C consolidated into G at Edwards AFB. During October-November 1960 Detachment G began flights over Cuba in connection with the planned counter-revolutionary invasion of Cuba. The missions were to map invasion sites and obtain information on Cuban air and land order of battle.

At the end of 1960, President Eisenhower ordered that Detachment G monitor Laos. That year saw fighting break out between the Royal Lao Army and the communist Pathet Lao. Laos was beginning to lose its historical neutrality. A new government of Laos appealed to the UN for help against invasion by China and/or North Vietnam. During 3-18 January 1961, five Detachment G pilots made seven flights from Cubi Point, Philippines to confirm or deny invasion preparations. The evidence did not support the Laotian claims, and the government of Laos retracted the charges.

During April 1961, Detachment G flew 15 missions covering the invasion of Cuba at the Bay of Pigs and the aftermath.

During May 1961, Lockheed modified several U-2s for aerial refueling and called the new model U-2F. All Detachment G pilots received training on aerial refueling. Two U-2s later crashed after their wings broke off when flying into wake turbulence from the tankers' wings.

In mid-August 1961 one Detachment G U-2 pilot flew Operation Ebony over North Vietnam.

Throughout the rest of 1961 and into 1962 Detachment G monitored Soviet arms buildup in Cuba, flying from Laughlin AFB, Texas or Edwards AFB California, using aerial refueling. During August 1962, Detachment G discovered eight SA-2 sites in Cuba.

On 5 September 1962, mission 3089 found three more SA-2 sites and a MiG-21 stationed in Cuba. CIA Director McCone argued that the SA-2 sites were likely there to protect high priority targets such as medium-range ballistic missile (MRBM) sites.

Complicating matters, Black Cat U-2 pilot Huai Chen was shot down 9 September 1962 over mainland China by a SA-2 missile. The Soviets were also protesting a USAF U-2 overflight of Sakhalin Island in what was supposed to have been a peripheral mission. It became politically harder, therefore, to receive permission for further overflights of Cuba at this point.

Weather also proved unfavorable through much of September. By early October 1962, the Committee on Overhead Reconnaissance (COMOR) began pressing for intensive coverage by U-2: "Ground observers have reported sightings of what they believe to be the SS-4 MRBM in Cuba. These reports must be confirmed or denied by photo coverage."

A national intelligence estimate had already incorrectly concluded that the Soviets would not install such missiles in Cuba.

On 14 October 1962 the mission flown by a SAC U-2 pilot over San Cristobal, Cuba came back with conclusive evidence of Soviet MRBMs in Cuba. President Kennedy gave command and control authority to the USAF for the rest of the Cuban U-2 operation, taking it away from Detachment G.

On 27 October 1962, a Russia SA-2 missile shot down Maj Rudolph Anderson's U-2 over Cuba, adding pressure towards nuclear war. At this time the System XII warning system for the U-2 that detected Fan Song radar was coming into use. Major Anderson's plane either did not have System XII or it may not have worked correctly.

Across the globe, from 1962 to 1964, Detachment G pilots were flying from Takhli, Thailand over Vietnam in 36 missions. The 'Trojan Horse' and 'Olympic Torch' sorties flew along the North Vietnam border mapping military and industrial targets. 'Senior Book' sorties involved flying along the Chinese border gathering COMINT.

From late November 1962 to January 1963 at the request of India, Detachment G flew from Takhli to cover the Sino-India border in the wake of the October 1962 China-India border war.

During April-May 1963 Detachment G covered border areas of Laos, Cambodia, and China.

Following the Southeast Asia missions, during December 1963 Detachment G flew from Ramey, Puerto Rico to cover portions of Venezuela and British Guiana. As 1964 began, Detachment G pilots were undergoing training to land on and take off from aircraft carriers. The first successful carrier landing was on 2 March 1964.

During May 1964 two Detachment G pilots flew operational missions from USS Ranger to monitor French A-tests. Also, during May 1964 and again during December 1964, Detachment G flew out of Charbatia, India to cover Sino-India border areas.

During 1966 the Agency's inventory of U-2s was down from the original 20 purchased to only six: four out of Edwards AFB and two used by Detachment H in Taoyuan, Taiwan. Twelve new versions called U-2R were ordered from Lockheed, six for the USAF and six for the CIA. Detachment G performed the lion's share of development to establish U-2R cockpit layout, adjustment to new S2010 pressure suit, flight tests of the new systems, rehearsal of mission flight profiles, and so on.

In the late 1960s, Detachment G operations over denied areas were beginning to wind down with the increased use of satellite reconnaissance. Detachment G undertook many flights over the continental US in response to requests by NASA, the US Geological Survey (for mapping), the Department of Commerce (for snowpack/hydrology surveys), and the Office of Emergency Preparedness regarding hurricane damage baseline and coverage of earthquake damage.

Detachment G Quick Deployment

Note that during the 1960s Detachment G had worked hard on the "quick response" approach to global power projection. The detachment repeatedly drilled and practiced rapid response to the forward deployment bases.

Quick deployment capability required keeping an inventory of forward bases, knowing how quickly to appoint a commander and operations officer for deployment, how to assemble a personnel roster on quick notice, arrange a transport aircraft, put together briefings, and launch the mission at any time.

The CIA arranged transport priority for the U-2 driver, flight planner, life-support personnel, and maintenance crew. The Operations Officer rode jump-seat in the transport, to keep close communication with the accompanying U-2. All radio communications were on secure, dedicated frequencies. The C-141 transport and U-2 traveled together.

The flights might land before the overseas deployment destination to refuel. If necessary, a KC-135 tanker might join for refueling requirements over the ocean.

In the meantime, the deployment commander would ensure that the forward deployment base was ready with staff cars to give mobility to the operations officer, transport crew, life support crew, maintenance team, and communications/security personnel when they arrived. A transport bus, billeting, power cart, fuel, tow vehicles, forklifts, fire truck, and so on, all had to be ready.

Consider now, the situation in early August 1970 when Secretary of State Kissinger offered the services of the US to monitor and verify the pullback areas along the Suez Canal in the wake of the Arab-Israeli conflict. Satellite coverage did not prove sufficient to verify items smaller than a jeep vehicle. When asked, the Air Force said its U-2s could not be ready for several weeks.

CIA Director Helms then told the National Security Council that the Detachment G U-2s could do the job right away. The Detachment G U-2 arrived 71 hours after authorization to deploy. Detachment G flew 29 verification missions there that came to be known as Project Even Steven. Its H camera could resolve objects on the ground to a size of ten to twelve inches.

In 1973 and 1974 Detachment G further monitored the Israeli-Egyptian and Israeli-Syrian disengagement areas.

Finally, in 1974 CIA Detachment G (as well as Detachment H, the Black Cat Squadron) disbanded. Also, OSA phased out, the inventory of remaining Agency U-2Rs turned over to the Air Force.

Tony LeVier

WHALE TALE - U-2 Carrier Operations

At one time, the CIA tried to extend the U-2's operating range and eliminate the need for foreign governmental approval for U-2 operations from USAF bases in foreign countries. There was also the matter of Soviet protest notes and pressure on these foreign governments.

As it was U-2s taking off from bases in Turkey and Pakistan for 'out and back missions' over the Soviet Union returned with fuel tanks mostly empty. Gimo Yang's mission for Detachment H during May 1963 faced the same fuel limitation. He had only ADF as a navigational aid, and weather was bad with a low ceiling, making it hard to find the landing strip in Korea. Yang's engine quit for lack of fuel when still rolling down the runway on landing. The CIA thus suggested operating the U-2 from aircraft carriers and approached the Navy about developing this ability.

During August 1963 the CIA's project WHALE TALE first tested a modified U-2C on the USS Kitty Hawk sailing out of San Diego. Famed test pilot Bob Schumacher had the stick as the insubstantial looking aircraft fired up its single jet engine and rolled to a graceful liftoff in only 321 feet without using the ship's catapults.

Following his successful launch, Schumacher made several landing approaches, proving that the U-2's slow approach speed and high excess power provided plenty of margin for error in case of a wave off. On his first attempt at an actual landing, however, one wingtip struck the deck, and Schumacher barely got the aircraft back in the air before it tumbled over the side.

Undaunted, Lockheed and the Navy modified three U-2As, adding stronger landing gear, an arresting hook, and wing spoilers to decrease lift during landing.

One of the initial problems with the U-2 was that it seemed not to want to land. On the aircraft's initial test flight, Lockheed pilot Tony LeVier tried unsuccessfully to land four times before finally bringing it back to earth on his fifth attempt. Understandably, carrier landings would be a further challenge.

While these modifications were taking place, Schumacher and several CIA pilots developed and honed their carrier landing skills flying T-2 Buckeye trainers from the USS Lexington. Schumacher landed the first modified U-2G on the USS Ranger on March 2, 1964. This landing had only one minor problem. When the arrestor hook engaged, it forced the plane's nose toward the deck and broke off the pitot tube. After repairs, Schumacher took off again successfully. During the next several days, Schumacher and the CIA

pilots received carrier qualifications from the Navy.

The U-2G only flew twice in operations. Both flights from Ranger occurred during May 1964, to observe France's development of an atomic bomb test range at Moruroa in French Polynesia.

The carrier-based U-2 missions in 1964 carried a new camera called the 112B, developed for use from satellites. The camera was able to resolve items on the ground of 10 to 12 inches in size. For Operation Fish Hawk monitoring French H-test preparation, the film was developed in the aircraft carrier's photo lab.

Future missions from carriers were considered to be impractical, given the expense of deploying carrier groups, as well as the rapid improvement of reconnaissance satellites.

Following the Gulf of Tonkin Resolution on August 1964 from a military aspect, tactical intelligence became more important than the CIA's

reconnaissance information. Consequently, SAC took over all the U-2 missions in Indochina.

In early 1964, SAC sent a detachment of U-2s from the 4080th to South Vietnam for high-altitude reconnaissance missions over North Vietnam. On 5 April 1965, U-2s from the 4028th Strategic Reconnaissance Squadron (SRS) took photos of SAM-2 sites near Hanoi and Haiphong harbor

On 11 February 1966, the 4080th Wing became the 100th Strategic Reconnaissance Wing (100 SRW) and moved to Davis-Monthan Air Force Base, Arizona. The detachment at Bien Hoa Air Base, South Vietnam was re-designated the 349th SRS.

U-2 carrier landing

The only loss of a U-2 during combat operations occurred on 8 October 1966, when Maj Leo Stewart, flying with the 349th Strategic Reconnaissance Squadron, developed mechanical problems high over North Vietnam. The U-2 managed to return to South Vietnam, where Stewart rejected. The U-2 crashed near its base at Bien Hoa. During July 1970, the 349th Strategic Reconnaissance Squadron at Bien Hoa moved to Thailand re-designated as the 99th Strategic Reconnaissance Squadron, remaining there until March 1976.

In 1969, the larger U-2Rs tested from the carrier America ended the U-2 carrier program afterward.

About the U-2 Effort - In Their Own Words

CIA Director George Tenet: "We desperately needed to know what Soviet intentions and capabilities were. In short, we were blind. The U-2 program gave us eyes to see inside the Iron Box. It may be one of the greatest achievements of any intelligence service of any nation."

CIA Director Allen Dulles when his reconnaissance chief, Richard Bissell told him what routes the U-2s had flown: "Oh my Lord. Do you think that was wise the first time?" "Allen," Bissell replied, "the first is the safest."

President Eisenhower when shown U-2 photos taken as British and French bombers attacked an Egyptian airfield during the Suez crisis, showing the smoking field: "Ten-minute reconnaissance — now that's a goal to shoot for!"

NRO Historian Outzen: Dwight Eisenhower is perhaps most important of all. Driven by the desire to avoid another Pearl Harbor, he provided presidential leadership that accelerated overhead reconnaissance efforts and protected them."

U-2 Pilot's View of Russia

Knutson: "I flew over Leningrad, and it blew my mind because Leningrad was my target as a SAC pilot and I spent two years training with maps and films, and here I was, coming in from the same direction as in the SAC battle plan, looking down on it through my sights. Only this time I was lining up for photos, not a bomb drop. Through my drift sight I saw fifteen Russian MiGs following me from about fifteen thousand feet below. The day before, Carmen Vito had followed the railroad tracks right into Moscow and saw two MiGs collide and crash while attempting to climb to his altitude. On most of the flights, '56-'57-'58, there was usually a constant stream of Russian fighters below you." - Rich, p. 147-148

Vito: "Those fields [small mosaic fields of the collective farms passing below] reminded me that Russian farmers tilled the land by hand. I couldn't get mad at people who work that hard. I was never mad at the Soviet people themselves, just at their government." On his way back to West Germany, the Canadian radar came on the air: "Lone Ranger, this is Tonto. Keep heading in the way you are going. You have twenty-two minutes to touchdown." Vito: "I knew they were talking to me because they had the best radar around and had picked us up on previous flights. But U-2 pilots had to observe complete silence on the radio. To this day, I wish I could have replied, 'Ti Ee Kemosabe.'" –
roadrunnersinternationale.com/vito.html

Powers: "There was no abrupt change in the topography, yet the moment you crossed the border, you sensed the difference. Of special interest were Soviet rocket launches. For some reason, and many of these occurred at night, and from the altitude at which we flew, they were often spectacular, lighting up the sky for hundreds of miles." - Powers p. 32, 39

Cherbonneaux: "Found and photographed the Soviet SAM and radar test site at Sary Shagan. Cherbonneaux flew over Semipalatinsk nuclear weapon assembly facility. Over the desert, he glanced at the drift sight and received a shock. He had flown many times over the US nuclear test area and recognized what he saw. Below, he spotted a large tower with a nuclear device 'shot cab' at its top. Cherbonneaux began to sweat and hyperventilate at the thought that the weapon might be detonated as he passed over. He shouted into his faceplate, 'Wait, goddamnit. Wait, will you? Let me pass and then light it!'" - Peebles p. 177-178

Chapter 18 – NASA and the U-2

NASA Finally Conducted Weather Research

During September 1956, the USAF's 4080th Strategic Reconnaissance Wing received five U-2A aircraft, modified for the High-Altitude Sampling Program (HASP) air sampling-including atmospheric gasses and particulate sampling devices.

During October 1957, the Air Force U-2 HASP flights began with the detachment of planes and crews to Ramey AFB, Puerto Rico and Plattsburgh AFB, New York. The following month, a CIA U-2 flew over "Typhoon Kitt" conducting high-altitude photographic surveillance of the storm to bolster their weather recon cover story and support the US Air Force's Air Weather Service's typhoon research.

The plane overflew the top of "Kitt," in the western Pacific Ocean, north of the Philippine Island of Luzon. The aircraft photographed the typhoon's cloud formations and inner eye dynamics—looking straight down from over 65,000 feet. This U-2 flight produced the first high-altitude, high-resolution images of the upper tropopause region of a tropical cyclone.

In 1958, the Formosa Strait crisis developed, and CIA Detachment C would fly missions to monitor a possible Chinese invasion of Taiwan. Also, Detachment C continued to fly weather missions on 14, 15, and 16 July high over Super Typhoon Winnie.

Winnie, a particularly powerful typhoon with winds more than 175 mph, struck the westernmost end of Taiwan, causing severe damage. The storm continued across the Strait and then impacted the southeast coast of mainland China.

Later, during September (1958), additional U-2 missions flew over the tops of Super Typhoons Ida and Helen, photographing spectacular cloud features and structures—looking down into the storms' eyes from the lower stratosphere.

The CIA's Detachment C flew these early Pacific Typhoon U-2 overflights under a fake cover designation as the USAF's 3rd Weather Reconnaissance Squadron-Provisional. The CIA publicly stated that the squadron was supporting the AF AWS typhoon research base at NAS Atsugi, Japan. The typhoon flights helped to bolster their "weather reconnaissance" cover story while providing area tactical reconnaissance coverage of the region during the "Offshore Island Crisis," where the armed confrontation was occurring between the People's Republic of China and Taiwan's Nationalists Chinese in the summer of 1958.

In 1961, the US Air Force began using the U-2 as a control research vehicle for Air Force's Cambridge Research Laboratory (AFCRL) flights. The flights were for high-altitude space particle (micrometeorite) collection, the US Weather Bureau's WSR-57 weather radar network development, and Ozone Research. The latter involved flying into Atlantic Hurricane Ginny, the first-ever high-altitude ozone research probe of a tropical cyclone.

While the CIA continued its U-2 reconnaissance flights, the US Air Force continued using its U-2s for research. The Agency moved from studying typhoons and hurricanes to High-altitude Clear Air Turbulence (HICAT) flight tests in the Lightning Research Project (in association with Project Rough Rider) and participated in the Jet Stream Cirrus (Cloud) Research Project.

The US Air Force also conducted clear air turbulence research flights. The US Air Force flew an AFCRL U-2 flight over Hurricane Isbell and with specialized ozone instruments, conducted ozone measurements, and cloud photography over the hurricane's eye.

AFCRL made research U-2 planes available to the Weather Bureau's hurricane research project (NHRP) and its component "Stormfury" experimental hurricane modification project. Over the next eight years, AFCRL research U-2 planes flew many Atlantic missions supporting direct and indirect hurricane research.

The US Air Force ended its hurricane scientific research flights with the September 1967 HICAT U-2 flight over and around the eye of Hurricane Beulah. The aircraft conducted hurricane surveillance and high-altitude aerial photography of the hurricane's eye using specialized thermal instruments. Beulah made landfall in southern Texas. During June 1971, the US Air Force ended its AFCRL U-2 scientific research flights, turning weather research over to

NASA as the CIA claimed in its cover story 15 years earlier.

During April 1971, NASA received approval for the use of U-2 spy planes for scientific research. Instead of CIA pilots flying "NASA weather research U-2 aircraft," NASA pilots began flying CIA U-2 aircraft. During June 1971, NASA received two CIA U-2C aircraft (#6681 / N708NA and #6682 / N709NA) for high-altitude research.

The end was in sight for the CIA reconnaissance. It started with President Kennedy ordering the CIA to allow the US Air Force to fly its better U-2A. The Agency gave its U-2Cs equipped with electronic countermeasures to the Air Force during the Cuban Crisis in 1962 and for all flights afterward there.

Now, NASA was also flying CIA aircraft. On 1 August 1974, the CIA's U-2 manned reconnaissance operations ended with the advent of improved satellite coverage. The CIA transferred all its U-2 aircraft, equipment, and logistic support parts directly to the US Air Force.

****

By 1968, the US Air Force had slowly privatized their organic research laboratories and decreased its direct support of scientific research within its operational commands.

At this point that NASA, cognizant of the capabilities and contributions the early Air Force Research U-2 provided over the years, lobbied the US government for the acquisition of its U-2 research aircraft to carry on similar research.

In 1971, the lobby succeeded with NASA granting authorization to operate U-2 aircraft for scientific research.

On June 3 and 4, 1971, the NASA Ames research facility received two (ex-CIA) U-2C aircraft on permanent loan from the US Air Force, as high-altitude scientific research aircraft.

Fifteen years after it said that it was the U-2 aircraft program manager and operating these new U-2 aircraft as scientific research tools, NASA finally received U-2 aircraft for that purpose.

Cuba

From October 1960 onward, U-2s operated over Cuba firstly in the build-up to the Bay of Pigs invasion, and then during the aftermath of the invasion's failure. During summer 1961, the CIA modified six of its U-2s to allow in-flight refueling. The resulting model was known as the U-2F.

The CIA continued to upgrade its U-2 fleet. During late summer/autumn 1962, flights over Cuba discovered SAM sites, followed during October 1962 with the discovery of IRBM sites. In 1963, the CIA modified several aircraft called the U-2G for carrier operations, and in 1966, placed an order for several new re-designed aircraft known as U-2R fitted with the Pratt & Whitney J75/P-13B engine, and capable of flying at 75,000 feet. Lockheed delivered the last aircraft on 11 December 1968. On 23 November 1966, the CIA ordered six more with the new camera, a modified version known as the H-camera developed for the A-12.

From October 1960, Detachment G made many overflights of Cuba from Laughlin Air Force Base, Texas. Many of these missions originated at the North Base at Edwards Air Force Base using the six Lockheed modified U-2F aircraft capable of aerial refueling. Pilot fatigue limited these flights to 10 hours.

At the Central Intelligence headquarters in Langley, Dick Bissell was concentrating on plans for the Bay of Pigs invasion of Cuba. Following a US-sponsored bombing run against the Cuban Air Force, a group of 1,500 armed exiles landed at the Bay of Pigs on 17 April 1961. As the invasion faltered, President John F. Kennedy called off promised air support, leaving the CIA and the Cuban exiles at the fate of Castro.

Cuba executed hundreds and sentenced around 1200 to 30 years in prison for treason (Cuba ransomed some 1100 prisoners in 1962 for $53 million in food and medicine). In August 1961, Russia and East Germany commenced construction of the Berlin Wall to divide Germany physically and ideologically.

The US Reaction to the Iron Curtain and Berlin Wall

"Brinksmanship" of the time is noted in this intelligence summary: "On 23 June 1959 the US Embassy in Moscow reported that Khrushchev, in private conversation with Ambassador Harriman on 23 June, took an uncompromising position on Berlin, boasted of Soviet military strength, and in effect, warned that the USSR was prepared to face a showdown over the Berlin issue. He bluntly asserted that the USSR was determined to liquidate the West's rights in Berlin and that if the West insisted on perpetuating or prolonging its rights in Berlin, this means war."

Thus, in early 1960, the United States feared the Soviet Union might decide to invade West Germany. The United States reacted by having the Nevada Atomic Proving Grounds next door to Area 51 deploy an atomic cannon to defend West Germany.

The US Army established an air defense and retaliatory ICBM missile sites across the United States. It deployed Hawk missile units to West Germany and along the beaches of Florida to deter open war with Russia. The military services discouraged military dependent travel for much of Europe. Wives and children of military personnel living in West Germany wore military dog tags, and every six months, practiced evacuation from West Germany to Paris, France, while the military member of the family remained behind on military alert status.

The Air Force Initiated Operation Chrome Dome, a continuous airborne alert with B-52 Stratofortress strategic bomber aircraft armed with thermonuclear weapons flying routes to points on the Soviet Union border.

At the cost of more than the Manhattan Project, the United States established the Semi-Automatic Ground Environment (SAGE) system of large computers, networking with radar sites across the United States. SAGE provided a single unified image of the airspace over a wide area. SAGE sent its tracking data directly to CIM-10 Bomarc missiles and some of the US Air Force's interceptor aircraft in flight to provide an intercept course.

The United States and Canada formed the Distant Early Warning Line, the DEW line. The DEW line established a system of radar stations in the far northern Arctic region of Canada, with additional stations along the North Coast and Aleutian Islands of Alaska. The line built to detect incoming Soviet bombers included the Faroe Islands, Greenland, and Iceland, giving both the United States and Canada an early warning of any sea-and-land invasion.

Inside Cheyenne Mountain in Colorado, the United States and Canada further established NORAD, the North American Air and Space Defense Command to provide air and space warning, air sovereignty, and defense for North America. Completely underground, a 25-ton nuclear blast door protected the entrance.

The onset of the Cold War called for the development of highly specialized and secretive strategic reconnaissance aircraft, or spy planes.

In a nationally broadcast speech, Castro declared he was a Marxist-Leninist and that Cuba was going to adopt Communism.

In the Sino-Albanian split, the People's Republic of China severed diplomatic ties with Albania.

This all came to a head when an August 1962 flight revealed Soviet SA-2 SAM sites on the Cuban Island, and later overflights found more sites and MiG-21 interceptors. The increasing number of SAMs caused the United States to plan its Cuban overflights more cautiously.

At the time, only the CIA was conducting overflights of Cuba. The CIA flying and not the Air Force bothered many officials. Many believed it better using military officers in the case of the Russian shooting down one of these U-2s conducting aerial reconnaissance. The Air Force cover story would be that of routine ferry mission to Ramey AFB in Puerto Rico, where one of its U-2 detachments operated.

Following one last Cuba overflight originating from Edwards and ending at the McCoy Air Force Base, Florida on 14 October 1962, all further U-2 operations over Cuba originated from McCoy.

The CIA did not give up to those advocating the US Air Force perform the surveillance of the Russians in Cuba. The CIA argued its planes better equipped to survive Russian defenses. President Kennedy said, "Fine. The US Air Force will use your aircraft." The next thing the CIA knew, on October 14, 1962, Air Force pilots appeared at the North Base and before the day was out, left in the CIA's planes repainted with USAF insignia.

Early Sunday morning, 14 October 1962, then-

Major Heyser climbed into the CIA U-2F, Article 342. The plane flown by Heyser was the second CIA U-2 modified for in-flight refueling. Flying this repainted as 'USAF 66675,' at Edwards Air Force Base, California, he underwent a brief qualification on the type of U-2 and departed on a Cuban overflight, Mission 3101, dubbed BRASS KNOB. He met the sun over the Gulf of Mexico and flew over the Yucatan Channel before turning north to penetrate denied territory. With a 25% cloud cover, he flew the maximum altitude profile to ascend the U-2F to 72,500 feet where the plane left no contrail. Heyser switched on the camera in flight over the Island for fewer than seven minutes.

Heyser's potential exposure to the two SAM sites lasted over 12 minutes. His preflight briefing cautioned him to scan the drift sight for Cuban fighters or, worse still, a SA-2 missile launched at him. If attacked, his instructions were for him to turn sharply toward the missile, and away from it, in an S-pattern to break the missile radar's lock. Heyser encountered no opposition from Cuba's air defenses. He coasted-out and headed for McCoy Air Force Base, Florida, landing there at 0920 EST after seven hours in the air.

The US Air Force flew the film Washington, DC to the National Photographic Intelligence Center for processing, and the first images trucked under armed guard to analysts at NPIC. By noon, the analysts had identified SS-4 missile transporters. The photo images and other evidence caught on the films set in motion the Cuban Missile Crisis. On 22 October 1962, the first of thirteen days marking the most dangerous period of the Cuban missile crisis, President Kennedy announced the photographs proving that the Soviet Union was building secret sites for nuclear missiles. The crisis ended after Soviet Premier Nikita Khrushchev ordered the missiles withdrawn from Cuba.

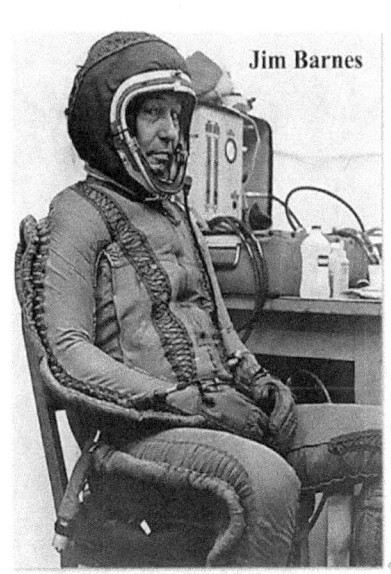
Jim Barnes

There is an unanswered question who was the first to spot the Soviet missiles in Cuba, the CIA's Jim Barnes, or the Air Force's Maj Richard Heyser. Both were flying at the same time.

Note, CIA pilot Barnes was also flying Cuban coverage. There is some question who was the first to spot the Soviet missiles in Cuba. Both were flying at the same time.

The Cuban incident was still not over. On 27 October, a SA-2 missile shot down a U-2C, killing Air Force Major Rudolf Anderson, the only person killed by enemy fire during the Cuban Missile Crisis. He posthumously received the first Air Force Cross, awarded for heroism.

On that fateful October 27, Anderson took off in a U-2F (AF Serial Number 56–6676, former CIA Article 343) from a forward operating location at McCoy Air Force Base in Orlando, Florida. A few hours into his mission, a Soviet-supplied S-75 Dvina (NATO designation SA-2 Guideline) surface-to-air missile system shot him down near Banes, Cuba.

The CIA report issued at 0200 hours the next morning stated, "The loss of the U-2 over Banes was probably caused by intercept by a SA-2 missile from the Banes site, or pilot hypoxia, with the former appearing more likely based on present information that shrapnel from the exploding proximity warhead on the SA-2 punctured Maj. Anderson's pressure suit and helmet, causing decompression and killing him almost immediately."

Major Anderson had trained for midair refueling with Majors Heyser and Captain John Campbell at Edwards AFB in early 1962. The Air Force needed Anderson and Heyser to ferry their U-2s across the Pacific for a precision ELINT mission carried out before the crisis begins to develop in Cuba.

Anderson was part of the 4028th Strategic Reconnaissance Weather Squadron; 4080th Strategic Reconnaissance Wing headquarters at Laughlin Air Force Base, Texas.

Fulfilling the CIA officials' fears of a USAF takeover, the CIA pilots never again flew over Cuba; the Strategic Air Command retained control over Cuban overflights, which continued until the 1970s under the code-name OLYMPIC FIRE.

At the same time as the Cuban crisis, Royal

Air Force English Electric Lightning planes of the Air Fighting Development Squadron made several practice interceptions against U-2s. Using ground-controlled interception and using energy climb profiles, the Lightning could intercept the U-2 at up to 65,000 feet.

Although Anderson was the only combat death of the crisis, three reconnaissance-variant Boeing RB-47 Stratojet planes of the 55th Strategic Reconnaissance Wing crashed between September 27 and November 11, 1962, killing 11 crewmembers.

Seven more airmen died when a Boeing C-135B Stratolifter delivering ammunition to Naval Base Guantanamo Bay stalled and crashed on approach during October 1962. The RB-47 missions of the Cuban missile crisis involved peripheral ELINT flights, weather reconnaissance in support of U-2 flights, and tracking of individual Soviet ships during the naval quarantine.

SOVIET SHIP POLTAVA ENROUTE TO CUBA
15 SEPTEMBER 1962

"The Cuban missile crisis ended on October 28, 1962, when Khrushchev agreed to remove the missiles from Cuba. This was enabled by a super-secret private agreement by Kennedy with Khrushchev to remove US Jupiter IRBMs, which had a range of 1,500 miles, from Turkey and Italy. By plan, the US did not carry out the removal until early 1983, done very quietly and never publicized."

"On October 29th Kennedy asked to meet with Major Steve Heyser personally. Before the meeting, his colonel asked Heyser what he planned to say. He said —'I'm going to tell him we are not getting enough money for per diem and the food is lousy!!'"

"Sixty-nine U-2 missions over Cuba were flown during November 1962 to monitor Russian withdrawal. This included 40 missile launchers plus the missiles and their nuclear warheads. On November 20th Kennedy publicly announced that they had been withdrawn, departed on Russian ships and that the US Navy had inspected the ships. The Cuban missile crisis was over."

– recalled by Sherm Mullin (via Roadrunners Internationale)

Chapter 19 – Epilogue of the CIA U-2

Early on, General Curtis LeMay had said that once the CIA brought the U-2 into operation, the Air Force would take it away from them. LeMay and President Kennedy did exactly that in the case of the Cuban crisis.

In 1979, the Air Force also took Area 51 from the CIA. They, however, failed to take away the proud legacy of the CIA and those serving and supporting it at Area 51.

The CIA would accomplish even improved feats at its Area 51 station in Nevada manned by the new Science and Technology Directorate. There the CIA would design and build America's first stealth plane, which is still the fastest and highest-flying manned air-breathing plane ever built.

It would turn Area 51 into an operating location for exploiting enemy aerial assets and technology, radar technology, stealth, and much more that remains classified.

The Science and Technology Directorate transformed Area 51 into a highly technical laboratory where military and aerospace could advance and test proof-of-concept models in total secrecy. The second and third books of this *CIA Area 51 Chronicles* series, **The Archangel** and **The Company Business** cover these advances.

The most enduring of the CIA's legacy is that it built one of the United States' most secret aerial reconnaissance platforms that more than half a century later, still flies some of the US military's most sensitive spy missions worldwide in the form of the current U-2S. The satellite legacy is not far behind it in terms of years of service.

U-2 overflights of the USSR ceased permanently during May 1960 when Russia shot down Francis Gary Powers. However, the CIA continued flying with Lockheed manufacturing models of the CIA's original aircraft 'A' through H, then R and S.

The Pratt & Whitney J57-P37 engine powered the A models. The 'B' model upgraded to the J57-P31 engine. The 'C' models contained the new J75 engine with enlarged air inlets to improve engine performance. Lockheed modified five aircraft ('D' models) with optical spectrometers used to scan IR wavelengths for missile plumes, as part of the warning system for Soviet ICBM launches. (This aircraft included a place for a second crew member.)

The 'E' model aircraft increased sensor capacity and, a year or so after the Gary Powers incident, several F Model U-2s received upgrades for in-flight refueling. Lockheed later restored them to 'C' models and in 1963, modified two 'C' model aircraft for carrier operations. (G model)

A restart of the production line in 1967 yielded a 40% larger 'R' model with double the range, four times the payload capability and much-improved subsystems. The extended size of the R model reduced the "coffin corner" effect of the flight profile, increasing the safety margin on either side of the critical Mach number. The R line restarted in 1981 for the Air Force and NASA with new sensors and added mission capabilities.

The US Air Force called its version the TR-1 from 1981 through 1992, at which time the US Air Force renamed them the U-2Rs. From 1994 to 1998, all the aircraft received new engines and a new name, becoming the U-2S, the most capable and reliable high-altitude intelligence, surveillance, and reconnaissance (ISR) platform compared to any system flying today.

The original 1955 A-model planes, smaller and cruder in many respects than those flying today, expected a short lifetime. In the form of the B model with its improved J57/ P-31 engine, the earliest U-2 flew in many parts of the world for diverse missions.

On 1 May 1956, the US Air Force activated the 4080th Strategic Reconnaissance Wing at Turner Air Force Base, Georgia to fly peripheral strategic missions of the Communist world.

During August 1956, the second CIA U-2 class completed their training. As the second class deployed, the third CIA training class arrived at Watertown. Among others, it included Frank G. Grace, Jr., and Bob Ericson. By early 1957 the CIA deployed the class to the Far East and the Air Force group was in training at the Ranch.

Richard Bissell, known as "Dick" to his friends, "Mr. B" at the Skunk Works and Area 51, was called "the mad story" by his aides. None of his limbs seemed connected exactly right, making him appear perpetually in motion. He drove old cars until they fell apart, once showing up with a distributor cap in the pocket of his suit. He was chronically late and—dangerously for the chief of the clandestine service—sloppy about keeping track of paper. Considered a poor manager, but a great leader, he imbued his staff with idealism, that they were doing something important.

Bissell was courteous, well-mannered and calm during crises, but would unleash a tantrum over a misplaced paperweight. Where other Ivy Leaguers learned to swear by being around military men during wartime, Bissell's profanity was limited to "Sweet Jesus" said with real emphasis: "Swe-e-et JEEus." Affixed with Fidel Castro, he declared that there would be no Communist government in Latin America while he was DD/P and head of the clandestine service. (Bissell did not like to use the word "assassination." He preferred "executive action.")

On 16 August 1956, following Soviet protest of U-2 overflights, Bissell conducted the first meeting on reducing the radar cross-section of the U-2. The effort evolved into Project RAINBOW, a bid to prolong the aircraft's operational life through a package of modifications called "Trapeze," that added wires, and painted impregnated with tiny iron ferrite beads and electronic countermeasure systems. The modified U-2s, called "Dirty Birds," failed to reduce the U-2's RCS. The failure led to the decision to develop a new aircraft with stealth characteristics.

On 5 October 1956, Col Jack A. Gibbs (then Deputy Project Director), advised Dick Bissell to consider the workforce. He felt they needed to see if operations planned to increase deep penetration of the USSR by next spring. If not, they should review the Headquarters personnel roster to initiate a reduction in force.

Gibbs believed the front office in project headquarters was sufficient, meaning for the present workload. Delineation of the responsibilities was excellent. Gibbs cautioned Bissell against the front office issuing similar action instructions to several different individuals, as all the staff was busy and fast-moving.

The project staff needed space badly. On 1 May 1955, the staff moved to separate quarters on the top floor of 2210 E. Street, NW, Washington, DC—joining the nucleus of the Agency photo interpretation staff already set up by Mr. Arthur C. Lundahl.

At the time, the AQUATONE headquarters office composed of administration, personnel, finance, logistics, contract management, and operations—including intelligence, weather, and photointerpretation. Security and Communications staff assigned to the project still worked from their offices. Additional staff joining throughout the summer of 1955 made more space needed by October.

The first project headquarters operated from CIA's Administration (East) Building, a small red brick building (the Briggs School) at 2430 E Streets, NW in Washington, DC. Continual growth forced the AQUATONE staff to move several times during its first two years.

On 3 October 1955, the headquarters moved to Quarters Eye, Wings A, and C, on Ohio Drive. Colonel Ritland joined the staff and took a more active part as Deputy Project Director.

Defense Secretary Charles Wilson visited Watertown for a briefing on the U-2 operation late in 1955. Ray Goudy made a flyby, and Captain Meierdierck gave a pilot's report from high-altitude over the facility.

By December 1955, following Defense Secretary Wilson visiting Watertown for this briefing on the U-2 operation, the staff realized the extent of quarters that would be required for the operational phase. They needed a restricted area in a fire-resistant building, with adequate facilities for an operations center and a communications center, and with a minimum of 9600 square feet. The project leased the fifth floor of the Matomic Building at 1717 H Street NW, in Washington, DC.

By the beginning of 1956, the Groom Lake test site received delivery of four U-2 aircraft. The Air Force IPs checked the CIA pilots in Det A out in the U-2 and put support planning in effect for overseas deployment with assistance by Colonel Shingler and Lt Col Arthur Lien.

On 25 February 1956, the project staff moved into their quarters on the fifth floor of the Matomic Building. The staff remained there for the next six years until it moved into the new CIA Headquarters building at Langley during March 1962.

By March 1956, the fleet at Watertown consisted of nine aircraft, and with Bruce Grant having to withdraw, six CIA pilots were undergoing flight training at the site. The SAC 4070th Instructor Pilots and the LAC test pilots continued flight testing.

The CIA made decisions regarding navigation procedures, use of cameras and other equipment, planned for overflight needs, and checked pilot proficiency. During the training of Detachment A, the number of trainee pilots fluctuated between six and seven. In early 1956 with Bruce Grant withdrawing from the program, the number of CIA pilots dropped from seven.

During March 1956, Col Landon McConnell became the base commander at Watertown, and CIA Director Allen Dulles visited Watertown to meet the first training class personally.

Detachment A deployed to Europe during May 1956. After September 1956 the detachment was back down to six pilots. Nine had trained at the Ranch. Grant withdrew, Rose killed in training during May 1956, and his replacement Carey lost over Germany flying for Detachment A on September 17, 1956.

The second group for CIA Detachment B arrived at the Ranch and started flying and classroom training. A group of four Greek and one Polish pilot came to Groom for familiarization in the U-2. The Greek pilots all washed out during initial training, and the CIA never allowed the Polish pilot to fly the U-2.

Instructor pilot Louis Setter checked out one of the Greek pilots, asking his student what the previous flying he had done. The pilot replied that he had flown a Spitfire. During cockpit orientation, he tried to move the U-2 wheel from side to side, like a Spitfire, instead of rotating it. Setter put him in the driver's seat of the chase station wagon and had him drive out on the lakebed until he learned how to at least steer by rotating the wheel. Setter then put him in an L-20 and gave him some flying lessons. The L-20 had a wheel for aileron control and had a tailwheel. The Greek pilot's first and only flight in the U-2 ended in him stalling it over the lakebed about 10 feet in the air. The plane hit heavily on the left wingtip. That was the end of his U-2 flying.

During August 1956, the second U-2 class,

Detachment B, completed their training, and the third U-2 training class for Detachment C arrived at Watertown.

Frank. Grace, a CIA pilot trainee for Detachment C, did not survive a night takeoff. His aircraft stalled and cartwheeled into the base of the control tower. During the ensuing fire, Hank Meierdierck in Mobile Control, and the base doctor in the ambulance drove into the flames to try and rescue Grace but unfortunately could not save him.

During December 1956, Bob Ericson was flying a U-2A (56–6690) at 35,000 feet when he suffered an oxygen failure. As he began to pass out, the aircraft went out of control. Ericson managed to open the canopy, exit the plane, and parachute to a safe landing on the Navajo Indian Reservation in Arizona.

This same month, Article 341 received a modification for a series of radar cross-section (RCS) tests called Project RAINBOW. The modification was an attempt by the LAC to reduce the RCS of the U-2 using radar-absorbent materials.

They strung a second U-2, Article 344, with piano wire of varying dipole lengths between the nose and wings of the aircraft to reduce the radar signature. These methods created extra drag with a resultant penalty in range and altitude. The U-2 planes modified under Project RAINBOW received the nickname "dirty birds" because they were not aerodynamically "clean."

During April, 1957, during a Project RAINBOW test flight, Article 341 suffered a flameout at 72,000 feet due to airframe heat build-up. Pilot Robert Sieker's pressure suit inflated, but his helmet faceplate failed, and he lost consciousness. The aircraft stalled at 65,000 feet and entered a flat spin. Sieker revived at low altitude and attempted to bail out, but without enough altitude for safe manual egress. The responders located Sieker's body near the wreck, with his parachute partially deployed.

USAF pilots soon finished training at Watertown. The SAC 4070th training unit departed, and as of 11 June, 1957 Commander Jack Nole of the 4028th SRW led flights of the SAC U-2s from Watertown to Laughlin AFB, Texas. Laughlin was now the main base for the

SAC U-2s, and Edwards AFB became the main U-2 base and flight test center for the CIA.

As described in Chapter 15, Area 51's Groom Lake facility was vacant and shuttered during the nearby atomic test series called Operation Plumbbob from around the end of May through October 1957. The 29 tests reportedly released more than twice as much radioactive Iodine-131 as any other continental A-test series. Effects on the health of civilians downwind are uncertain. A study by the National Cancer Institute estimated the expectation that such radiation release could become associated with as many as 1900 deaths due to thyroid cancer.

An AEC information booklet called "Background Information on Nevada Nuclear Tests" published in 1957 gave a cover story for the Watertown operation. It stated that the National Advisory Committee for Aeronautics (NACA) was operating U-2 aircraft at the Groom Lake site "with logistical and technical support [from] the Air Weather Service of the US Air Force to make weather observations at heights unattainable by most aircraft." At that time, the aircraft was unpainted except for NACA markings in case the CIA lost one of them off-site.

The final move came during January 1968, when the project staff known as the Office of Special Activities moved again to the new office complex at Langley, Virginia.

Bissell reported to the DCI, who relied on the DDCI, Gen Charles Pearre Cabell, the official more involved in the day-to-day affairs of the overhead reconnaissance project. Cabell held an extensive background in Air Force overhead reconnaissance, making him ideally qualified to oversee the U-2 project, which meant attending the frequent White House meetings on the U-2 for the DCI.

The Air Force U-2 program became organized along similar lines to the CIA's, with one headquarters base plus three overseas or global detachments. Among other locations, the Air Force U-2s deployed to Alaska, and Ramey, Puerto Rico. One of the SAC Foreign Field Bases was at Buenos Aires, Argentina and another at Bien Hoa in Vietnam.

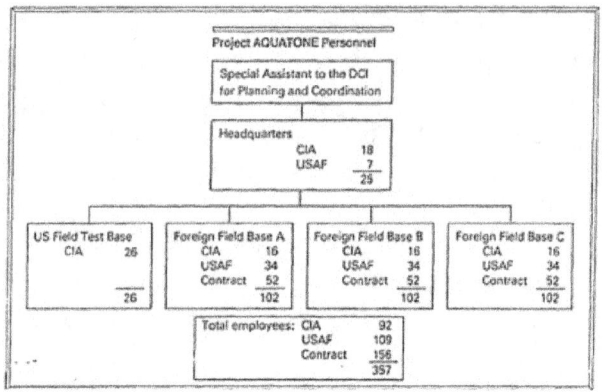

****

*Stanley Olson, James Wood, & John Crilley
playing tourist in Turkey near Saint Paul's birthplace*

*Dunnigan, Wolverton, Clendinning, Andrews
(via Jim Wood)*

Marty Knutson's Ford Thunderbird in Adana, Turkey

Marcussin & Franklin en route from Adana to Germany

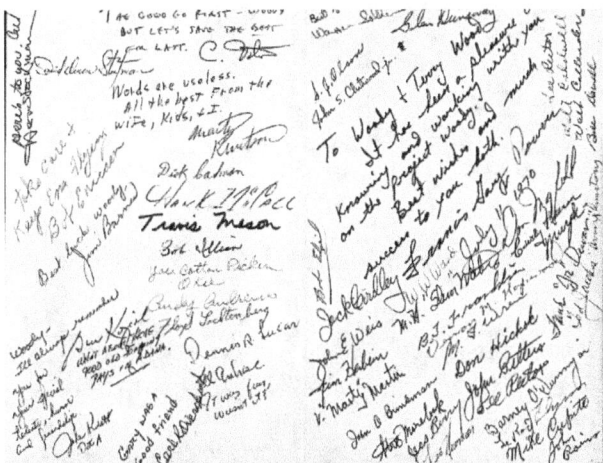

Agency pilot and support personnel signatures
(via Jim Wood)

Marty Knutson's maid with one of the pilots' daughters
in front of Knutson's Thunderbird at Incirlik in Adana, Turkey

Pilots' kids playing on Wood's BMW 500

Francis Gary Powers and Bob Ericson

Lockheed's Bob Murphy at Incirlik in Adana, Turkey 1958

Francis Gary Powers

Photo by Hervey Stockman of four Detachment A U-2s.

This is undoubtedly the first and only four-aircraft formation flown in U-2s.

Lead aircraft flown by Glendon "Glen" Dunaway, right wing flown by

Jacob "Jake" Kratt, left wing flown by Carl Overstreet, slot, Carmine Vito.

Agency U-2 pilots Vito & Knutson in 1956 (via Jim Wood)

The years 1970–2000

During August 1970, the National Reconnaissance Office (NRO) deployed two U-2Rs to cover the Israeli-Egypt conflict under the code-name EVEN STEVEN.

During June 1976, the USAF transferred the U-2s of the 100 SRW to the Ninth Strategic Reconnaissance Wing (9 SRW) at Beale Air Force Base, California, and merged with SR-71 aircraft operations there. When SAC disbanded in 1992, the wing transferred to the new Air Combat Command (ACC) and was re-designated the 9th Reconnaissance Wing (9 RW).

In 1977, a U-2R received a retrofitting to give it an upward-looking window for high-altitude astronomical observations of the cosmic microwave background (CMB). This experiment was the first to measure the motion of the galaxy about the CMB and to establish an upper limit on the rotation of the universe.

In 1984, during a major NATO exercise, Royal Air Force Flight Lieutenant Mike Hale intercepted a U-2 at the height of 66,000 feet (20,100 m), a height considered safe from interception. Hale climbed to 88,000 feet (26,800 m) in his Lightning F3.

In 1989, a U-2R of the 9th Reconnaissance Wing, flying out of Patrick Air Force Base, Florida photographed a space shuttle launched for NASA to assist in identifying the cause of tile loss during launches in the initial post-Challenger missions.

On 19 November 1998, a NASA ER-2 research aircraft set a world record for the altitude of 20,479 meters (67,190') in horizontal flight in the 12,000-16,000 kg (26,000-30,000 lbs.) weight class.

Replacements Proposed, But the U-2 Remains Flying

Sixty years after its first flight, the U-2's incredible technological and operational capabilities were enabling missions from natural disaster support to intelligence-gathering.

The U-2, in a span of over 55 years, evolved into many variations of Lockheed's original A Model U-2, all still referred to as U-2s: U-2A, U-2B, U-2C, U-2D, U-2E, U-2F, U-2CT, U-2G, U-2H, U-2R, U-2RT, U-2EPX, WU-2, TR-1A, TR-1B, ER-2, U-2S, TU-2S, U-2E/F/H, U-2R/S, and ER-2.

The plane's development and evolution have not been the only change to the program that the CIA began. Eight women have now flown the U-2, and Merryl Tengesdal was the first African American woman to do so.

Today, over 60 years since its maiden flight, the U-2 remains the premier aerial intelligence platform, by far. No other matches its altitude, power, and payload capacity. No other plane has matched the U-2's ability to swap in and out technologies that make it adaptable for whatever the mission.

The Dragon Lady, with a cruise speed of 475 miles per hour and a wingspan of 103 feet, loves to fly. When it comes to capturing data, flying at 70,000 feet provides a sweet spot for an ISR platform. The U-2 can fly over nearly any inclement weather condition.

Since 1955, the mission of the U-2 has remained the same, providing essential data and information to make informed decisions in near real-time. When it comes to precise targeting and accurate damage assessment, the U-2 sets the bar—and sets it high.

Superior Surveillance

In 1955, the CIA introduced a high-altitude ISR asset that is still operating today. No other ISR in use—or in development —can accomplish the daily peacetime Strategic Reconnaissance Operations of the U-2S or compete with its future capabilities. U-2S flies more than 10,000 feet higher, 100 mph faster, and has larger bandwidth links than any other high-altitude ISR platform. Flying 24/7 around the world at record-high operating rates, U-2S collects critical targets as no other platform can.

Only the U-2

Stands out from other high-altitude ISR platforms with its ability to field a 10-band imaging sensor in theater – producing strong intelligence to help the warfighter assess and anticipate threats. This capability proved able to deliver real-time information to commanders on the ground.

Has maintained a mission success rate above 95% for the past decade – significantly ahead of other high-altitude ISR platforms.

Can routinely collect more than double the amount of intelligence within the same amount of on-station time as any other high-altitude ISR platform.

Can collect intelligence at 70,000-plus feet, giving it a deep look at targets from 10,000 feet higher than any other high-altitude ISR platform.

Can be airborne within three hours of an emergent crisis – much faster than other ISR platforms. The Dragon Lady offers a proven rapid response for the warfighter.

Can rapidly adjust to changing conditions and runways, flying through snow, wind, and ice – setting it apart from other high-altitude ISR platforms that are more limited by weather.

Can collect intelligence going 100 mph faster than any other high-altitude ISR platform.

Has nearly twice the generator power of other high-altitude ISR platforms, giving it the power to collect intelligence from extreme distances.

Can defend itself from threats while collecting valuable intelligence from 70,000 feet.

Offers a unique design that allows for adding new sensors and capabilities cost-effectively. It could make a quantum leap in capabilities for 1/10 of the costs to upgrade other high-altitude ISR platforms.

Has twice the life of any other high-altitude

ISR platform. The Dragon Lady can deliver unparalleled performance for potentially decades to come.

Needs just one airframe to collect multiple types of intelligence, while other high-altitude ISR platforms need three.

Recent Use and Plans for Retirement

The USAF planned to retire its entire U-2 fleet by 2019 and upgrade its Global Hawks to match the capability of the venerable Dragon Lady's sensors. Northrop Grumman during February 2016 flew the first demonstration of a Universal Payload Adapter, a 17-point system on the belly of a Global Hawk that permitted the remotely operated aircraft to carry multiple sensors, such as the Senior Year Electro-Optical Reconnaissance System. Later in the year 2016, the U-2's Optical Bar Camera and an MS-177 multispectral sensor flew on the Global Hawk.

Now, in a 60-year program, the U-2 has evolved since its first flight in 1955 as an A model. In the late 1960s, Lockheed redesigned the U-2 and went into production with the U-2R designation. The 12 R models produced 40 percent larger than the 1950s U-2A and U-2C models carried a modular payload, and a larger cockpit accommodated a pilot's full pressure suit.

In the 1980s, the TR-1, structurally identical to the U-2R, went into production. The TR-1 included provisions for more advanced sensors. With the stand-up of Air Combat Command, all U-2s updated to a standard configuration and designated U-2R.

In the mid-1990s, the U-2R converted to the U-2S, receiving the GE F118 turbofan engine. The F118 engine provides higher thrust yet is efficient enough for the U-2S to fly 7000 miles on its tank of fuel—needing no refueling. The U-2S Block 10 electrical system upgrade replaced legacy wiring with advanced fiber-optic technology and lowered the overall electronic noise signature to provide a quieter platform for the newest generation of sensors. The cost of each new plane is currently classified.

The U-2 outlasted its Mach 3 A-12 replacement, retired in 1968, and the SR-71 replacement retired in 1998. On December 2005, the Pentagon approved a classified budget document calling for the U-2s termination no earlier than 2012, with some aircraft retiring by 2007. During January 2006, Secretary of Defense Rumsfeld announced the U-2s pending retirement as a cost-cutting measure during a larger reorganization and redefinition of the USAF's mission. Rumsfeld said this would not impair the USAF's ability to gather intelligence, which satellites, and a growing supply of unmanned RQ-4 Global Hawk reconnaissance aircraft would do.

In 2009, the USAF stated it planned to extend the U-2 retirement from 2012 until 2014 or later to allow more time to field the RQ-4. In 2010, the RQ-170 Sentinel replaced the U-2s operating from Osan Air Base, South Korea. Upgrades late in the War in Afghanistan gave the U-2 greater reconnaissance and threat-detection capability. By early 2010, U-2s from the 99th Expeditionary Reconnaissance Squadron had flown over 200 missions in support of Operations Iraqi Freedom and Enduring Freedom, and the Combined Joint Task Force-Horn of Africa.

A U-2 station in Cyprus during March 2011 helped in the enforcement of the no-fly zone over Libya, and a U-2 station at Osan Air Base in South Korea provided imagery of the Japanese nuclear reactor damage from the 11 March 2011 earthquake and tsunami.

From March 2011 until 2015, the US Air Force operated a fleet of 32 U-2s. However, in 2014, Lockheed Martin determined the U-2S fleet having used only one-fifth of its design service life, and it remained one of the youngest fleets within the USAF.

In 2011, the USAF stated its intent to replace the U-2 with the RQ-4 before the fiscal year 2015. The proposed legislation required any replacement to have lower operating costs.

During January 2012, the USAF announced plans to end the RQ-4 Block 30 program and extended the U-2s service life until 2023. The US Air Force changed its plans and kept the RQ-4 Block 30 in service because of political pressure despite its objections.

The US Air Force argued the U-2 costs $2,380 per flight hour compared to the RQ-4's $6,710 as of early 2014. Critics point out the RQ-4's cameras and sensors having less capability and lacking all-weather operations capability even with some of the U-2's sensors installed on the RQ-4. A decrease in the RQ-4's per flying hour costs and its matching the U-2's capabilities motivated proposals to replace the U-2s by FY 2016.

During May 2014, the U-2 faced accusations of causing an air traffic disruption in the Western US because of an apparent ERAM software glitch. The USAF stated the U-2 did not cause the problem as it did not emit any electronic signals capable of scrambling the control center's computers. The FAA later determined the cause of a flight plan entry error overwhelming the air traffic system's memory capacity.

The US Air Force calculated the U-2s retirement would save $2.2 billion. It proposed spending $1.77 billion over ten years to enhance the RQ-4, including $500 million on a universal payload adapter to attach U-2 sensors onto the RQ-4. USAF officials feared it was retiring the U-2 amid RQ-4 upgrades creating a capability gap.

The Air Force proposed using other high-altitude ISR platforms to substitute for the U-2 and RQ-4 during the interim. Such proposal included using satellites and the secretive RQ-170 and RQ-180 UAVs. The House Arms Services Committee's markup of the FY 2015 budget included language prohibiting the use of funds to retire or store the U-2; it requested a report outlining the transition capabilities from the U-2 to the RQ-4 Block 30.

In late 2014, Lockheed Martin proposed an unmanned U-2 version with greater payload capability. However, the concept did not gain traction with the USAF. In 2015, the USAF received a directive to restart modest funding for the U-2 for operations and research, development, and procurement through to FY 2018.

The former head of the USAF Air Combat Command, Gen Mike Hostage, helped to extend the U-2S to ensure commanders received enough intelligence, surveillance, and reconnaissance (ISR) coverage. He stated, "It would take eight years before the RQ-4 Global Hawk fleet could support 90% of the coverage of the U-2 fleet." Although the US Air Force intended for the RQ-4 to replace the U-2 by 2019, Lockheed claims it could remain viable until 2050.

On 23 May 2017, it became official that the U-2 was not retiring. The US Air Force Fiscal Year 2018 funded both the Global Hawk and the U-2 to meet the demand for ISR. It funded the upgrade to the ASARS-2B radar and the stellar tracking initiative, the star tracker that provided an alternative means of navigation in case of GPS jamming. Also funded is the enhancements to the optics and focal planes of the SYERS-2C imaging sensor; the SIGINT system; and the defensive electronic warfare system. There's even some funding for a "technical refresh" of the good old wet-film Optical Bar Camera (OBC).

There's more money to continue development and testing of the U-2's ability to act as an airborne communications node. These include the Link 16, the F-22's In-Flight Data Link (IFDL) and the F-35's Multifunction Advanced Data Link (MADL).

Also, in the works: upgrades to the pilot helmet and pressure suit, a fresh look at the ejection system, and the installation of a flight data recorder.

You can't keep a Good Lady down!

What Comes Next?

Like the A-12, the Corona satellites, and the SR-71 expect the unmanned ISR platform to have stiff competition emerging from the black world of Area 51. Don't be surprised someday to learn of a next-generation high-altitude ISR platform, a stealthy TR-X design incorporating the best features of the U-2: high-altitude flight, ISR, and rapid deployment.

Summary – OSA Overhead Reconnaissance Milestones

1954 Prologue: US reconnaissance lacked hard information on the secretive Soviet Union. Uninformed guesses could lead to "bankrupt" policies, as Herbert Miller put it. Also, missions carried out by warplanes such as B-47s were provocative, carrying a 'military silhouette.' Unable to operate much higher than 40,000 feet, they were increasingly vulnerable. Richard Leghorn observed: "We knew Americans were dying on these missions. It created a real sense of urgency to come up with new aircraft and to think about moving into space."

-March 1954 Killian Land Committee formed to advise the President. Urges high altitude reconnaissance by a civilian agency, CIA

- 24 November 1954 U-2 Project approved by Eisenhower.

- 10 June 1955 USAF/CIA Agreement on Project AQUATONE.

- 6 August 1955 First flight of U-2, eight months after contract let.

- 7 May 1956 U-2 Detachment A deployed to England.

- 20 June 1956 First U-2 overflight of Soviet Bloc countries.

- 4 July 1956 First U-2 flight over Russia.

- 20 August 1956 U-2 Detachment B deployed to Turkey.

- February 1957 U-2 Detachment C deployed to Japan.

- November 1957 first GUSTO stealth study reported to President Eisenhower's Board of Consultants.

- March 1958 Project CORONA satellite program started.

- 1 September 1958 Land Panel approved GUSTO effort leading to

OXCART; first GUSTO contract let.

- 21 January 1959 First CORONA launch.

- June 1959 Presidential approval for OXCART advanced study

- 4 September 1959 Letter of intent with Lockheed for OXCART.

- November 1959 Presidential approval to proceed with OXCART.

- 8 February 1960 Production contract for OXCART let.

- 25 August 1960 CIA's and military's satellites placed under NRO

- October 1960 U-2 Detachment in Turkey returned to the US.

- 26 October 1960 First U-2 flight over Cuba from the US.

- 5 October 1962 Last CIA U-2 flight over Cuba (50 flown in all)

- 29 February 1964 A-11 surfaced by President Johnson

- 23 November 1965 OXCART declared operationally ready

- 1974 OSA phased out

OSA Epilogue

Former National Reconnaissance Office (NRO) Director Dr. Hans Mark looked back at the role of the U-2 starting in 1954: "The U-2 was intended as an interim between the ground zero, which was March '54 when the Killian Land Committee was set up, and the time that satellites would be available. Interestingly, Gary Powers was shot down during May of 1960, and the first successful Corona flew during August of 1960."

CIA's first Deputy Director of Science and Technology, Albert "Bud" Wheelon, recalled the deployment of overhead reconnaissance against the USSR, "it was as if a floodlight had been turned on in a darkened warehouse." The 'bomber gap' and the 'missile gap' disappeared, and American leaders no longer had to err on the side of survival by thinking, 'worst case.'

"Tonight, we know how many missiles the enemy had," President Lyndon Johnson declared in 1967, "and, it turned out, our guesses were way off. We were doing things we didn't need to do. We were building things we didn't need to build. We were harboring fears we didn't need to harbor." The U-2 and the satellites (especially the latter, after both sides had them) largely allayed fears of a surprise attack and permitted diplomacy and arms control to move ahead.

Pilot transition from Groom Lake

Watertown flight line

Preparing to close canopy and launch

Particulate sampling payload

Taxiing without pogos
CIA pilots became adept at taxiing without pogos

Chapter 20-Project GUSTO

The CIA never expected the U-2 to last over 18 months before being shot down. Even before the U-2 went operational, the CIA had focused its attention on building a jet that could fly at extremely high speeds and altitudes while incorporating state-of-the-art techniques in radar absorption or deflection. Codenamed GUSTO, in the fall of 1957, U-2 project manager Richard Bissell established an advisory committee to help select a design for the U-2's successor. The selection boiled down to Lockheed's A-12 and Convair's Kingfish designs.

On 29 August 1959, the selection panel voted for the A-12. Project GUSTO was terminated, and "by a sort of inspired perversity," OXCART was selected from a random list of code names to designate this R&D and all later work on the A-12.

From Area 51, the CIA focused on the intelligence-gathering aspect of the Cold War. It trained its pilots and had them overflying Russia from 1956 to 1960.

Where it was difficult for American pilots to fly over mainland China, the CIA trained ROCAF pilots and crews to conduct clandestine flights. While the CIA Detachment H Black Cats flew over mainland China, American Detachment G pilots continued covering many other areas of the world.

The CIA accomplished much: It confirmed through U-2 missions that the feared bomber gap did not exist. Next, through Corona, it confirmed to President Kennedy that a missile gap also did not exist. In the process, the CIA proved the value of stealth and of overhead reconnaissance that continues today.

The position of Acting Director for OSA remained vacant for several months until filled as one of the recommendations of the Inspector General's survey of the spring of 1962. The first incumbent, Col (later brigadier general) Jack C. Ledford, served from September 1962 to August 1966.

(DD/S&T General Order No. 37 dated 27 July 1965, changed the CIA title of the Acting Director for Special Activities to the Director of Special Activities.)

During September 1962, Ledford received an assignment as the air commander with the 1040th US Air Force Field Activity Squadron at Bolling Air Force Base, Washington, DC, and became the Director of Special Projects, Headquarters US Air Force. He was Director of the Office of Special Activities, D/S&T for Project OXCART at Area 51. For his service in this position, he received the Distinguished Service Medal.

The CIA approved two increments of personnel for OSA during the latter part of 1962, for the A-12 program, bringing the T/O back up over 500. In 1963, the CIA requested an additional 217 slots with the approval of 121, making the total strength 629 instead of the 725 considered essential by June 1964, the delivery date for the last A-12 to Area 51.

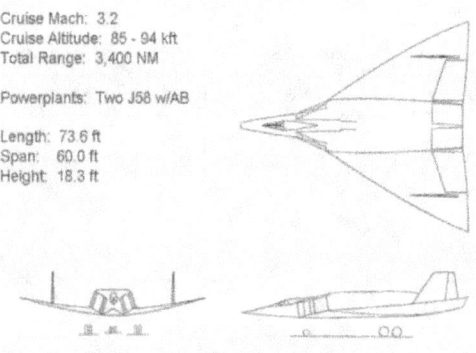

CONVAIR KINGFISH
JULY 1959

Cruise Mach: 3.2
Cruise Altitude: 85 - 94 kft
Total Range: 3,400 NM

Powerplants: Two J58 w/AB

Length: 73.6 ft
Span: 60.0 ft
Height: 18.3 ft

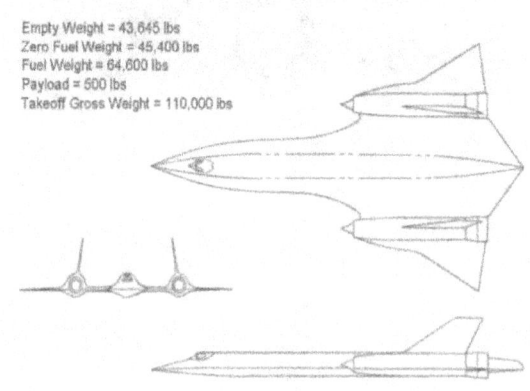

A-12 INITIAL CONFIGURATION 3-VIEW

Empty Weight = 43,645 lbs
Zero Fuel Weight = 45,400 lbs
Fuel Weight = 64,600 lbs
Payload = 500 lbs
Takeoff Gross Weight = 110,000 lbs

22% Increase in Empty Weight Compared to A-11 - "Cost of Stealth"

Glossary

AEC The United States Atomic Energy Commission (AEC) was an agency of the United States government established after World War II by Congress to foster and control the peacetime development of atomic science and technology. President Harry S. Truman signed the McMahon/Atomic Energy Act on August 1, 1946, transferring the control of atomic energy from military to civilian hands, effective from January 1, 1947. The Energy Reorganization Act of 1974 abolished the agency and assigned its functions to two new agencies: The Energy Research and Development Administration and the Nuclear Regulatory Commission.

CAT	Civil Air Transport
Comp	Abbreviation for "complimentary
COMSEC	Communications Security
CONUS	Continental United States
CRYPTO	Cryptographic
CW	Continuous Wave
CWAR	Continuous Wave Acquisition Radar
DPRK	North Korea
DREAMLAND	Name of the Groom Lake facility in Nevada during the USAF stealth Project
DTS	Data Transmission System
DYCOMS	Dynamic Coherent Measurement System used for RCS
ECCM	Electronic Counter-Countermeasures
ECM	Electronic Counter Measures
EG&G	Edgerton, Germeshausen, and Grier, Inc. EG&G is a United States national defense contractor and provider of management and technical services.
EMT	Emergency Medical Technician
ESPIONAGE	a covert act or acts of spying on others—often a systematic use of spies by a government to discover Military, Political, Economic secrets
FAA	Federal Aviation Administration
FCF	Functional Check Flight
FOUO	For Official Use Only
IFF	Identification Friend or Foe
Isinglass	CIA/General Dynamics Mach 4/5 rocket-powered, air-launched recon vehicle study to replace Oxcart, capable of reaching orbital velocities, not completed/never flown
LVCP	Landing Craft, Vehicle, Personnel. A Higgins boat
MOS	Military Occupational Specialty
NASA	National Aeronautics and Space Administration
NASA High Range	A 400-mile high speed corridor extending from NASA/Dryden at Edwards AFB to Ely, Nevada with Beatty tracking station being in the middle of the corridor.
NERVA	Nuclear Engine for Rocket Vehicle Application
Nevada Proving Ground	The site, established on 11 January 1951, for the testing of nuclear devices
NOFORN	Not Releasable to Foreign Nationals - restricted to US citizens
NRDS	Nuclear Rocket Development Station located at Jackass Flats within the AEC' atomic testing grounds
OCS	Officer Candidate School
Operation BLACK SHIELD	The operational phase of Project Oxcart flying the A-12 over North Vietnam and North Korea
PARADISE RANCH	Name of the Groom Lake facility in Nevada during the CIA A-12 Project

OXCART

- Project AQUATONE — Initial program name for CIA-sponsored U-2 Reconnaissance plane
- Project AURORA — USAF, classified program, most likely for B-2A procurement
- Project BALD EAGLE — USAF counterpart to the CIA's Aquatone; big-wing B-57 Canberra, became RB-57D
- Project GUSTO — A 1957 advisory committee selected by the CIA to select a successor to the U-2
- Project HAVE BLUE — The Lockheed XST ultra-secret stealth prototype that developed into the F-117 Nighthawk
- Project HAVE DOUGHNUT — One MiG-21F-13, used for Air Combat Training at Groom Lake, USAF/USN joint project, predating Have Idea (1968)
- Project HAVE DRILL — Two ex-Syrian MiG-17F from Israel, used for Air Combat Training at Groom Lake, USAF/USN joint project, predating Have Idea (1969)
- Project HAVE FERRY — Exploitation of a MiG 17.
- Project HAVE GARDEN — Evaluation of Soviet R-13-300 turbojet engine (used in MiG-21) blade profiling and compressor maps by Pratt & Whitney (1978)
- Project HAVE IDEA — Air Combat Training with various MiG planes, to Constant Peg
- Project Have RADIO — Exploitation of a foreign radar system.
- Project OXCART: — The CIA project at Groom Lake to design, build, and operate the A-12 Blackbird reconnaissance plane to replace the U-2. Originally classified Top-secret, OXCART was UNCLASSIFIED according to Senior Crown Security Class Guide dated 11/01/89. The CIA declassified the identity of the personnel during September 2007, and the restriction on using the name Area 51 lifted September 2010.
- PX — Post Exchange
- Pylon — The Special Projects team preferred to call the pylon a pole or pylon identified the tie point under aircraft wings where they fastened external loads.
- RatScat — Radar Target Scatter Site
- Reconnaissance — An overt act of reconnaissance in the field. A search made to produce useful military information to inspect, observe, or survey for enemy positions, strengths, and intent, etc. No longer just military purposes in this blended Technology Age
- ROK — South Korea
- SCR — Short for Signal Corps Radio # 584
- SPY — A person employed by a Gov to obtain secret info or Intel on another Gov Alt – Any person who clandestinely seeks info on people or projects for a profit
- Tank Battery — An installation of identical or nearly identical oil storage tanks
- TO&E — Table of Organization and Equipment
- TOC — Tactical Operations Center
- UCMJ — Uniform Code of Military Justice
- USO — United Service Organizations
- USOM — Office of Rural Affairs (Counterinsurgency)
- USS Pueblo — American spy ship captured by the North Korean navy in 1968
- WATERTOWN — Name of the Groom Lake facility in Nevada during the CIA U-2 Project AQUATONE
- WNINTEL — Warning Notice - Intelligence Sources and Methods Involved
- X-15 — The North American X-15 rocket-powered aircraft/spaceplane member of the X-series of experimental aircraft
- ZULU — Zulu Time is the world time used by international shortwave broadcasters, ham radio operators, shortwave listeners, the military, plane and ship navigation, and utility radio services — also known as UT or UTC (Universal Time [Coordinated]). The entire planet is on the

same time. There are no time zones for UTC. UTC also has no Daylight-Saving Time 0800 hours at the Beatty Tracking station was 1600 hrs. ZULU time.

Bibliography

Bissell, Richard M., Jr., with Jonathan E. Lewis and Frances T. Pudlo. ***Reflections of a Cold Warrior: From Yalta to the Bay of Pigs***. New Haven, CT: Yale University Press, 1996.

Beschloss, Michael R. ***Mayday: Eisenhower, Khrushchev and the U-2 Affair***. New York, Harper & Row, 1986.

Brugioni, Dino A. ***Eyeball to Eyeball: The Inside Story of the Cuban Missile Crisis.*** Random House. 1992.

Brugioni, Dino A. ***Eyes in the Sky: Eisenhower, the CIA, and the Cold War Aerial Espionage***. Naval Institute Press, 2010.

Darlington, David. ***Area 51: The Dreamland Chronicles***. NY: Henry Holt & Co, 1997

Eisenhower, Dwight D. ***Waging Peace, 1956-1961***. Garden City, NY: Doubleday 1965.

Fursenko, Aleksander and Timothy Naftali. ***Khrushchev's Cold War: the Inside Story of an American Adversary***. NY: Norton, 2006.

Hopkins, Robert S. III. ***Spyflights and Overflights: US Strategic Aerial Reconnaissance, Vol. 1 1945-1960***. Manchester, England: Hikoki Publications, 2016

Jacobsen, Annie. ***Area 51: An Uncensored History***. NY: Little, Brown & Co., 2011.

Jenkins, Dennis R. ***Dressing for Altitude: US Aviation Pressure Suits***. NASA, 2012

Johnson, Clarence L. with Maggie Smith. ***Kelly: More than My Share of It All***. Washington D.C.: Smithsonian Institute Press, 1985.

Kleyla, Helen H., and Robert D. O'Hern. ***History of the Office of Special Activities, DS&T***. CIA Directorate of Science and Technology Historical Series, 1974.

Leary, William M. ***Perilous Missions: Civil Air Transport and CIA Covert Operations in Asia.*** University of Alabama Press, 1984.

McIlmoyle, Gerald R., Brigadier General (Ret) with Linda Rios Bromley. ***Remembering Dragon Lady: Memoirs of the Men Who Experienced the Legend of the U-2 Spy Plane***. West Midlands, England: Helion & Co., 2011

Merlin, Peter W. ***Area 51***. (*Images of Aviation*), Charleston, SC: Arcadia Publishing, 2011.

Merlin, Peter W. ***Unlimited Horizons: Design and Development of the U-2***. NASA 2015.

Miller, Jay. ***Lockheed U-2***. (*Aerograph 3*), Austin, Texas: Aerofax, 1983.

Mikesh, Robert C. ***Martin B-57 Canberra: The Complete Record.*** Atglen PA: Schiffer, 1995.

Pedlow, Gregory & Welzenbach, Donald. ***The Central Intelligence Agency and Overhead Reconnaissance: The U-2 and OXCART Programs, 1954-1974.*** New York, NY: Skyhorse

Publishing, 2016. (2013 declassification, new typeset; earlier PDF is on CIA website)

Peebles, Curtis. *Shadow Flights: America's Secret Air War Against the Soviet Union*. Novato, CA: Presidio Press, 2000.

Pocock, Chris. *Dragon Lady: The History of the U-2 Spyplane*. Shrewsbury, England: Airlife Publishing, 1987

Pocock, Chris. *The U-2 Spyplane: Toward the Unknown – A New History of the Early Years*. Atglen, PA: Schiffer Publishing, 2000.

Pocock, Chris. *50 Years of the U-2.* Atglen, PA: Schiffer Publishing, 2005.

Pocock, Chris. *The Black Bats: CIA Spy Flights Over China From Taiwan 1951-1969.* Atglen, PA: Schiffer Publishing, 2010.

Polmar, Norman. *Spyplane: The U-2 History*. Osceola, WI: MBI Publishing Co., 2001.

Powers, Francis Gary with Curt Gentry. *Operation Overflight*: *a Memoir of the U-2 Incident*. Washington, D.C.: Potomac Books, 2004.

Rich, Ben R. with Leo Janos. *Skunk Works.* Boston: Little, Brown & Co., 1994.

Suhler, Paul. *From Rainbow to Gusto: Stealth and the Design of the Lockheed Blackbird*. American Institute of Aeronautics & Astronautics, 2009

Taubman, Philip. *Secret Empire.* NY: Simon & Schuster, 2003

US Army, *History of Strategic Air and Ballistic Missile Defense, Vol. 1: 1945– 1955* Amazon

Military History Quarterly, Hall, R. Cargill. "The Truth About Overflights." Volume 9 No. 3

https://www.nro.gov/Portals/65/documents/foia/declass/ForAll/050918/F-2018-00003c.pdf

Pinetreeline.org, Collections hosted by the Military Communications and Electronics Museum: "Metz, France – 1 Air Division" (see 'History' tab)

http://c-and-e-museum.org/Pinetreeline/metz/metz.html (Starting point to search for following:)

US News & World Report. "The Secret War: How US Airmen Risked Their Lives to Spy on The Soviet Union (special edition)." By D. Stanglin, S. Headden, and P. Cary. Summer 2003

Video

American Experience: Spy in the Sky. WGBH Boston Video. Originally aired in 1996. DVD produced in 2003. Now listed for sale by PBS, Amazon, Walmart, and others.

The Inquisitive Angel: The True Story of the U-2. Declassified in 2006, originally written and filmed for the CIA by Don Downie & Jim Jarboe of Hycon Manufacturing Co. A conversion to DVD by Creative Fission Productions may become available through Amazon).

Note: an abridged version of this video is released to the CIA's official Youtube channel—

Angels in Paradise: The Development of the U-2 at Area 51 (CIA Youtube release, about 21 minutes long). See: https://www.youtube.com/watch?v=nQnBJrj_-l8&t=404s

Wings of Russia: Spies – Watching From Above. Wings Air. 2008, Episode 18 of 18

https://www.youtube.com/watch?v=dFBPpcRhFug

Government Documents

History of Air Force Atomic Cloud Sampling 1963. PDF via George Washington University:

https://nsarchive2.gwu.edu//nukevault/ebb249/doc07.pdf

Unlimited Horizons: Design and Development of the U-2 (NASA Astronautics Book Series). Merlin, Peter, 2015. Available in Kindle, EPUB, and PDF formats, free downloads:

https://www.nasa.gov/connect/ebooks/unlimited_horizons_detail.html

Dressing for Altitude: US Aviation Pressure Suits—Wiley Post to the Space Shuttle (NASA) Jenkins, Dennis R., 2012. Available in Kindle, EPUB, and PDF, free downloads:

https://www.nasa.gov/connect/ebooks/dress_for_altitude_detail.html

CIA Declassified Documents cited in this book:

"National Intelligence Estimate 18" (The Probability of Soviet Employment of BW and CW in the Event of Attacks Upon the US) 10 January 1951

https://www.cia.gov/library/readingroom/docs/CIA-RDP79R01012A000400030002-1.pdf

"Future Plans for Project AQUATONE/OILSTONE" 29 July 1957

https://www.cia.gov/library/readingroom/docs/DOC_0000743239.pdf

"The Central Intelligence Agency and Overhead Reconnaissance: The U-2 and OXCART Programs, 1954-1974" by CIA Historians Pedlow and Welzenbach, 2002 (41 MB PDF file)

https://www.cia.gov/library/readingroom/docs/2002-07-16.pdf

"National Intelligence Estimate Number 11-5-55" (Air Defense of the Sino-Soviet Bloc 1955-60)

https://www.cia.gov/library/readingroom/docs/CIA-RDP79R01012A005100040007-6.pdf

National Reconnaissance Office Declassified Documents cited:

"Heart Throb Pilot: Overflights in the European Theater." Cooke, Gerald. Released in 2018

https://www.nro.gov/Portals/65/documents/foia/declass/ForAll/050918/F-2018-00039a.pdf

"The Truth About Overflights." Hall, R. Cargill Released in 2018

https://www.nro.gov/Portals/65/documents/foia/declass/ForAll/050918/F-2018-00003c.pdf

"A Daytime Overflight of Soviet Siberia." Hall, R. Cargill & Hillman, Donald. Released 2018

https://www.nro.gov/Portals/65/documents/foia/declass/ForAll/050918/F-2018-00039a.pdf

"Project Home Run Operations." Brown, George A. Released in 2018

https://www.nro.gov/Portals/65/documents/foia/declass/ForAll/050918/F-2018-00039i.pdf

Other Government Publication cited:

History of Strategic Air and Ballistic Missile Defense, Vol. 1: 1945 – 1955. US Army 1975

https://history.army.mil/html/books/bmd/BMDV1.pdf (hard copy reprint at Amazon)

Appendix 1

Lockheed Employees at Area 51 During Project AQUATONE

ADP Engineering - Permanent Party - Salaried

Deal, Paul	Flight Test Engineer
Fulkerson, Glen	Flight Test Supervisor
Goudey, Ray	Pilot
Klinger, Robert	Flight Test Engineer
LeVier, Tony	Pilot
Matye, Robert	Pilot
Reedy, Jack	Flight Test Engineer
Schumacher, Bob	Pilot
Sieker, Robert	Pilot
Yoshii, Mich light	Test Engineer

ADP Support – Salaried

Caravan, Fred

Johnson, C. L. (Kelly)

ADP Permanent Party - Managers & Supervisors - Salaried

Carney, Jerry	Crew Chief Supervisor– Ship 2
Frye, Fritz	Crew Chief Supervisor– Ship 1
Harvey, Frank	Supervisor
Kammerer, Dorsey	Manager

ADP Permanent Party - Hourly

Buckner, Vernon	Electrician – Ship 1
Christman, Pop	Shop
Cuthbert, Gene	Stock Material
Esley, Jack	Mechanic – Ship 2
Flynn, Leroy	Mechanic – Ship 1
Harvey, Frank	Painter
Herman, Carl	Shop
Huff, Dick	Utility Electrician
Hutchings, Mervin	Electrician – Ship 2
Johnson, Bob	Shop
King, Bob	Mechanic – Ship 2
Murphy, Bob	Mechanic – Ship 1
Paget, Dick	Inspector – Ship 2
Smith, Paul	Clerk and Photographer
Wilkerson, Pete	Inspector – Ship 1

ADP Chase Pilots Beechcraft & Cessna

All 5 of the U-2A ADP Pilots

ADP Transportation C-47

All 5 of the U-2A ADP Pilots

Bridan, _____ Pilot

Other

Alex Chiniaeff   Inspection Manager

Ron Harris

Gene Reynolds

Area 51 Humor in the Words of Instructor Pilot Louis Setter

"While Detachment A was in training, it snowed about 4 to 5 inches. Our lakebed was covered with a smooth layer of the white stuff, and U-2 flying canceled. I had an L-20 Beaver, a single-engine bush plane that we used for various jobs, so I flew it with no trouble. It has large tires. Col Stan Beerli, the detachment commander, was a skier, so he asked me to give him a tow. I agreed, so we tied a rope to the landing gear strut and off we went. I taxied at just below takeoff speed—at about 45 miles per hours, and Stan had a great ride. However, he complained about my prop blowing snow in his face. At that speed, it stings! So, I said, "I can fix that," and I took off, with him still on the end of the tow rope. I cruised at about 20 feet in the air, doing a big circle over the lakebed, and eventually got up to over 90 miles per hour. Stan had a ball. He must have been pulling a couple of Gs at the end of that rope. He later told me the probably broke the world speed record for being towed. The Germans did it using a car, with the skier alongside in the ditch"

Note: Above individuals pictured in this book:

Ray Goudey pictured throughout this book; Clendinning and Wood, pictured in Chapter 19; Raines pictured in Chapter 16

See http://area51specialprojects.com/u-2_deployment_orders.html

Quotes on Soviet Militarism

"Our new state, *now being born*, is also a state, for we too need detachments of armed men; we too need the *strictest order*, and must *ruthlessly* and forcibly crush all attempts at either a tsarist or a Guchkov-bourgeois counterrevolution."—Lenin *On the Tasks of the Proletariat in Our Revolution*, 1917, - quoted in: Garvey p. 54

"'Red militarism' is an apt term for many aspects of Soviet life in the 1930s . The style of political and economic leadership was military: the economic system is often called a 'command economy.' 'There is no fortress,' it was said, 'that the Bolsheviks cannot storm.'" - Holloway p. 9

After World War I, "the Red Army was part of a developing 'warfare state.' Stalin's USSR was a society organized for violence with a steady erosion of distinctions and barriers between military and civilian spheres. Moderation in defense planning was literally criminal. The Soviet system's legitimating ideology of Marxism-Leninism defined war as a science."

"' Possessed by a unique sense of its own destiny,' the former Soviet military economist Vitaliy Shlykov wrote, 'and hoping to establish a new world order "free of capitalistic exploitation," the communist leadership of the country saw a military confrontation with the rest of the world as inevitable. To be prepared for that confrontation, the Soviet State had to be turned into a permanent military camp.'" Rhodes 2007 p. 15

Soviet soldier: "Our system, it's a military system, essentially; we'll never have proper asphalt and manicured lawns. But there'll always be plenty of heroes." Rhodes 2007 p. 19

"Stalin realized that, in order to get his Bomb, his people would have to freeze and starve. In some places, electricity was cut off so that nuclear facilities could get all the power they needed." - DeGroot p. 137

"The civilian population continued to suffer for years as billions of rubles were diverted to the military and to the crash program to build the atomic bomb and stockpile a nuclear arsenal." - Rhodes 2007 p. 39

During November 1945, Molotov said the Soviet Union "must equal the achievements of contemporary world technology. We will have atomic energy and much else." - Holloway p. 27

"By 1945, with the beginning of the first rumblings of the Cold War, Stalin made up for his prior mistakes and rapidly took the offensive in accelerating the Soviet Union's ability not only to develop both the atomic and hydrogen bombs in rapid succession but also to recoup the gains that had been made on rockets in the 1930s. As early as 13 May 1946, the Council of Ministers of the USSR dedicated a meeting to the 'matters of the rocket weapons,' establishing the basic organizational structure for indigenous development of these weapons. From the outset, Stalin's primary goal was that the Soviet Union develop a long-range missile. In 1947, at a Communist Party Politburo meeting, he stressed that priority should be given to 'rocketry' and that the Soviet Union needed to develop 'transatlantic missiles.' No limits should be placed on the funding available to achieve that goal (Rhea, p. 287). On 14 April 1947, Stalin summoned Korolev to a key meeting of a special commission that called for the development of a long-range plan for rocket technology development." - Schultz p. 12

"By the summer of 1946, the basic institutional framework had been created for developing nuclear weapons, long-range rockets, radar, and jet propulsion. Special bodies were set up in the Party, the government, the secret police, and the Armed Forces to direct these programs." - Holloway p. 21

"'We never had a moral problem with what we were doing,' Nikitin recalls. 'It was a sacred thing.'" - DeGroot p. 141

"In 1956 the Polish economist Lange described the Soviet economy as a 'war economy'" - Holloway p. 9

Yuriy Rufov, Director of Biomedpreparat, Stepnogorsk, Russia: "No government could afford this today,' said Rufov. 'Of course, 90 percent of the Soviet industry connected with the military. That's what led to the collapse of the Soviet Union.'" - *N. Geographic*

Sources

DeGroot =   *The Bomb*, Gerard DeGroot, Harvard 2005
Garvey =   *Marxist-Leninist China*, James Garvey,
                    Exposition Press 1960

Holloway =   *The Soviet Union and the Arms Race*,
                    David Holloway, Yale 1983
N. Geographic =   "Weapons of Mass Destruction" *National
                    Geographic*, Nov. 2002
Rhodes =   *Arsenals of Folly*, Richard Rhodes, Knopf
            2007
Schultz =   "US National Security and the Geopolitical
                    Setting" Susan Schultz, NRO
                    National Reconnaissance, Spring 2012

Showalter = *On the Road to Modern War: 2*
                    "Processing Cataltrophe: Russia,"
                    Dennis Showalter, *International
                    Encyclopedia of the First World War*
                    http://encyclopedia.1914-1918-
                    online.net/article/on_the_road_to_modern_war

These few quotes are not intended to assign all blame for the Cold War, but to provide insight—especially for younger readers—into what US leaders faced during the period leading up to U-2 development

About the Author

Thornton D. "TD" Barnes, a multifaceted individual with a background in military intelligence, surface-to-air missile and radar electronics, and aerospace, was born in Dalhart, Texas, and raised on a ranch near Clayton, New Mexico, and Dalhart, Texas. His childhood during World War II instilled a passion for technology exploration, which he carried into adulthood. After completing high school in Oklahoma, 17-year-old Barnes embarked on a ten-year military career, beginning with service in Korea as an intelligence specialist. During his time in the Army, he honed his missile and radar electronics skills, focusing on countering Soviet threats. He also attended the Artillery Officer Candidate School before a military injury altered his career path.

Transitioning to aerospace pursuits, Barnes became involved in significant projects at NASA's High Range in Nevada, contributing to the X-15 program, atomic bomb tests at the Atomic Energy Commission's Nevada Proving Grounds, and the NERVA nuclear rocket project. He furthered his involvement in secretive projects by participating in the CIA's Mach 3 A-12 Project OXCART and stealth initiatives at Area 51.

Beyond his aerospace endeavors, Barnes founded and led an oil and gas exploration company for over four decades, delving into uranium and gold mining ventures. He has dedicated himself to preserving the history of Area 51, serving as president of Roadrunners Internationale and as the Nevada Aerospace Hall of Fame Director Emeritus. His contributions have been featured in documentaries on major networks like the National Geographic Channel, the Discovery Channel, the Fox News Channel, and the History Channel.

Barnes is also an accomplished author, with notable works about the Cold War, including "The Secret Genesis of Area 51," "The CIA Area 51 Chronicles," and " CIA Station D - Area 51. Currently residing in Henderson, Nevada, he continues to exert influence in aerospace, exploration, and literature, focusing particularly on the formerly highly classified aspects of the CIA's era at Area 51.

Other Books by Author:

Fiction
The EMP Series:
Nuclear Winter - Book 1
Nuclear Spring - Book 2
Nuclear Summer - Book 3
By the Sword - Book 4
The Hanson War Series
First Strike - Book 1
POW - Book 2
The Border - Book 3

The Unit
Beyond Earth – The Odyssey to Numerus X

Non-fiction
Soaring with the Eagles
The Secret Genesis of Area 51
CIA Project Oxcart
MiGs Over Nevada
The CIA Area 51 Chronicles (The CIA at Area 51 1955–1979)
Book 1 - The Angels
Book 2 - The Archangels
Book 3 - The Company Business
CIA Station D - Area 51
MiGs at Area 51
Black Ops Science & Technology
Project HAVE PAD
Nonfiction Books scheduled for publication 2024 by Begell House Publishing
Against the Tide
Secret Skies
Wizards of Area 51

www.ingramcontent.com/pod-product-compliance
Lightning Source LLC
Chambersburg PA
CBHW080008210526

45170CB00015B/1915

* 9 7 8 1 5 4 7 0 1 2 9 3 0 *